D0762123

Elements of Ocean Engineering

Cover photo credits:

Augur tension leg platform by Jenifer Tule
South Lake Worth, Florida, inlet by Dr. Billy L. Edge
Two-man, human powered submarine by Dr. Robert Randall

Elements of Ocean Engineering

Robert E. Randall

Published by
The Society of Naval Architects
and Marine Engineers
601 Pavonia Avenue
Jersey City, New Jersey 07306

Library of Congress Cataloging-in-Publication Data

Randall, Robert E., 1940-
Elements of Ocean Engineering / Robert E. Randall. -- 1st ed.
p. cm.
Includes bibliographical references and index.
ISBN 0-939773-24-4 (alk. paper)
1. Ocean engineering. I. Title.
TC1645.R36 1997
620' .4162--dc21

97-5912
CIP

THIS BOOK IS DEDICATED TO:

BARBARA, MY WIFE

BRIAN AND NEIL, MY SONS

PEG AND JACK, MY PARENTS

CONTENTS

CHAPTER 3: OFFSHORE STRUCTURES 47

CHAPTER 4: COASTAL PROCESSES AND STRUCTURES 79

PREFACE

This text is intended for use in a first course for ocean engineering students and to serve as an overview of the ocean engineering field and its applications. Students are expected to have completed one to two years of engineering college studies including at least two semesters of engineering calculus, a semester of engineering physics (mechanics) and a semester of engineering statistics.

Although engineers have been working in the oceans ever since man first ventured into the sea, the discipline of ocean engineering didn't develop until the late 1960's when there was a need to recover resources beneath the sea floor, protect coastlines, maintain and create waterways, and better understand the ocean environment. At the same time curricula in ocean engineering began to appear in U.S. universities located in states with shorelines, and in some cases, some long-standing marine programs changed their name to ocean engineering, and similar name changes were also observed in other countries. Private industry experienced the expansion of the offshore industry, dredging industry, and consulting firms specializing in coastal and offshore applications; and the defense industry developed a focus on undersea defenses and related ocean engineering applications.

The author defines ocean engineering as the application of engineering principles to the ocean environment. This discipline is nearly thirty years young, and it is still establishing its identity. Although many of the first generation of ocean engineers are only now completing their careers, the question "What is an ocean engineer?" is still commonly encountered, and it is hoped that this text will help clarify this question.

Ocean engineering is a very exciting and challenging field addressing man's use of the ocean frontier that covers approximately 79% of the earth's surface, and ocean engineers are involved in developing this vast and harsh frontier while at the same time striving to protect the same ocean environment. This field is especially challenging because it requires innovation and solutions to problems that have never been attempted before. The resources of the ocean are vast and ocean engineers must lead the way in its resource development in an environmentally acceptable way.

ACKNOWLEDGMENTS

Since 1975, I have had the pleasure of teaching undergraduate and graduate students at Texas A&M University in the discipline of ocean engineering. It is the interest of these students and those who will follow that have made it possible for this textbook to be published.

Many people at Texas A&M University, in the marine industry, and at the Society of Naval Architects and Marine Engineers (SNAME) contributed to the content and appearance of this book. These acknowledgments provide the opportunity to thank some of those people individually.

The assistance of Aubrey L. Anderson, Daniel T. Cox, William P. Fife, Richard B. Griffin, John B. Herbich, Jack Y. K. Lou, and Jun Zhang in reviewing and providing comments on various chapters of the text is greatly appreciated. Bruce L. Crager of Oceaneering Production Systems also provided comments. Several ocean engineering students: Billy D. Ambrose, Allan M. Breed, Ryan B. Cantor, David E. Cobb, Michael S. Ellis, and Christopher L. Ross have also contributed by making drawings and developing the problem solutions. Appearing on the front cover is the photograph of the Auger tension leg platform furnished by Jenifer Tule, and appearing on the back cover is the photograph of the South Lake Worth, Florida, inlet provided by Billy L. Edge. Joan Pope and Yen-hsi Chu of the Coastal Engineering Research Center in Vicksburg, MS, provided a number of photographs used to depict coastal structures and processes. The ocean engineering students enrolled in the course "Introduction to Ocean Engineering" at Texas A&M University provided comments and feedback to the class notes that were used to develop this text. The Ocean Engineering Program and Civil Engineering Department at Texas A&M University are also acknowledged for their financial assistance and encouragement. Both Melanie Estes and Celeste Phillips, with the Texas A&M Printing Center, were extremely helpful in producing the final photographs used for publication. I am grateful to Jose Femenia, chair of the SNAME control committee, for his encouragement and review of the text.

Many thanks go to Jaime Horowitz, Associate Director for Publications and Technical Activities for SNAME for her guidance, her assistance in obtaining the necessary copyright permissions for the many of the figures and tables, and her contributions to both the overall appearance of the text and the book's cover.

I appreciate the assistance provided by Joyce Hyden, secretary for the Ocean Engineering Program at Texas A&M University, in the preparation of the final manuscript. Last, I want to express gratitude to my wife, Barbara, for her dedication in editing and contributing to the text's readability, and for her encouragement, understanding, and patience during the preparation of this book.

List of Figures

List Of Tables

Elements of Ocean Engineering

CHAPTER 1: OVERVIEW OF OCEAN ENGINEERING

BACKGROUND

Ocean engineering is a relatively new engineering discipline whose future is linked to mankind's need to use natural energy and mineral resources beneath the sea surface, provide a food source, accommodate recreational activities, transport goods and people, provide alternative space for living quarters and facilities, further understand oceanic processes and develop engineering concepts for protecting the land from various ocean meteorological processes. Ocean engineering may be defined as the application of engineering principles to the analysis, design, development and management of systems that must function in a water environment such as oceans, lakes, estuaries, and rivers. This definition is similar to that for aerospace engineering which applies engineering principles to systems operating in space and the upper atmosphere.

There are other related disciplines that are applied to the ocean environment such as coastal engineering, marine engineering, naval architecture, naval engineering and offshore engineering. Coastal engineering typically applies engineering principles to systems operating in the coastal zone. Marine engineering applies to energy and mechanical systems used on marine vehicles, and naval architecture refers to the design of the hull structure and propulsion systems for marine vehicles. Offshore engineering indicates the application of engineering principles to systems in the offshore zone which implies deeper water outside the coastal zone. The application of engineering laws to naval systems or ships is often termed as naval engineering. In the author's view, ocean engineering encompasses both coastal and offshore engineering and overlaps with the areas of naval architecture, marine and naval engineering.

Since ocean engineering is a relatively new engineering field, it is only now that authoritative books and references are available for use by practitioners in industry and for academic instruction. Several texts are now published addressing the area of offshore structures such as Graff (1981), Dawson (1983), Gerwick (1986), McClelland and Reifel (1986), Patel (1989), Barltrop (1991), and Mather (1995) and evaluation of wave forces including Sarpkaya and Issacson (1981) and Chakrabarti (1994). Water wave theories are described in texts by Dean and Dalrymple (1984) and Kinsman (1984) and Mei (1992). The area of coastal processes and protection is addressed by the US Army Corps of Engineers' Shore Protection Manual (USACE, 1984). Other references related to coastal processes include Horikawa (1988), Ippen (1966), Fischer et al. (1979) and Wiegel (1965). Ocean engineering handbooks have been written by Herbich (1990, 1992) and Meyers et al. (1969). Underwater systems references include the National Oceanic and Atmospheric Administration Diving Manual (NOAA 1991), Allmendinger (1990) and the US Navy Diving Manual (USN 1985). Lewis (1988) and Berteaux (1991) have written texts for the related areas of naval architecture and buoy engineering, respectively.

EDUCATIONAL INSTITUTIONS

Educational curricula at the graduate and undergraduate level have been developed at many United States academic institutions located in states with coastlines. Table 1-1 illustrates the institutional names, location and degree programs in the United States, and Table 1-2 lists some institutions located abroad. Although these tables are believed to be fairly complete, some institutions may have inadvertently been excluded.

Table 1-1. List of selected academic institutions in the United States offering ocean engineering, naval architecture, marine engineering and other related degree programs.

Name of Institution	Location	Degree Program	Degree Types
California Maritime Academy	Vallejo, California	Marine Engineering	Undergraduate
California State Polytechnic University	Pomona, California	Ocean Engineering minor	Undergraduate
Florida Atlantic University	Boca Raton, Florida	Ocean Engineering	Undergraduate Graduate
Florida Institute of Technology	Melbourne, Florida	Ocean Engineering	Undergraduate Graduate
Great Lakes Maritime		Marine Engineering	Undergraduate
Maine Maritime	Castine, Maine	Marine Engineering	Undergraduate
Massachusetts Institute of Technology	Cambridge, Massachusetts	Ocean Engineering	Undergraduate Graduate
Massachusetts Maritime	Cape Cod, Massachusetts	Marine Engineering	Undergraduate
Oregon State University	Corvallis, Oregon	Ocean Engineering	Graduate
State University of New York Maritime College	Fort Schuyler, New York	Naval Architecture & Marine Engineering	Undergraduate
Stevens Institute of Technology	Hoboken, New Jersey	Coastal and Ocean Engineering	Graduate
Texas A&M University	College Station, Texas Galveston, Texas	Ocean Engineering Maritime Systems Engineering & Marine Engineering	Undergraduate Graduate Undergraduate
University of California at Berkeley	Berkeley, California	Naval Architecture and Offshore Engineering	Graduate
University of Delaware	Newark, Delaware	Coastal Engineering	Graduate
University of Florida	Gainesville, Florida	Coastal & Oceanographic Engineering	Graduate
University of Hawaii	Honolulu, Hawaii	Ocean Engineering	Graduate
University of Michigan	Ann Arbor, Michigan	Naval Architecture & Marine Engineering	Undergraduate Graduate
University of New Hampshire	Durham, New Hampshire	Ocean Engineering	Graduate
University of New Orleans	New Orleans, Louisiana	Naval Architecture & Marine Engineering	Undergraduate Graduate
University of Rhode Island	Kingston, Rhode Island	Ocean Engineering	Undergraduate Graduate
University of Washington	Seattle, Washington	Ocean Engineering	Graduate
US Coast Guard Academy	New London, Connecticut	Naval Architecture & Marine Engineering	Undergraduate
US Merchant Marine Academy	Kings Point, New York	Marine Engineering	Undergraduate
US Naval Academy	Annapolis, Maryland	Ocean Engineering	Undergraduate
Virginia Polytechnic Institute	Blacksburg, Virginia	Aerospace & Ocean Engineering	Undergraduate Graduate
Webb Institute of Naval Architecture	Glen Cove , New York	Naval Architecture & Marine Engineering	Undergraduate

Table 1-2. List of selected universities outside the United States offering education in ocean engineering or related fields of study.

Name of Institution	Location	Degree Program	Degree Types
Delft University of Technology	The Netherlands	Coastal Engineering	Undergraduate Graduate
University of Trondheim	Norway	Ocean Engineering	Undergraduate Graduate
Dalian University	Peoples Republic of China	Ocean Engineering	Undergraduate Graduate
Danish Hydraulic Institute	Denmark	Coastal Engineering	Undergraduate Graduate
Memorial University of Newfoundland	Newfoundland	Ocean Engineering	Undergraduate Graduate
Cranfield Institute of Technology	United Kingdom	Ocean Engineering	Undergraduate Graduate
Seoul National University	Korea	Naval Architecture	Undergraduate Graduate
Shanghai Jiao Tong University	Peoples Republic of China	Ocean Engineering	Undergraduate Graduate
University of Tokyo	Japan	Naval Architecture	Graduate
University of Buenos Aires	Argentina	Naval Architecture	Graduate
University of New South Wales	Australia	Naval Architecture Coastal Engineering	Graduate
Tianjin University	Peoples Republic of China	Offshore Engineering Coastal Engineering	Graduate
Technical University of Berlin	Germany	Ocean Engineering	Graduate
University of Hamburg	Germany	Naval Architecture	Graduate
Ecole Nationale Super. de Tech. Avancees	France	Offshore Engineering Naval Architecture	Graduate
University College Cork	Ireland	Coastal Engineering Offshore Engineering	Graduate
Yokohama National University	Japan	Ocean Engineering Naval Architecture	Graduate
Nihon University	Japan	Naval Architecture	Graduate
Kagoshima University	Japan	Ocean Engineering	Graduate
Inha University	Korea	Naval Architecture	Graduate
Royal Institute of Technology	Sweden	Naval Architecture	Graduate
University of Glasgow	United Kingdom	Ocean Engineering Naval Architecture	Graduate
University College London	United Kingdom	Ocean Engineering	Graduate
University of Strathclyde	United Kingdom	Coastal Engineering Marine Technology	Graduate
University of Auckland	New Zealand	Coastal Engineering Offshore Engineering	Graduate
India Institute of Technology	India	Naval Architecture	Undergraduate Graduate

BRIEF HISTORY

Although engineers have been engaged with engineering applications in the ocean since before the beginning of this century, the academic discipline of ocean engineering only surfaced at some universities in the late 1960's and early 1970's. As a consequence, engineers educated in ocean engineering are relatively new. The development of ocean engineering was fueled by exploration of the underwater environment, development of offshore gas and oil, and the continued need for coastal protection and port expansion. The US Navy, J. Y. Cousteau and E. A. Link pioneered the development of underwater habitats (e.g.. Sea Lab I-III, Conshelf and Hydrolab) and manned submersibles (e.g. Aluminaut, Ben Franklin, Deep Diver, Deep Submergence Rescue Vehicle, Deepstar, Johnson-Sea-Link I & II, Star I - III, and Trieste) that have provided platforms to explore and develop ocean resources. The development of offshore oil and gas fields by the various oil and gas companies (e.g. Amoco, Arco, British Petroleum Chevron, Conoco, Esso, Exxon, Mobil, Shell, Statoil, and Texaco) in the Gulf of Mexico, North Sea and Persian Gulf has been tremendous, and opportunities for ocean engineering applications have prospered at the same time. Other offshore development has occurred offshore Alaska, Canada, Brazil, Mexico, China, Africa, and Indonesia. Several major US ports are undergoing deepening, widening, and modernization of cargo handling facilities. Examples are the Los Angeles 2020 project in Los Angeles, California and the Houston ship channel deepening and widening project in Houston, Texas. Contaminated sediments in ports has created new engineering problems related to maintenance dredging in ports and the related disposal of dredged materials that is necessary to allow ships to continue accessing the facilities. The development of the nation's coastlines and ports as centers of trade and recreation continues to expand.

Coastal

Protection of coastlines and beaches from erosion and flooding has been a concern of engineers from the beginning of time. In 1950, the beach erosion board was first established in the United States to protect the nation's coastlines. A major activity that occurs world wide is beach nourishment which is placing beach material back on beaches after severe erosion over many years or due to severe storms. These beach nourishment projects, such as occurred at Miami Beach, Florida in the USA (Figure 1-1), are necessary to protect the land from flooding and wave action, provide beaches for recreation, and protect wetlands where a diverse marine habitat exists. Coastlines are protected by many different man-made coastal protection structures that include seawalls, breakwaters, revetments, groins, and submerged berms.

In the 1960's, port, harbor, and marina development rose sharply. Large ports contribute to a strong economy and increased commerce and trade vital to all nations. Safe and navigable entrance channels are critical to ports and harbors, and the construction and maintenance of channel jetties and breakwaters have provided safe passage to these important trading ports. Recreational boating and fishing that occur along coastlines and in coastal bays and estuaries, inland lakes, and rivers also require the development of small boat marinas to support water recreation, sport fishing, and pleasure boat activities. Commercial fishing and the seafood industry require a port and harbor infrastructure to support these very important activities.

Development of ports and harbors requires dredging of the bottom sediments and that led to the initiation of the Dredging and Dredge Material Disposal research program spearheaded by the US Army Corps of Engineers in the U.S. The need for dredging is world wide so that the ports and harbors can remain open for commercial and military ships, submarines, and other water borne craft. New ships have greater drafts, and consequently, it is necessary to further deepen the entrance channels to the ports (e.g. 17 m or 55 ft). Large dredges, such as cutterhead, hopper, and bucket wheel, are used to deepen the channels and subsequently maintain the channel depths. The disposal of the dredged material is also an important engineering activity. In some cases, the dredged material can be beneficially used (e.g. beach nourishment, land and wetland development, and providing wildlife habitats) or placed in an environmentally safe manner in specific disposal areas upland or offshore.

Figure 1-1. Before (left) and after (right) beach restoration in Miami, Florida. (Reprinted with permission from USACE, 1984, *Shore Protection Manual.* (Full citing in references).

The federal agency in the United States that is most responsible for the maintenance of all navigable waterways and coastal protection is the U.S. Army Corps of Engineers. In the late 1970's, the Coastal Engineering Research Center moved to the U.S. Army Engineer Waterways Experiment Station to consolidate the Corps of Engineers laboratories and researchers in a central location and establish itself as the nation's leading coastal research and physical modeling facility in the United States and one of the leader's in the international community. In 1987 the Dredging Research Program was initiated by the Corps of Engineers to improve efficiency of dredging operations and in 1993 the Coastal Inlets Program was begun to study the nations inlets that are very dynamic and provide the access to the nation's ports and inland waterways.

Offshore

Trends in the ocean engineering field have paralleled the Offshore Industry whose center in the United States is in Houston, Texas. However, the first offshore exploration for oil was in 1887 off the coast of California in a few feet of water. In 1910, an oil well was drilled in Ferry

Lake, Louisiana. Internationally, the first wells were drilled in Lake Maricaibo, Venezuela in 1929, and the Gulf of Mexico followed with the development of the Creole field in 4.3 m (14 ft) of water off the coast of Louisiana. Shallow water wells continued the slow development, and in 1959 Shell installed a platform in 30.5 m (100 ft) of water off Grand Isle, Louisiana. The Persian Gulf and the North Sea experienced oil finds and subsequent offshore platform development starting in 1960. Development in the 1970's was explosive, and offshore platforms and drilling advanced into deeper water at a rapid rate. In 1973, the North Sea was the site of the first concrete gravity platform which is a concrete structure built on land, floated to the site, and sunk to the bottom. The Hondo platform was installed by Exxon in 259.1 m (850 ft) of water off the California coast near Santa Barbara, and in 1978, Shell placed their Cognac platform in 312.5 m (1025 ft) of water in the Gulf of Mexico. Exxon installed the first guyed tower, Lena, in the Gulf of Mexico during 1983. A guyed tower is a slender, bottom supported tower that is laterally braced by cables (guy wires). The following year, 1984, Conoco placed the first tension leg platform in the North Sea in 147.9 m (485 ft) of water. In 1988, Shell installed the Bullwinkle fixed platform in 548.8 m (1800 ft), and five years later (1993), Shell installed the Auger tension leg platform in a water depth of 852 m (2795 ft) in the Gulf of Mexico. A brief history of offshore platform installations is illustrated in Figure 1-2. As of 1984 there were 16 gravity structures in the North Sea in depths ranging from 230 to 500 ft water depth, and other gravity platforms were installed offshore Brazil and the Baltic Sea. McClelland and Reifel (1986) report that over 3500 offshore structures have been placed in offshore waters of over 35 nations and nearly 98% of them are steel structures supported by piles driven into the seafloor. The 1990's are experiencing the push to deeper waters (> 2000 ft or 610 m) with the installation of tension leg platforms, and floating production systems are being used to produce oil in marginal fields (2-6 year production life). New platform concepts continue to be proposed with the goal to reduce the cost of production and to be able to work in greater and greater water depths since large oil reserves have been found in very deep water depths (> 6000 ft or 1829 m).

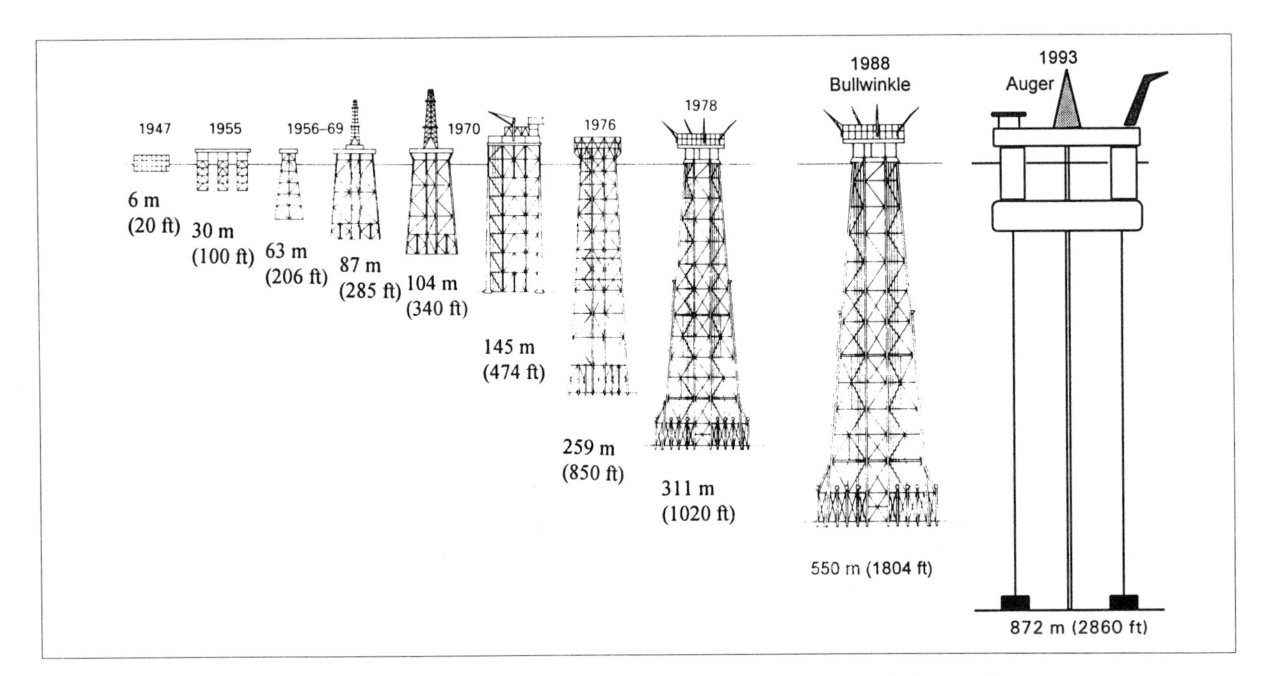

Figure 1-2. Historical development of deep water production platforms and their installed water depth.

Underwater Systems

Underwater habitats, diving equipment, submarines and subsea completion equipment are examples of underwater systems that ocean engineers are researching, developing, designing, and operating to advance man's use of the ocean environment. The first manned underwater habitat for saturation diving, Man-in-the-Sea I, was developed and tested in 1962 by E. A. Link. Since then over 65 underwater habitats have been built worldwide (NOAA 1991), and some of the more well known habitats are Conshelf I, II and III developed by J. Y. Cousteau of France, Helgoland I and II built in Germany in 1969 and 1971, Chernomor I and II operated by Russia in the Black Sea. SeaLab I, II and III, Tektite, LaChalupa, and Aegir are habitats built and operated in the United States waters and the Caribbean. A more complete description of underwater habitats may be found in NOAA (1991) and Miller and Koblich (1984). The most used underwater habitat was Hydrolab which hosted over 700 scientist/aquanauts. It was decommissioned in 1985 and is now displayed at the Smithsonian Institute in Washington, D.C. NOAA has recently constructed a new habitat, named Aquarius, for scientific missions in the Caribbean Sea.

Divers use many different types of breathing equipment to assist them in their exploration and work in the underwater environment, and a history of the development of this equipment is described by Bachrach et al. (1988). Prior to the use of compressed air, breath-hold divers developed goggles, snorkels, and fins to improve their diving efficiency. Self contained underwater breathing apparatus (SCUBA) have been around since the 1500's, but the double hose Cousteau-Caglan aqualung that was developed in 1943 and sold world wide over the subsequent ten years started the common use of SCUBA for research and recreational diving. The SCUBA breathing apparatus consists of a compressed air cylinder and demand regulator. The cylinder is normally carried on the divers back and the demand regulator is inserted into the diver's mouth and supplies air when the diver inhales. The need to conserve the breathing gas used for mixed gas breathing (e.g. helium/oxygen mixtures) resulted in the development of the semi-closed breathing apparatus which recirculates the exhaled breathing gas through a carbon dioxide absorber. This system was followed by the development of the closed circuit breathing apparatus (NOAA 1991) that totally contains the breathing gas (i.e. no breathing gas leaves the system).

Surface supplied diving equipment supplies the breathing gas to the diver through a flexible hose (umbilical) to the diver's helmet. The location of the gas supply can be from the surface, habitat, personnel transfer capsule (diving bell), or lock-out submersible. The helmets can be free flow or can have a demand regulator, and in some cases the helmet is equipped to remove carbon dioxide. One atmosphere diving suits date back to the 1700's, but the most widely used and successful system was developed in 1935 and called the JIM system. The JIM systems have been rated as deep as 610 m (2000 ft) and have actually worked in depths as deep as 543 m (1780 ft). A similar one atmosphere system (WASP) has been developed by Oceaneering International for mid-water work on offshore structures, and it uses small thrusters and crane to position the diver at the work location.

Submarines are important military undersea vehicles, and they were first used in World War I and II. The U-boats of Germany were formidable weapons in the sea that patrolled the shipping lanes and disrupted shipping and supply routes. The German and Allied submarines were typically 91.5 m (300 ft) long and could work to depths of near 122 m (400 ft), and they were powered by diesel engines on the surface and electrical batteries underwater. In the late 1960's and early 1970's, larger and faster submarines were constructed that used nuclear power, inertial navigation systems, and oxygen generating equipment so that the submarines could stay beneath the water for nearly unlimited times and could travel under ice caps without having to surface. The USS Nautilus was the world's first nuclear powered submarine (Allmendinger 1990).

Submersibles are usually small submarines that are manned with only a few people, and their purpose is to allow exploration of the ocean depths in a one atmosphere environment while observers view the undersea environment through windows and video cameras. Mechanical or electro-hydraulic manipulators are used to assist in observations and collection of samples. Some submersibles have a lock-out chamber that allows divers to exit and return to the submersible. The first submersible, "bathysphere" was built in 1930, and in 1934, it was used to reach a depth of 934 m (3028 ft). Since then submersibles and the lock-out submersibles have developed rapidly, and some of the well known submersibles include Alvin, Deepstar, Johnson-Sea-Link, Pisces series, and the Perry built PC series.

A remotely operated vehicle (ROV) is an unmanned underwater system consisting typically of a propulsion device, closed circuit television, and mechanical or electro-hydraulic manipulator. The vehicle is controlled from a surface vessel through an umbilical, and video pictures and data are also transmitted through the umbilical and viewed on the surface vessel. The first ROV to gain fame was CURV that was developed by the US Navy and used to recover a hydrogen bomb resting on the sea floor at a depth of 869 m (2851 ft) off the coast of Spain in 1966. As the offshore industry moved into deeper and more hostile waters such as the North Sea, the development and use of ROV's grew tremendously. More than a thousand ROV's are available ranging in size from that of a basketball to a large truck, and a few are capable of working in the deepest depths of the ocean. These vehicles are generally classified as tethered (free swimming), towed, bottom reliant, structure reliant, untethered (autonomous) or hybrid (MTS 1984). ROV's are being used in the offshore industry, military applications, and scientific investigations. In September 1985, a Woods Hole Oceanographic Institute team of scientists used a towed ROV, Argo and Jason, to locate and video tape the Titanic which sank in the Atlantic Ocean in 1912. Autonomous ROV's are the newest development and can be preprogrammed for a specific task without using umbilical. The future of remotely operated vehicles is very bright, and these vehicles are a valuable tool for the ocean engineer in a wide variety of underwater applications.

Generally, the equipment associated with subsea systems is necessary for the production of oil and gas from subsea wells. When the oil and gas fields are marginal, subsea production technology is more economical than conventional platform production techniques (Goodfellow 1990). Marginal fields apply to oil reserves of 30-50 million barrels and are typically in shallow water depths of less than 160 m (525 ft). As subsea equipment and technology advance, then its use in deeper waters is anticipated. Subsea equipment includes subsea wells, blowout preventers, templates, flow lines, well testing equipment, production risers, subsea trees, manifolds, controls and chokes (Figure 1-3).

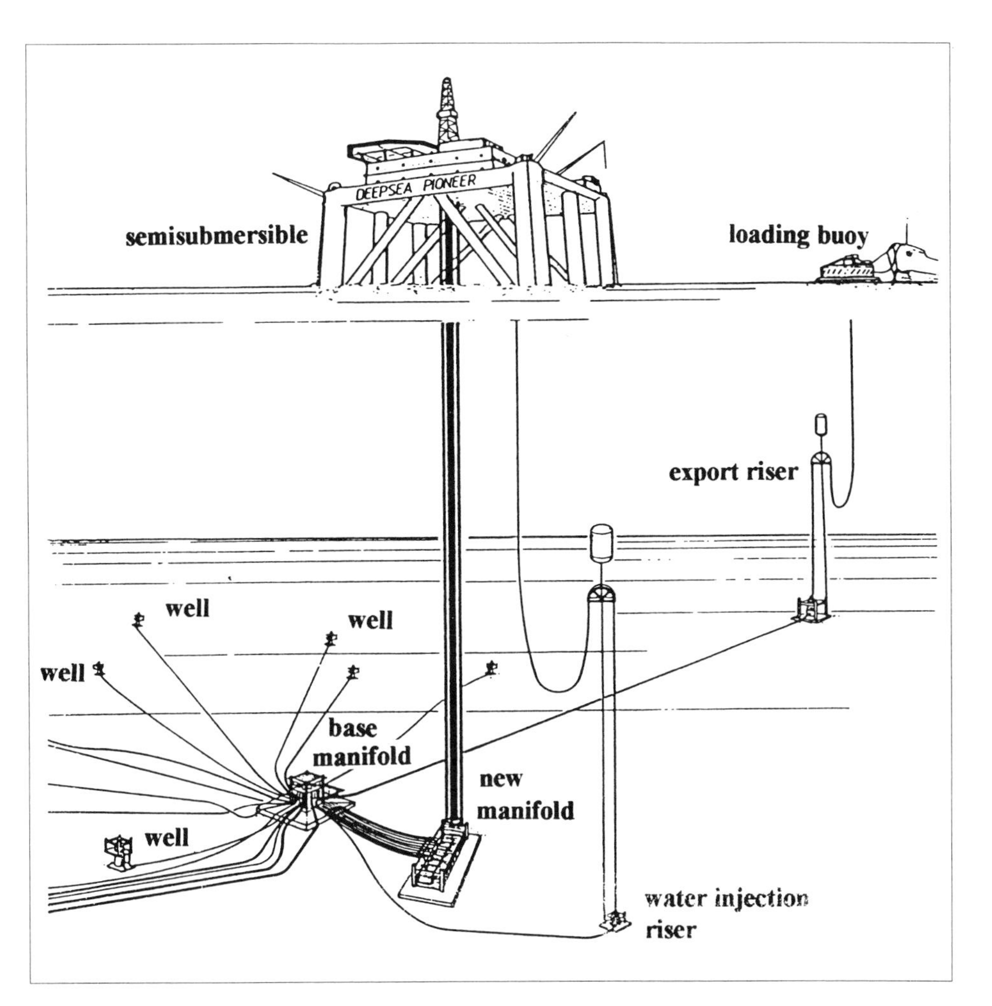

Figure 1-3. Production riser system for Argyll field in the North Sea. Reprinted with permission from Goodfellow,1990, *Applications of Subsea Systems*. (Full citing in references.)

APPLICATION AREAS IN OCEAN ENGINEERING

Ocean engineering is an interdisciplinary field, and consequently there are a large number of application areas. A list of some of these application areas are tabulated below:

- Coastal protection and erosion control (breakwaters, groins, seawalls, jetties, offshore berms, beach nourishment)
- Dredging and dredged material disposal (navigation channel maintenance, port and harbor development and maintenance, mechanical and hydraulic dredges)
- Marine hydrodynamics (floating and moored body motions)
- Marine foundations (seafloor support for ocean structures)
- Monitoring ocean environment (environmental monitoring)
- Mooring systems (taut-wired, catenary, multi-point, single point)
- Naval architecture (vessel stability, hull structure, resistance and propulsion)
- Numerical modeling (structures, fluids, interactions)
- Ocean energy (thermal, currents, waves, and tides)

- Ocean instrumentation (wave gauges, current meters, conductivity-temperature-depth-dissolved oxygen (CTD/DO), water samplers, tide gauges, transmissometers)
- Ocean mining (manganese nodules, placers)
- Ocean waves (wave theories, wave kinematics, wave forces, wave forecasting and hindcasting, wave refraction, reflection and diffraction, wave spectral distribution)
- Offshore disposal (clean dredged material, contaminated dredged material, hazardous waste)
- Offshore pipelines (oil and gas)
- Offshore structures (fixed, gravity, compliant, oil and gas platforms, jack-up rigs)
- Physical modeling (wave tank, wave basin, circulating water tunnel, towing tank, open channel recirculating flumes)
- Ports, harbors, and marinas (commercial and recreational, construction, maintenance and operation)
- Search and salvage (locating and recovering sunken objects and treasure)
- Submarines and floating structures (military submarines, semisubmersible drilling rigs, floating production systems)
- Submersible vehicles (small manned submersibles, remotely operated vehicles, autonomous underwater vehicles
- Underwater acoustics (SONAR, side scan, sub-bottom profiler, depth sounder, seismic exploration)
- Underwater systems (habitats, diving bells, remotely operated vehicles)

EMPLOYERS OF OCEAN ENGINEERS

Ocean engineers are employed by private industry, offshore industry, defense industry, federal agencies and laboratories, dredging industry, and other organizations involved in developing engineering systems that operate in the oceans, estuaries, lakes, and rivers of the world. With offshore exploration and development expanding off coastlines of other nations, there are opportunities for ocean engineers in industries abroad. Typical 1997 starting salaries for entry level ocean engineers range from $30,000 to $43,000. Although organizations employing ocean engineers are very dynamic, a summary of employers known to the author are:

Private Industry

- Offshore Oil (AMOCO, British Petroleum, Chevron, Conoco, ESSO, Exxon, Shell, Statoil, Texaco, independents, etc.)
- Offshore Construction (Brown & Root, Bechtel, McDermott, FMC, Gulf Aker Marine, LeTourneau, etc.)
- Shipyards (Houston, Ingalls, Mare Island, Newport News, Norfolk Navy, etc.)
- Offshore Consulting Firms(Acker Maritime, MPC Inc., INTEC, Noble Denton & Associates, Oceaneering International, Mustang Engineering, McClure & Associates, Petromarine, etc.)
- Coastal and Dredging Engineering Consulting Firms (Moffat & Nichols, Black & Veatch, Rosenblatt, Greenhorne & O'Mara, Gahagan & Bryant, HartCrowser, Hartman Consulting, etc.)

- Dredging Contractors (Bean Dredging, Ellicott, Great Lakes Dredge and Dock, Dutra, T. L. James, Weeks, etc.)
- Defense Contractors (General Dynamics, Lockheed, McDonnell Douglas, Rockwell)
- Instrument and Data Analysis (Endeco/YSI, General Oceanics, Sea Bird, Hydrolab, EG&G, Benthos, InterOcean, etc.)
- Diving, Submersibles and ROV's (Oceaneering Technologies, SONSUB, Subsea, Perry Tritech, etc.)
- Drilling Companies (Rowan, Reading Bates, Sonat, Global, etc.)
- Seismic and Hydrographic Surveying Companies (John Chance & Associates, Chris Ransome and Associates, etc.)
- Offshore Service (Schlumberger, Diamond Offshore, etc.)

Government

- US Navy and US Coast Guard
- Navy Civilian Laboratories (Coastal Systems Station; Naval Civil Engineering Laboratory; Naval Command, Control & Surveillance Center; Naval Research Laboratory; Naval Undersea Warfare Center)
- National Oceanic and Atmospheric Administration (NOAA)
- US Army Corps of Engineers [Districts, Divisions, and Waterways Experiment Station (WES), Cold Regions Research and Engineering Laboratory (CRREL)]
- Regulatory Agencies (Environmental Protection Agency, EPA; American Bureau of Ships, ABS; Det Norske Veritas, DNV)

PROFESSIONAL ORGANIZATIONS

The professional organizations to which ocean engineers generally belong are normally related to professional societies associated with the major science and engineering areas. There are also several divisions within these major societies that are devoted to ocean engineering. These professional societies or organizations are:

- American Geophysical Union (AGU)
- American Institute of Aeronautics and Astronautics (AIAA), Hydronautics Division
- American Society for Engineering Education (ASEE), Ocean and Marine Engineering Division
- American Society of Civil Engineers (ASCE), Waterway, Ports, Coastal and Ocean Engineering Division
- American Society of Mechanical Engineers (ASME), Ocean Engineering Division
- Institute of Electrical and Electronics Engineers (IEEE), Ocean Instrumentation Division
- Marine Technology Society (MTS)
- Offshore Mechanics and Arctic Engineering (OMAE)
- Shore and Beach Association
- Society of Naval Architects and Marine Engineers (SNAME)

- Western Dredging Association (WEDA)
- The Coastal Society
- International Society of Offshore and Polar Engineering (ISOPE)

JOURNALS AND MAGAZINES

A source of information for ocean engineers is located in technical journals and trade magazines, and a short list of titles for the current publications are:

Journals

- Coastal Engineering
- Journal of Applied Ocean Research
- Journal of Engineering Education
- Journal of Coastal Engineering Research
- Journal of Fluid Mechanics (JFM)
- Journal of Geophysical Research (JGR)
- Journal of the International Society of Offshore and Polar Engineering (ISOPE)
- Journal of Physical Oceanography (JPO)
- Journal of Waterways, Port, Coastal and Ocean Engineering (ASCE)
- Marine Technology Journal (SNAME)
- Marine Technology Society Journal (MTS)
- Ocean Engineering, An International Journal
- Offshore Mechanics and Arctic Engineering Journal (OMAE)
- Terra et Aqua

Magazines

- Oil & Gas Journal
- Sea Technology
- Oceans
- Offshore Magazine
- Shore and Beach
- Underwater Systems Design
- World Dredging, Mining and Construction
- Dredging and Port Construction

CONFERENCES

The exchange of basic and applied research results is accomplished through annual, biannual and specialty conferences in the broad areas of ocean engineering, coastal engineering, offshore technology, dredging, naval architecture, and marine engineering. There is also an annual educational conference that includes education and teaching of ocean and marine engineering that is sponsored by the American Society for Engineering Education. Some of the more well established conferences for ocean engineers are:

- Oceans Conference (MTS, IEEE, annual)
- International Conference on Coastal Engineering (ICCE, every two years)
- Offshore Mechanics and Arctic Engineering Conference (OMAE, annual)
- Society of Naval Architects and Marine Engineers Conference(SNAME, annual)
- Specialty Conferences (ROV, Marine Instrumentation., Civil Engineering in the Oceans, Coastal Sediments, Coastal Practices, Dredging, etc.)
- World Dredging Congress (WODCON, every three years)
- American Society for Engineering Education Conference (ASEE, annual)
- Offshore Technology Conference (OTC, annual)
- Western Dredging Association Technical Conference and Texas A&M Dredging Seminar (WEDA, TAMU, annual)
- International Society of Offshore and Polar Engineers Conference (ISOPE, annual)
- Underwater Intervention (MTS, ADC, annual)

PROFESSIONAL REGISTRATION

Each state in the United States has a Board of Registration for Professional Engineers that reviews and approves applications for practicing engineers to become registered Professional Engineers. In most states engineering students take the Engineering in Training (EIT) or Fundamentals in Engineering (FE) examination during their senior year or after graduating from an engineering curriculum accredited by the Accreditation Bureau for Engineering and Technology (ABET). The engineer is then normally required to obtain four years of experience working for a professional engineer before sitting for the Professional Engineer Exam (PE). On satisfactory completion of the PE exam the engineer is given a professional license. Details and timing of exams varies slightly depending on the state in which the engineer is practicing. Engineers are also commonly registered in more than one state and reciprocity agreements exist between states.

The Engineering in Training or Fundamentals in Engineering exam is common to all engineering disciplines and generally covers the subjects of mathematics, physical and chemical sciences, basic engineering sciences and engineering economy. Examination review material is available in university bookstores, and review sessions are generally administered through the various engineering colleges. The professional engineering exam is discipline-related such as civil, mechanical, electrical, or other engineering disciplines. Ocean engineers usually must prepare for exams in civil or mechanical engineering. Similar to the EIT, review material is available through university bookstores and the State Board of Registration. Most states use examinations that are prepared by the National Council of Engineering Examiners (NCEE). Recently, there have been discussions concerning the development of examinations for naval architecture and ocean engineering, but currently there are no exams in these disciplines. All ocean engineering students are encouraged to become professional engineers.

EXAMPLE OCEAN ENGINEERING CURRICULUM

As an example, the Ocean Engineering degree program leading to a Bachelor of Science degree at Texas A&M University at College Station, Texas is illustrated in Table 1-3. Curricula and credit hours at other universities will differ to some degree. In the illustrated 137 semester credit hour curriculum, the first two years are typical of a basic engineering program. The third year begins with basic fluid dynamics and structural theory, and these are followed by ocean-related courses of ocean wave mechanics, physical oceanography, and advanced hydromechanics. The fourth year further emphasizes ocean engineering topics with courses related to the dynamics of offshore structures, coastal engineering, and underwater acoustics. The final semester includes a project (capstone) design course intended to provide the opportunity for the students to apply what they have learned to a real world ocean engineering design project with input from industry representatives. The final project results are orally presented, and a final design project report is submitted.

Table 1-3. Ocean Engineering curriculum at Texas A&M University from 1995-96 undergraduate catalog.

Course Title	Semester Credit Hours	Course Title	Semester Credit Hours
FIRST YEAR			
Fundamental of Chemistry II	3	Engr. Problem Solving & Computing	3
Fundamentals of Chemistry Lab	1	Engineering Mathematics II	4
Engineering Graphics	2	Engineering Physics (Mechanics)	4
Composition & Rhetoric	3	US History elective	3
US History elective	3	Social Science elective	3
Engineering Mathematics I	4	Physical Education	1
Physical Education	1		
SECOND YEAR			
Engineering Mathematics III	3	Engineering Mechanics of Materials	3
Engineering Mechanics I (statics)	3	Engineering Mechanics II (dynamics)	3
Introduction to Oceanography	3	Differential Equations	3
Engineering Physics (electricity & optics)	4	Introduction to Ocean Engineering	2
Political Science (US)	3	Political Science (Texas)	3
Physical Education	1	Social Science elective	3
		Physical Education	1
THIRD YEAR			
Fluid Dynamics	3	Computer Applications in Engineering	2
Fluid Dynamics Laboratory	1	Intro. to Geotechnical Engineering	3
Theory of Structures	3	Technical Writing	3
Geology for Civil Engineers	3	Ocean Engineering Wave Mechanics	3
Thermodynamics	3	Hydromechanics	3
Humanities elective	3	Introduction to Physical Oceanography	3
FOURTH YEAR			
Electrical Circuits & Instrumentation	4	Materials Engineering	3
Dynamics of Offshore Structures	3	Design of Ocean Engineering Facilities	4
Underwater Acoustics	3	Ocean Engineering Laboratory	1
Basic Coastal Engineering	3	Humanities elective	3
Seminar	1	Technical Electives	3
Technical Elective*	3	Technical Electives	3
*Selected from approved list of technical electives Total Semester Credit Hours = 137			

REFERENCES

Allmendinger, E. E., Editor. *Submersible Vehicle Systems Design.* Jersey City: The Society of Naval Architects and Marine Engineers, 1990.

Bachrach, A. J., B. M. Desiderati, and M. M. Matzen. *A Pictorial History of Diving.* San Pedro: Best Publishing Co., 1988.

Barltrop, N. D. P., and A. J. Adams. *Dynamics of Fixed Marine Structures.* Third Edition, Oxford: Butterworth-Heinemann, 1991.

Berteaux, H. O. *Coastal and Oceanic Buoy Engineering.* Woods Hole: H. O. Berteaux, 1991.

Chakrabarti, S. K. *Hydrodynamics of Offshore Structures.* Boston: Computational Mechanics Publications, 1994.

Dawson, T. H. *Offshore Structural Engineering.* Englewood Cliffs: Prentice-Hall, 1983.

Dean, R. G., and R. A. Dalrymple. *Water Wave Mechanics for Engineers and Scientists.* New Jersey: World Scientific, 1991.

Fischer, H. B., E. J. List, R. C. Y. Koh, J. Imberger, and N. H. Brooks. *Mixing in Inland and Coastal Waters.* New York: Academic Press, 1979.

Gerwick, B. C. *Construction of Offshore Structures.* New York: John Wiley & Sons, Inc., 1986.

Goodfellow & Associates. *Applications of Subsea Systems.* Tulsa: PennWell Publishing Co., 1990. Figure 1-3 reprinted with permission: "Source: Ron Goodfellow=C9s "Applications of Subsea Systems" Copyright PennWell Publishing, 1990."

Graff, W. J. *Introduction to Offshore Structures: Design, Fabrication, Installation.* Houston: Gulf Publishing Co., 1981.

Herbich, J. B. *Handbook of Coastal and Ocean Engineering.* Vol. I, II, and III. Houston: Gulf Publishing Co., 1990, 1990, and 1992.

Horikawa, K. *Nearshore Dynamics and Coastal Processes.* Tokyo: University of Tokyo Press, 1988.

Ippen, A. T. *Estuary and Coastline Hydrodynamics.* Iowa City: Iowa Institute of Hydraulic Research, 1966.

Kinsman, B. *Wind Waves: Their Generation and Propagation on the Ocean Surface.* New York: Dover Publications Inc., 1984.

Lewis, E. V. *Principles of Naval Architecture.* Second Revision, Vol. I, II, III. Jersey City: The Society of Naval Architects and Marine Engineers, 1988.

Marine Technology Society (MTS). *Operational Guidelines for Remotely Operated Vehicles.* Washington: Marine Technology Society, 1984.

Mather, A. *Offshore Engineering: An Introduction.* London: Witherby & Company Limited, 1995.

McClelland, B. and M. D. Reifel, *Planning and Design of Fixed Offshore Platforms.* New York: Van Nostrand Reinhold Company, 1986.

Mei, C. C. *The Applied Dynamics of Ocean Surface Waves.* Second Printing, New Jersey: World Scientific, 1992.

Meyers, J. *Handbook of Ocean and Underwater Engineering.* New York: McGraw-Hill, 1969.

Miller, J. W., and I. G. Koblick, *Living and Working in the Sea.* New York: Van Nostrand Reinhold Co., 1984.

National Oceanic and Atmospheric Administration (NOAA). *NOAA Diving Manual.* Washington: National Oceanic and Atmospheric Administration, October 1991.

Patel, M. H. *Dynamics of Offshore Structures*. London: Butterworths, 1989.

Sarpkaya, T., and M. Isaacson. *Mechanics of Wave Forces on Offshore Structures*. New York: Van Nostrand Reinhold, 1981.

U.S. Army Corps of Engineers (USACE). *Shore Protection Manual*. Vol. I & II. Washington: U.S. Government Printing Office, 1984. Figure 1-1 reprinted with permission: "Source: *Shore Protection Manual*. U.S. Army Corps of Engineers (USACE). Vol. I & II. Copyright U.S. Government Printing Office, 1984."

U.S. Navy (USN). *US Navy Diving Manual*. Washington: U.S. Government Printing Office, 1985.

Wiegel, R. L. *Oceanographical Engineering*. Englewood Cliffs: Prentice-Hall, Inc., 1965.

CHAPTER 2: THE OCEAN ENVIRONMENT

GENERAL

Almost three-fourths, 71 percent, of the Earth's surface is covered by oceans and the remainder, 29 percent, is land. The major ocean areas are: 1) Southern Ocean, 2) Atlantic Ocean, 3) Pacific Ocean, 4) Indian Ocean and 5) Arctic Ocean. If the Southern Ocean is considered as part of the Pacific Ocean, then the world ocean area is subdivided into the Pacific Ocean (46 %), Atlantic Ocean (23 %), Indian Ocean (20 %) and the others account for the remaining 11 %. Some examples of smaller bodies of water are: 1) Mediterranean Sea, 2) Caribbean Sea, 3) Sea of Japan, 4) Bering Sea, 5) North Sea, 6) Gulf of Mexico, and 7) Baltic Sea.

The average ocean depth is 3800 m (12,500 ft), and the maximum depth of 11,524 m (37,800 ft) occurs in the Mindanao Trench in the Pacific Ocean. In comparison, the average land elevation is 840 m (2760 ft) and the highest elevation is 8840 m (29,000 ft) at the top of Mount Everest. The distributions of depths in the oceans and elevations on land are illustrated in Figure 2-1 (Pickard and Emery 1990). The 8.4 % of the oceans that are shallower than 1000 m (3280 ft) and water depths shallower than 200 m (656 ft) consist of 5.3 % of the worlds ocean where a majority of offshore oil and gas platforms are found. In general, the land elevations are not as high as the elevations below sea level. Approximately, 52 % of the ocean depths are between 2000 and 6000 m (6560 and 19,680 ft).

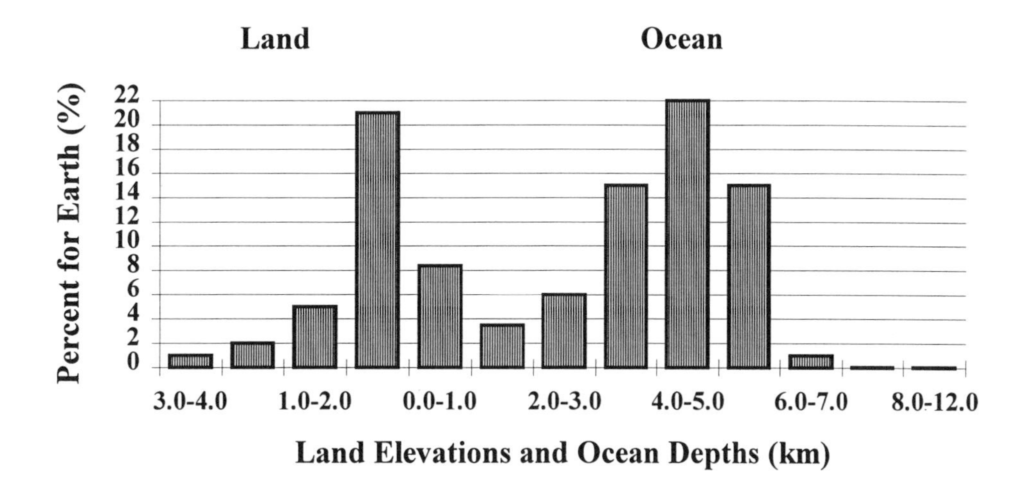

Figure 2-1. Distribution of depths in the ocean and elevations on land.

OCEAN FLOOR

There are mountains, valleys, and plains on the ocean bottom much like those found on land. The horizontal dimensions of the ocean are much larger than the vertical (depth). As a consequence, cross-sectional views of the ocean usually use a distorted scale. For example, the horizontal dimension is often scaled as 1 cm equals 100 km, and the vertical dimension is scaled as 1 cm equals 100 m. Using this procedure the slopes are greatly exaggerated. Figure 2-2 is an illustrative cross-section of the ocean floor whose main divisions are the shore, continental shelf, continental slope and rise, and the deep-sea floor. Definitions of these divisions or principle features are:

- Shore is the land mass close to the sea that is modified by sea action.
- Beach is the seaward limit of shore and extends from the highest to lowest tide levels.
- Continental Shelf extends seaward from shore with an average slope of 1:500. The outer limit occurs where slope is 1:20.
- Continental Slope extends seaward from the shelf edge to deep sea basin. Average slope is 1:20, and the average vertical dimension is 4000 m.
- Continental Rise is the lower portion of slope where it grades into the deep sea bottom.
- Deep sea bottom depths are 3000 to 6000 m and are in 74% of ocean basins, only 1% is deeper.
- Mid-Ocean Ridge runs near center of oceans.
- Seamounts are mountains in the oceans.
- Sills are ridges separating different water basins, e.g. fjord
- Trenches are the deepest features of the ocean floor that are very deep and narrow, e.g. Mariana Trench in the Pacific Ocean, Puerto Rico Trench in the Atlantic Ocean and Sunda Trench in the Indian Ocean.

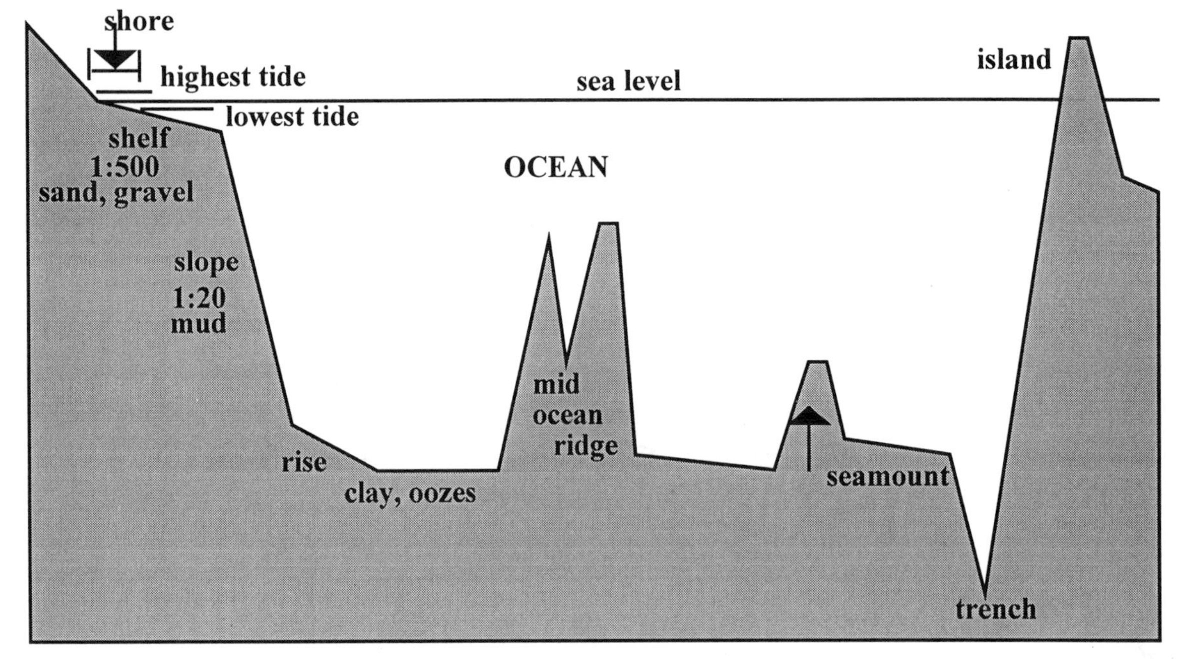

Figure 2-2. Principal features of the ocean floor.

Ocean bottom material on the continental shelf and slope generally comes from land via rivers or blown by wind. It generally consists of sands and gravel on the shelf and mud on the slope. Inorganic red clay and organic oozes are the two primary deep ocean sediments. The inorganic clays originate from land consisting of fine and volcanic material. Oozes come from the remains of living organisms, namely plankton. Calcareous oozes contain a large percentage of calcium carbonate originating from shells of animal plankton (zooplankton), and siliceous oozes have a large percentage of silica. Sediment deposition in the deep ocean is very slow with an average rate of 0.1 to 10 mm over 1000 years. Samples of deep ocean sediments are obtained using corers which are long steel pipes (2-30 m) that are dropped from the surface and penetrate the sea bottom. The core barrel is retrieved and the sediments are analyzed to determine engineering properties and reveal history of the deep ocean.

PHYSICAL PROPERTIES OF SEAWATER

Seawater contains a majority of the known elements, but the primary seawater elements are chlorine (55%), sulfate (7.7%), sodium (30.6%), magnesium (3.7%), and potassium (1.1%). The total amount of dissolved material in seawater is called the salinity, and it is defined as the total amount of solid materials in grams contained in a kilogram of seawater. The average salinity of ocean water is 35 grams of salts per kilogram of seawater. It is usually written as 35 o/oo or 35 ppt (35 part per thousand). Salinity is usually determined by measuring the conductivity of seawater which is related to salinity.

Temperature is the primary parameter in determining the density of seawater, and it is normally expressed in degrees C. Density ($\rho_{s,t,p}$) is the mass per volume of seawater and is a function of temperature, salinity, and pressure (depth). Open ocean values range from 1021.00 to 1070.00 kg/m^3. The density of seawater with a salinity of 35 o/oo and temperature of 10° C at atmospheric pressure is 1026.95 kg/m^3 ($\rho_{35,10,0}$ = 1026.95 kg/m^3). In oceanography, it is common practice to use the last four digits of the density and name it the quantity $\sigma_{s,t,p}$ that is defined as

$$\sigma_{s,t.p} = \rho_{s,t,p} - 1000 \qquad \textbf{2-1}$$

In most cases, the effect of pressure on density can be ignored and the quantity,

$$\text{sigma} - \text{t} = \sigma_t = \sigma_{s,t,0} \qquad \textbf{2-2}$$

is used. For example, the sigma-t value for the seawater density of 1026.95 is 26.95 ($\sigma_t = 26.95$). Values of σ_t are determined by knowing values of the seawater salinity and temperature and finding the associated sigma-t value in nomographs and tables published by the US Navy Oceanographic Office (1952). Empirical polynomial equations that are a function of salinity and temperature are also used to determine sigma-t. Millero and Poisson (1981) presented a new equation for the density of seawater as a function of salinity and temperature at standard atmospheric pressure, and this expression (Equation 2-3) for density is also given in Pond and Pickard (1983) as

$$\begin{aligned}\rho_{s,t.,0} = {} & 999.842594 + 6.793952\text{x}10^{-2}\,\text{T} - 9.095290\text{x}10^{-3}\,\text{T}^2 + 1.001685\text{x}10^{-4}\,\text{T}^3 \\ & - 1.120083\text{x}10^{-6}\,\text{T}^4 + 6.536322\text{x}10^{-9}\,\text{T}^5 + 8.24493\text{x}10^{-1}\,\text{S} - 4.0899\text{x}10^{-3}\,\text{TS} \\ & + 7.6438\text{x}10^{-5}\,\text{T}^2\,\text{S} - 8.2467\text{x}10^{-7}\,\text{T}^3\,\text{S} + 5.3875\text{x}10^{-9}\,\text{T}^4\,\text{S} - 5.72466\text{x}10^{-3}\,\text{S}^{3/2} \\ & + 1.0227\text{x}\,10^{-4}\,\text{TS}^{3/2} - 1.6546\text{x}10^{-6}\,\text{T}^2\,\text{S}^{3/2} + 4.8314\text{x}10^{-4}\,\text{S}^2\end{aligned} \qquad \textbf{2-3}$$

In oceanography and ocean engineering, lines of constant properties are called isoplethes. The common isoplethes are isotherms (constant temperature), isohalines (constant salinity), isopycnals (constant density) and isobaths (constant depth). Hydrographic measurements and surveys consist of measuring temperature, salinity, and depth, and these data are used to compute density. Results of these surveys are often illustrated as cross-sections of isotherms, isohalines, and isopycnals to show the physical characteristics of the water column.

Dissolved oxygen and nutrients (phosphate, nitrate, silicate, etc. ions) are also important seawater characteristics. However, these properties are non conservative which means biological processes may change their concentrations as the water masses move from one location to another. In contrast, the properties of salinity, temperature, and density are considered to be conservative below the sea surface which means that these quantities do not change, and as a consequence, water masses can be traced using these properties.

Sound velocity is another important property of seawater, and it is a function of salinity, temperature, and depth. Since visibility in the ocean is very restricted and electromagnetic waves are absorbed very quickly, sound waves are used extensively to "see" below the ocean surface and detect the sea bottom and other ocean features. Depth sounders, side scan sonars, search sonars, and seismic surveys are based on the transmission of sound waves through the ocean. The speed of sound in seawater with a salinity of 34.85 o/oo and 0 ^oC is 1445 m/s (4740 ft/s). A more detail discussion of underwater sound is contained in Chapter 8.

Visible light is also absorbed quickly by ocean waters, and it is the major source of heat energy that raises the water temperature. In the clearest ocean waters, only 22 % of the light is visible at 10 m (33 ft) depth and a mere 0.5 % is visible at 100 m (330 ft). In the case of turbid coastal waters, only 8 % is visible at 2 m (6.6 ft) depth, and no light is visible at 10 m (33 ft). Blue and green light penetrates the deepest in clear ocean water, and red and yellow light are absorbed quickest. In turbid coastal waters, the green and yellow light penetrate deepest.

OCEAN CURRENTS

Currents in the ocean are an important contributor to many physical, chemical, and biological processes that occur in the ocean environment. In Ocean Engineering, currents create forces on structures, vehicles, shorelines, and other systems which must be designed to withstand these forces. Newton's second law and the conservation of mass are important physical laws used to describe the movement of ocean waters. A first step in discussing the ocean currents is to describe the forces and accelerations included in Newton's second law, also known as the equations of motion. In vector form, Newton's second law is

$$\sum \vec{F} = m\vec{a} \qquad \textbf{2-4}$$

On the left side of Equation 2-4, the principal forces acting on ocean waters are due to pressure gradient (F_{pg}), coriolis (F_{cf}), gravity (F_g), friction (F_f), and centrifugal (F_c). The equation may be expressed in terms of force per unit mass and the rate of change of velocity as

$$\frac{\vec{F}_{pg}}{m} + \frac{\vec{F}_{cf}}{m} + \frac{\vec{F}_g}{m} + \frac{\vec{F}_f}{m} + \frac{\vec{F}_c}{m} = \frac{d\vec{V}}{dt} \qquad \textbf{2-5}$$

In component form, the x-direction is positive to the east, y-direction is positive to the north and z-direction is positive downward. The component velocities are u, v, and w respectively. The acceleration term is the sum of the local $(\partial/\partial t)$ and convective $(u\partial/\partial x + v\partial/\partial y + w\partial/\partial z)$ accelerations that is written in component form as

$$
\begin{aligned}
\text{x-component: } & \frac{du}{dt} = \frac{\partial u}{\partial t} + u\frac{\partial u}{\partial x} + v\frac{\partial u}{\partial y} + w\frac{\partial u}{\partial z} \\
\text{y-component: } & \frac{dv}{dt} = \frac{\partial v}{\partial t} + u\frac{\partial v}{\partial x} + v\frac{\partial v}{\partial y} + w\frac{\partial v}{\partial z} \\
\text{z-component: } & \frac{dw}{dt} = \frac{\partial w}{\partial t} + u\frac{\partial w}{\partial x} + v\frac{\partial w}{\partial y} + w\frac{\partial w}{\partial z}
\end{aligned}
\qquad \textbf{2-6}
$$

The pressure gradient term is the result of horizontal changes in pressure related to the high and low pressure areas in the ocean. The flow of ocean water due to these pressure differences is generally from a region of high pressure to low pressure and the magnitude of the current is related to the magnitude of the pressure gradient. The three components of the pressure gradient force per unit mass are

$$\left(\frac{F_{pg}}{m}\right)_x = -\frac{1}{\rho}\frac{\partial p}{\partial x}; \quad \left(\frac{F_{pg}}{m}\right)_y = -\frac{1}{\rho}\frac{\partial p}{\partial y}; \quad \left(\frac{F_{pg}}{m}\right)_z = -\frac{1}{\rho}\frac{\partial p}{\partial z} \qquad \textbf{2-7}$$

where the negative sign shows the flow is toward the low pressure area and ρ is the fluid density.

The Coriolis force is caused by the Earth's rotation, and it causes ocean waters to deflect to the right in the northern hemisphere and to the left in the southern hemisphere. The amount of deflection is a function of the flow speed and latitude location. The Coriolis force diminishes to zero when the fluid is at rest or at the equator, and the maximum deflection occurs at the north and south poles. This force is important in describing the large scale circulation in the oceans and atmosphere. The component form of the Coriolis force is given as

$$\left(\frac{F_{cf}}{m}\right)_x = fv; \quad \left(\frac{F_{cf}}{m}\right)_y = -fu; \quad \left(\frac{F_{cf}}{m}\right)_z = 0 \qquad \textbf{2-8}$$

where $f = 2\Omega\sin\phi$, Ω is the Earth's angular velocity (7.29 x 10^{-5} rad/s), and ϕ is the latitude location in degrees. Ocean waters are also affected by gravitational attraction, and this gravitational force per unit mass acts in the z direction only and is expressed as

$$\left(\frac{F_g}{m}\right)_z = g = 9.81 \text{ m/s}^2 \qquad \textbf{2-9}$$

The Earth's rotation also results in a centrifugal force that acts outward from the axis of rotation of the earth. The vertical component is included in the gravity (z-component) term, and the horizontal component is neglected.

Friction tends to retard the movement of ocean waters and is shown as a stress term in the equations of motion. Turbulent eddies in the ocean are superposed on the mean flow and are the mechanism for spreading frictional effects through the fluid. The instantaneous velocities are a combination of the time averaged component and a turbulent component. Incorporating these velocities in the equations of motion results in additional terms known as Reynold's stresses. The details of the development of Reynold's stresses are discussed in many fluids and oceanographic references (e.g. Bishop 1984). These Reynold's stresses are related to an eddy viscosity coefficient (K) by Prandtl (1952), and the component frictional force per unit mass is expressed as

$$\left(\frac{F_f}{m}\right)_x = -\frac{1}{\rho}\left[K_x \frac{\partial^2 \bar{u}}{\partial x^2} + K_y \frac{\partial^2 \bar{u}}{\partial y^2} + K_z \frac{\partial^2 \bar{u}}{\partial z^2}\right]$$

$$\left(\frac{F_f}{m}\right)_y = -\frac{1}{\rho}\left[K_x \frac{\partial^2 \bar{v}}{\partial x^2} + K_y \frac{\partial^2 \bar{v}}{\partial y^2} + K_z \frac{\partial^2 \bar{v}}{\partial z^2}\right] \quad \textbf{2-10}$$

$$\left(\frac{F_f}{m}\right)_z = -\frac{1}{\rho}\left[K_x \frac{\partial^2 \bar{w}}{\partial x^2} + K_y \frac{\partial^2 \bar{w}}{\partial y^2} + K_z \frac{\partial^2 \bar{w}}{\partial z^2}\right]$$

The entire three components of the equations of motion are written as

$$\frac{\partial \bar{u}}{\partial t} + \bar{u}\frac{\partial \bar{u}}{\partial x} + \bar{v}\frac{\partial \bar{u}}{\partial y} + \bar{w}\frac{\partial \bar{u}}{\partial z} = -\frac{1}{\rho}\frac{\partial \bar{p}}{\partial x} + f\bar{v} + \frac{K_x}{\rho}\frac{\partial^2 \bar{u}}{\partial x^2} + \frac{K_y}{\rho}\frac{\partial^2 \bar{u}}{\partial y^2} + \frac{K_z}{\rho}\frac{\partial^2 \bar{u}}{\partial z^2}$$

$$\frac{\partial \bar{v}}{\partial t} + \bar{u}\frac{\partial \bar{v}}{\partial x} + \bar{v}\frac{\partial \bar{v}}{\partial y} + \bar{w}\frac{\partial \bar{v}}{\partial z} = -\frac{1}{\rho}\frac{\partial \bar{p}}{\partial y} - f\bar{u} + \frac{K_x}{\rho}\frac{\partial^2 \bar{v}}{\partial x^2} + \frac{K_y}{\rho}\frac{\partial^2 \bar{v}}{\partial y^2} + \frac{K_z}{\rho}\frac{\partial^2 \bar{v}}{\partial z^2} \quad \textbf{2-11}$$

$$\frac{\partial \bar{w}}{\partial t} + \bar{u}\frac{\partial \bar{w}}{\partial x} + \bar{v}\frac{\partial \bar{w}}{\partial y} + \bar{w}\frac{\partial \bar{w}}{\partial z} = -\frac{1}{\rho}\frac{\partial \bar{p}}{\partial z} + g + \frac{K_x}{\rho}\frac{\partial^2 \bar{w}}{\partial x^2} + \frac{K_y}{\rho}\frac{\partial^2 \bar{w}}{\partial y^2} + \frac{K_z}{\rho}\frac{\partial^2 \bar{w}}{\partial z^2}$$

The continuity equation is another important equation that is used in combination with the previous equations of motion to describe fluid motion of the oceans. The continuity equation written in three dimensions is

$$-\frac{1}{\rho}\frac{d\rho}{dt} = \frac{\partial u}{\partial x} + \frac{\partial v}{\partial y} + \frac{\partial w}{\partial z} \quad \textbf{2-12}$$

and for an incompressible fluid such as ocean water, the continuity equation reduces to

$$\frac{\partial u}{\partial x} + \frac{\partial v}{\partial y} + \frac{\partial w}{\partial z} = 0 \quad \textbf{2-13}$$

Geostrophic Current

Simplified forms of the equations of motion and continuity are commonly used to obtain some practical and useful results where the simplifying assumptions are valid. An example is to assume that the ocean waters are not accelerated and are frictionless. These assumptions result in a geostrophic current, and for this case the component equations of motion are

$$fv = \frac{1}{\rho}\frac{\partial p}{\partial x}; \qquad fu = \frac{1}{\rho}\frac{\partial p}{\partial y}; \qquad g = \frac{1}{\rho}\frac{\partial p}{\partial z} \quad \textbf{2-14}$$

The horizontal components are a balance between the pressure gradient and the Coriolis force. These equations are commonly used to estimate ocean currents based on horizontal pressure distributions. As an illustration, consider a small volume of ocean water that is at rest in the northern hemisphere (Figure 2-3). The water volume accelerates as it moves from a high pressure region to a low pressure region, and it deflects to right. Finally, it is parallel to the lines of constant pressure (isobars) that define the horizontal pressure field. The vertical component is a balance between gravity and the vertical pressure gradient and is called hydrostatic equilibrium. When fluid acceleration and friction are relatively small, the geostrophic and hydrostatic equations are valuable tools for determining large scale ocean currents.

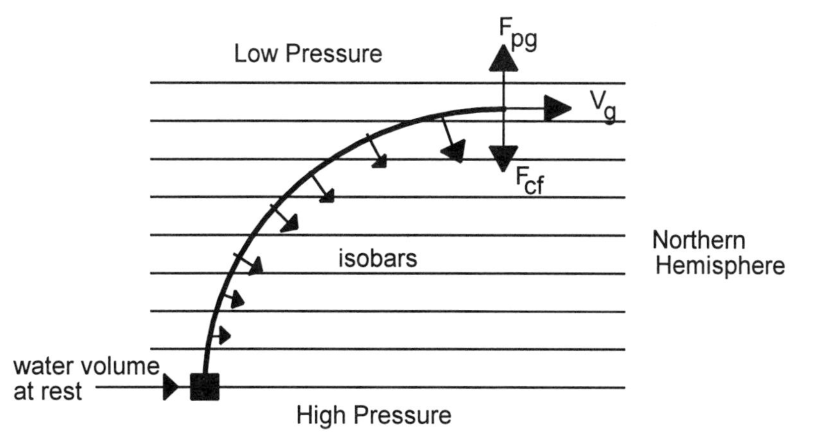

Figure 2-3. Illustration of geostrophic current (V_g) in the northern hemisphere.

Large scale ocean currents such as the Gulf Stream current can be evaluated when the horizontal pressure field is known. The pressure field may be determined from shipboard temperature and salinity measurements or by remote sensing techniques. For example, if the pressure change (Δp) across the Gulf Stream is the result of an increase in sea level (Δz) of 1 m over a distance across the Gulf Stream of 100 km (Δx), then the geostrophic current can be estimated as follows. The hydrostatic equation is

$$\Delta p = \rho g \Delta z \qquad \textbf{2-15}$$

and the geostrophic current (V_g) is determined from the geostrophic equation

$$V_g = \frac{1}{\rho f}\frac{\Delta p}{\Delta x} \qquad \textbf{2-16}$$

Substitution of Equation 2-15 into 2-16 and using appropriate values for Δx, Δz, g, and f yields

$$V_g = \frac{g}{f}\frac{\Delta z}{\Delta x} = \frac{(980\ \text{cm/s})\left(10^2\ \text{cm}\right)}{\left(10^{-4}\ \text{rad/s}\right)\left(10^7\ \text{cm}\right)} \cong 100\ \text{cm/s} \qquad \textbf{2-17}$$

Thus, the Gulf stream current in this example is approximately 100 cm/s and is in the positive y-direction (north) parallel to the lines of constant pressure.

Ekman Current

The Ekman current is based on the studies of upwelling by Ekman (1905), and it is the current resulting from a balance between Coriolis and surface wind frictional drag forces. Ekman current decays exponentially with depth and turns to the right with increasing depth in the northern hemisphere. Assuming infinitely deep and homogeneous water, the component form of the equations of motion developed by Ekman are

$$-fv = \frac{K_z}{\rho}\frac{d^2u}{dz^2}; \qquad fu = \frac{K_z}{\rho}\frac{d^2v}{dz^2} \qquad \textbf{2-18}$$

Two boundary conditions are used to solve this system of second order differential equations. First, the surface wind stress components (τ_x and τ_y) at the surface (z = 0) are assumed as

$$\tau_x = -K_z\frac{du}{dz}; \qquad \tau_y = -K_z\frac{dv}{dz} \qquad \textbf{2-19}$$

and the current is assumed to vanish (u = v = 0) at the bottom ($z = \infty$). The solution for these equations is named the Ekman spiral because the currents form a decaying spiral with increasing depth. Bishop (1984) shows the solution as

$$u = U_o\, e^{\left(-\frac{\pi z}{D}\right)} \cos\left(45^o - \frac{\pi z}{D}\right)$$
$$v = U_o\, e^{\left(-\frac{\pi z}{D}\right)} \sin\left(45^o - \frac{\pi z}{D}\right) \qquad \textbf{2-20}$$

Therefore, the Ekman current decays exponentially from its surface value (U_o) and rotates to the right as the depth increases in the northern hemisphere. In the southern hemisphere, the current rotates to the left. The surface current is 45° to the right of the surface wind direction in the northern hemisphere. The surface current magnitude is computed by

$$U_o = \frac{\tau_y}{\sqrt{\rho K_z f}} \qquad \textbf{2-21}$$

where the coordinate axis is oriented such that the wind is in the y-direction. The depth of frictional influence is computed from

$$D = \pi\sqrt{\frac{2K_z}{\rho f}} \qquad \textbf{2-22}$$

TIDES

The periodic rising and falling of the water that results from the gravitational attraction of the moon and sun and other astronomical bodies acting upon the rotating Earth is known as the tide. An ebb tide is the period of tide between high water and succeeding low water (falling tide), and it has an associated ebb current that is a tidal current away from the shore. A rising tide is a flood tide that is the period between low water and succeeding high water, and the associated flood current is a tidal current toward the shore. The depths on most navigational charts usually show the values for mean low water (MLW), but other reference datums may be used, so it is wise to check navigational charts to determine their reference datum. The difference in height between consecutive high and low waters is the tidal range. The tidal period is the interval of time between two consecutive like phases of the tide.

The earth has unequal diameters measured between the poles and at the equator with the equatorial diameter being the larger, and it completes one elliptical orbit around the sun in one year (365 days). The moon has an elliptical orbit about the earth with the closest point of approach being the perigee and the farthest point being the apogee. Additionally, the plane of the moon's orbit is inclined to that of the earth, and the axis of rotation for the earth is also inclined to the plane of the earth's solar orbit. These factors are part of over 150 complicated interacting factors that are needed to predict the tides precisely. However, many of these factors are small and can be neglected for many common purposes.

Tides are classified as semi-diurnal, diurnal, and mixed as shown in Figure 2-4. A semi-diurnal tide has two high and two low waters in a tidal day, and a diurnal tide has one high and one low water in a tidal day. The mixed tide is a combination of diurnal and semi-diurnal tides and is characterized by a large inequality in either the high or low water heights, with two high and two low waters usually occurring each tidal day. A tidal day is the time of the rotation of the Earth with respect to the Moon, approximately 24.84 hours and is also called the lunar day.

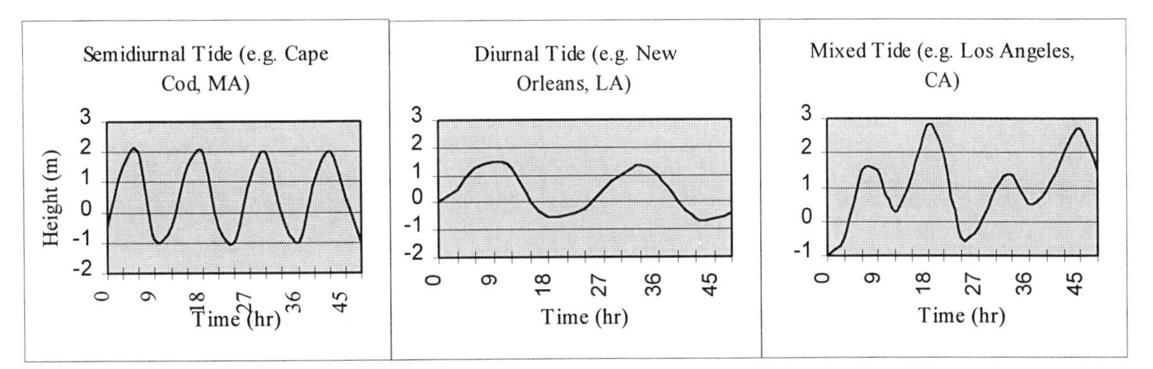

Figure 2-4. Examples of semi-diurnal, diurnal and mixed tides.

In 1687, Sir Isaac Newton proposed the theory of universal gravitation that indicated the tides are bulges of land and water resulting from the gravitational pull of the moon and sun on the earth. In Newton's equilibrium tide theory, water is assumed to cover the earth entirely at a uniform depth. The gravitational attraction of the moon and sun vary directly with their mass and inversely with the square of the distance from earth. Water bulges on the side of the sun due to gravity, and a similar bulge occurs on the opposite side of the earth as a result of the balancing force that keeps the earth in orbit around the sun. This balancing force is called the centrifugal force, and it is equal in magnitude and opposite in direction to the force of gravity. This equilibrium tide concept is illustrated in Figure 2-5 which shows the effect of the position of the earth, sun, and moon on ocean tides.

When tides occur in rivers, bays, and coastal inlets, the tidal flow is restricted by natural or manmade boundaries. In these locations the tidal currents are reversing which means that the tidal currents are in one direction, come to slack, and then reverse their direction in one tidal period. The speed of the flow varies between zero (slack) to a maximum speed at a time approximately half way between slack conditions. Tidal currents can be very strong, and examples are currents near 3.1 m/s (6 kts) found near San Francisco and as high as 5.2 m/s (10 kts) in other places. Ships entering these areas must be able to adjust their course in order to counteract the effects of these currents. Tidal currents are found to be rotary in nature in wide coastal or bay areas where the water movement is not restricted by barriers. In a tidal period these rotary tidal currents move through all compass directions and vary in magnitude with two maximum speeds in nearly opposite directions.

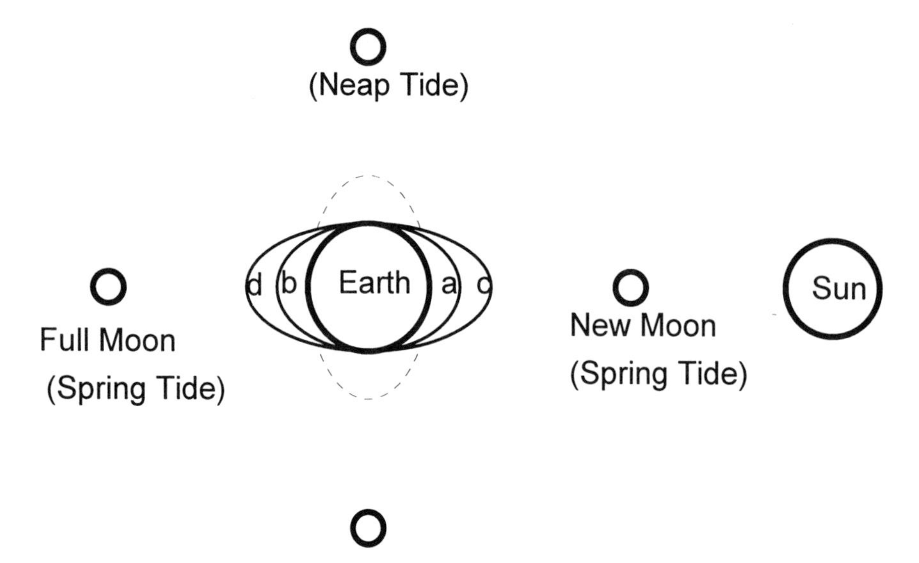

Figure 2-5. Equilibrium tide theory. Solar tides are gravitational (a) and centrifugal (b) and lunar tides are also gravitational (c) and centrifugal (d).

The equilibrium theory of tides gives a simplified explanation of tides, but the actual tides are much more complicated. In addition to the sun and moon, the tides are affected by other planets. Tidal predictions are based upon tidal measurements around the world and the National Ocean Survey (NOS) publishes the tide tables (NOS 1996) from which predictions of tides at various US locations can be determined. These tables are published each year.

OCEAN WAVES

Engineering systems that are designed to operate in the ocean must withstand the forces exerted by ocean waves, and consequently, ocean engineers must understand the physical processes and theories describing wave motion. The forces related to wind, currents, storm surges and ice are often less important than those due to waves. As waves propagate over the ocean surface, they eventually impact offshore platforms, subsea systems, coastal protection structures and the shoreline that must absorb, reflect, or dissipate the wave energy. Ocean engineers must design systems such that the wave forces do not cause it to fail.

The distribution of ocean surface wave energy, or wave energy spectrum, is illustrated in Figure 2-6 for the range of wave periods and frequencies found in the ocean. The primary wave generating forces (wind, storms, seismic, moon, and sun) and restoring forces (gravity, Coriolis, and surface tension) for the different ranges (period, frequency) of the spectrum are also indicated. The wind generated waves for periods spanning 1 to 30 s are the most important for ocean engineering design. These wind generated waves are classified as sea when they are under the influence of wind and as swell when the wind effect is very small.

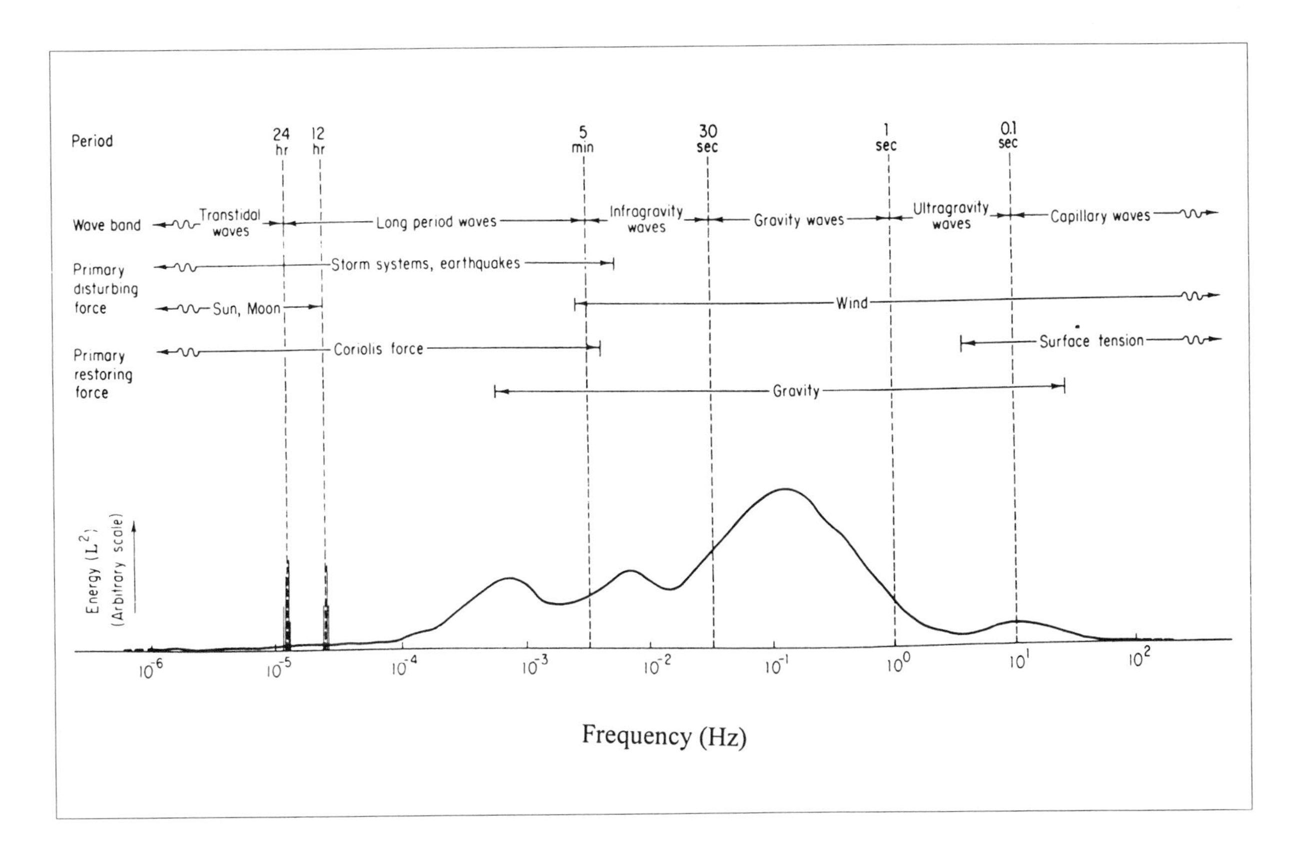

Figure 2-6. Approximate distribution of ocean surface wave energy. Reprinted with permission from Kinsman, 1965, *Wind Waves: Their Generation and Propagation on the Ocean Surface*, (Full citing in references).

Ocean waves are very complex and many wave periods may be present at a given location and time. Shorter waves are commonly found superimposed on longer waves. In addition, the waves from different directions interact and cause wave conditions that are very difficult to describe mathematically. An important tool for the ocean engineer is a simple wave theory that describes a wave propagating at a uniform period and height over a constant depth bottom. Such a theory is called the small amplitude (linear, Airy) wave theory, and it was first presented by Airy (1845). A brief derivation, useful equations, and important properties of linear waves are described in this chapter. More complete treatments of linear wave theory and more advanced wave theories, such as Stokes, stream function, and Cnoidal, are found in the texts of Wiegel (1964), Kinsman (1965), Ippen (1966), Le Mehaute (1969), McCormick (1973), Sorenson (1978), Sarpkaya and Isaacson (1981), Dean and Dalrymple (1984), USACE (1984), Horikawa (1988) and Chakrabarti (1990).

Linear Wave Theory

The simplest ocean waves are two dimensional, and linear wave theory has been developed using a linearized free surface boundary condition, bottom boundary condition, potential flow theory, and irrotational flow. Subsequently, wave characteristics such as celerity, particle displacement, velocity, acceleration, and pressure are determined from the velocity potential and used to evaluate the effects of waves on structures that are fixed and floating in the ocean. The major assumptions made in the derivation of linear wave theory are:

- Fluid is homogeneous and incompressible,
- Coriolis effect is neglected,
- Fluid is inviscid (no viscosity),
- Surface tension forces are negligible,
- Flow is irrotational,
- Bottom is horizontal, stationary and impermeable,
- Waves are two dimensional (long crested),
- Waves do not interact with other water motions,
- Surface pressure is constant, and
- Wave amplitude is small compared to the wavelength and water depth.

Figure 2-7 illustrates the characteristics of a typical sinusoidal progressive waveform. A rectangular coordinate system (x, z) is used with the origin at the still water level (SWL). The wave moves to the right with a celerity (C), period (T), and length (L) in a water depth (d). The wave height (H) is defined as the vertical distance between the wave crest and trough and is twice the amplitude of the wave. The water particles move in a clockwise orbit as the wave progresses from left to right, and their position at any instant is given by the horizontal and vertical coordinates (ξ, ε) referenced to the center of the orbit. Similarly, the horizontal and vertical components of water particle velocity (u, w) are defined, and the water surface elevation above the SWL at any point is defined as η. The water surface elevation profile is a function of space (x) and time (t).

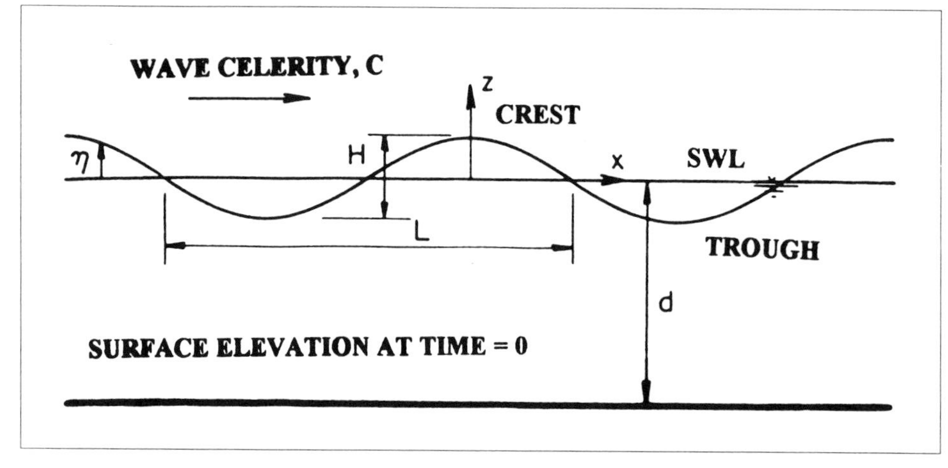

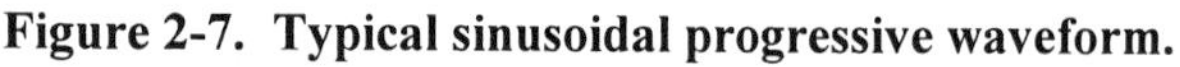

Figure 2-7. Typical sinusoidal progressive waveform.

For an ideal irrotational fluid, the velocity potential (ϕ) must satisfy the Laplace equation

$$\frac{\partial^2 \phi}{\partial x^2} + \frac{\partial^2 \phi}{\partial z^2} = 0 \qquad \textbf{2-23}$$

and the water surface elevation (η) is assumed to be a function of position (x) and time (t)

$$\eta = \frac{H}{2}\cos\left(kx - \omega t\right) \qquad \textbf{2-24}$$

where k is the wave number (k = 2π/L) and ω is the wave angular frequency (ω = 2π/T).

The bottom boundary condition requires no flow normal to the bottom and is written as

$$w = \frac{\partial \phi}{\partial z} = 0 \qquad \textbf{2-25}$$

at z = 0. The Bernoulli equation for irrotational flow may be expressed as

$$\frac{1}{2}\left(u^2 + w^2\right) + gz + \frac{p}{\rho} + \frac{\partial \phi}{\partial t} = 0 \qquad \textbf{2-26}$$

where g is gravitational acceleration (32.2 ft/s^2 or 9.81 m/s^2), ρ is fluid density, and p is pressure. Equation 2-26 is then applied at the free surface where the gauge pressure is zero and then linearized by neglecting the nonlinear velocity squared. The result is

$$z = \eta = -\frac{1}{g}\frac{\partial \phi}{\partial t} \qquad \textbf{2-27}$$

which is the water surface boundary condition. If the wave amplitude is small, which is assumed for linear wave theory, then Equation 2-27 is approximately the same at the still water level and is expressed as

$$\eta = -\frac{1}{g}\frac{\partial \phi}{\partial t} \qquad \textbf{2-28}$$

The velocity potential is assumed to be a sinusoidal function of position and time in the form

$$\phi = A\sin(kx - \omega t) \qquad \textbf{2-29}$$

where A is a function of z only. Substituting Equation 2-29 into the Laplace equation (Equation 2-23) and solving the resulting partial differential equation yields

$$\phi = \frac{H}{2}\frac{g\cosh k\left(z + d\right)}{\omega \cosh kd}\sin\left(kx - \omega t\right) \qquad \textbf{2-30}$$

Considering the water particle at the water surface, the vertical component of velocity (w) on the surface is expressed as $w = \partial \eta / \partial t$ and η is given by Equation 2-28. Therefore,

$$w = -\frac{1}{g}\frac{\partial^2 \phi}{\partial t^2} \qquad \textbf{2-31}$$

and from the definition of velocity potential, the vertical velocity is

$$w = \frac{\partial \phi}{\partial z} \qquad \textbf{2-32}$$

Combining Equations 2-31 and 2-32 yields

$$\frac{\partial^2 \phi}{\partial t^2} + g\frac{\partial \phi}{\partial z} = 0 \qquad \textbf{2-33}$$

Using the derived expression for velocity potential (Equation 2-30) and solving Equation 2-33 yield the linear dispersion relationship

$$\omega^2 = gk\tanh(kd) \tag{2-34}$$

Recalling the definitions of ω and k yield

$$\frac{\omega}{k} = \frac{L}{T} = C \tag{2-35}$$

and substituting into Equation 2-12 gives

$$C = \sqrt{\frac{gL}{2\pi}\tanh\frac{2\pi d}{L}} \tag{2-36}$$

This equation is the fundamental relationship between celerity, wave length, and water depth, and it should be observed that the celerity is not a function of wave height according to linear wave theory. In terms of the wave period, Equation 2-36 is written as

$$C = \frac{gT}{2\pi}\tanh\left(\frac{2\pi d}{L}\right) \tag{2-37}$$

and the wave length can be expressed as

$$L = \frac{gT^2}{2\pi}\tanh\left(\frac{2\pi d}{L}\right) \tag{2-38}$$

The above equation presents some difficulty in its solution because the wave length (L) appears on both sides of the equation. An iterative solution is one way to solve the equation. However, Eckart (1952) developed an approximate relationship

$$L \approx \frac{gT^2}{2\pi}\sqrt{\tanh\left(\frac{4\pi^2}{T^2}\frac{d}{g}\right)} \tag{2-39}$$

that is within 5 percent and considered sufficient for many engineering applications. The maximum error of 5 % occurs when $2\pi d/L \approx 1$.

Water waves are classified as deep, intermediate, or shallow water depending on the relative depth (d/L) and other dimensionless ratios such as $2\pi d/L$, d/gT^2, and tanh $(2\pi d/L)$ that are tabulated in Table 2-1. For deep water waves, tanh $(2\pi d/L) \cong 1$, and therefore, the deep water wave length (L_o) and wave celerity (C_o) from Equations 2-14 and 2-15 are written as

$$C_o = \frac{gT}{2\pi} = 1.56T(\text{m/s}) \text{ or } 5.12T(\text{ft/s}) \tag{2-40}$$

$$L_o = \frac{gT^2}{2\pi} = 1.56T^2(\text{m}) \text{ or } 5.12T^2(\text{ft}) \tag{2-41}$$

In the case of shallow water waves, tanh $(2\pi d/L) \cong 2\pi d/L$, and the wave celerity and wave length are

$$C = \sqrt{gd} \tag{2-42}$$

$$L = T\sqrt{gd} = CT \qquad \textbf{2-43}$$

Table 2-1. Classification of water waves.

Class	Dimensionless Ratio			
	$\frac{d}{L}$ (relative depth)	$\frac{2\pi d}{L}$	$\frac{d}{gT^2}$	$\tanh\left(\frac{2\pi d}{L}\right)$
Deep	> 1/2 (0.5)	> π	> 0.08	≅ 1
Intermediate	1/25(0.04) to 1/2 (0.5) or 1/20 (0.05) to 1/2 0.5)	1/4 (0.25) to π (3.14)	0.0025 to 0.08	$\tanh\left(\frac{2\pi d}{L}\right)$
Shallow	< 1/25 (0.04) or < 1/20 (0.05)	< 1/4 (0.25)	< 0.0025	$\cong \frac{2\pi d}{L}$

The difficulty in determining wave length for intermediate class waves led to the development of tables of values of d/L as a function of d/L_0 (Appendix Tables A-5 and A-6, Wiegel 1964, USACE 1984). Dividing Equation 2-37 by 2-41 or 2-38 by 2-41 yields

$$\frac{C}{C_o} = \frac{L}{L_o} = \tanh\left(\frac{2\pi d}{L}\right) \qquad \textbf{2-44}$$

Using the expression relating wave length to deep water wave length and multiplying by the water depth gives a useful relationship

$$\frac{d}{L_o} = \frac{d}{L}\tanh\left(\frac{2\pi d}{L}\right) \qquad \textbf{2-45}$$

that facilitates calculation of the wave length in any water depth when the deep water wave length is known. Tabulated values of d/L as a function of d/L_0 are contained in Appendix A (Tables A-5 and A-6), Wiegel (1964), and the Shore Protection Manual (USACE 1984).

Water particles under waves travel in orbits that are circular in deepwater and elliptical in intermediate and shallow water (Figure 2-8). Near the bottom in shallow and intermediate water, the elliptical orbits become very flat and the velocity is mostly horizontal.

Wave kinematics refers to the velocity and acceleration of the water particles under waves. The horizontal and vertical components (u, w) of water particle velocity are determined from the velocity potential which gives $u = \partial\phi/\partial x$ and $w = \partial\phi/\partial z$. Taking the partial derivative of Equation 2-30 with respect to x and z and using the dispersion relationship (Equation 2-34) yield the linear wave theory expressions for horizontal and vertical velocity components as

$$u = \frac{\pi H}{T}\frac{\cosh k(d+z)}{\sinh kd}\cos(kx - \omega t) \qquad \textbf{2-46}$$

and

$$w = \frac{\pi H}{T}\frac{\sinh k(d+z)}{\sinh kd}\sin(kx - \omega t) \qquad \textbf{2-47}$$

These equations show the particle velocities consist of a particle speed (πH/T) term, hyperbolic decay term depending on depth, and a phase term that is a function of position and time.

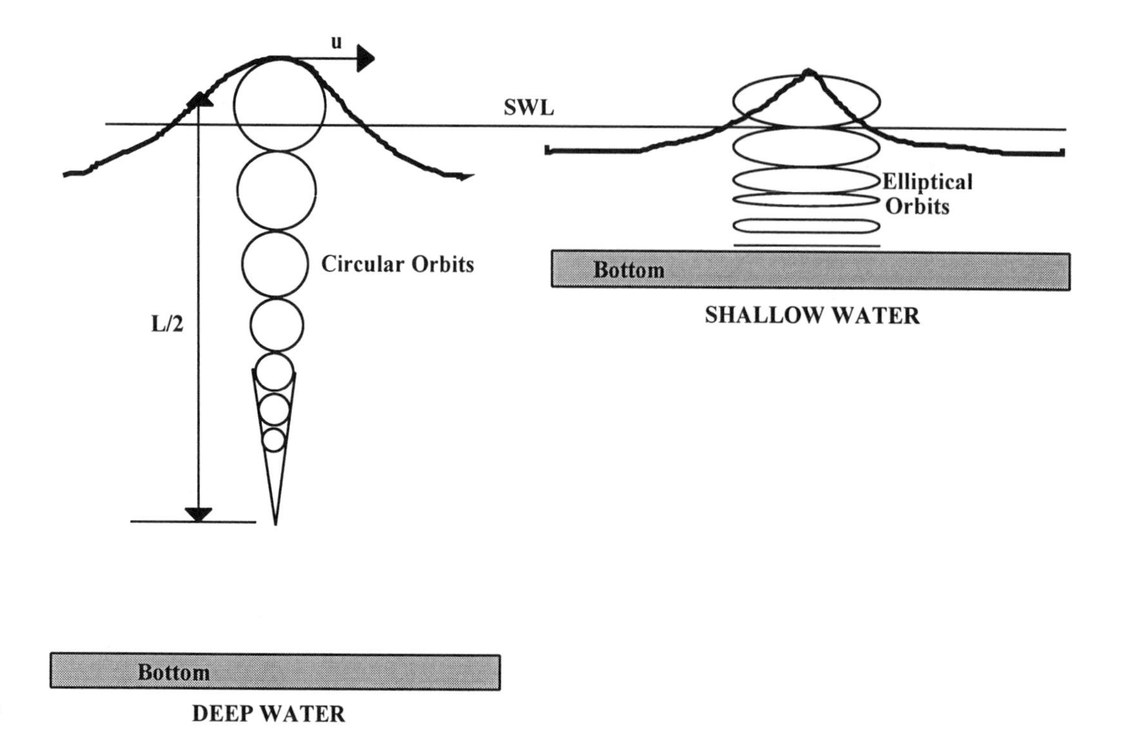

Figure 2-8. Water particle orbits in deep and shallow water.

The acceleration of a particle consists of local and convective acceleration terms, and for the horizontal component the equation is given as

$$a_x = u\frac{\partial u}{\partial x} + w\frac{\partial u}{\partial z} + \frac{\partial u}{\partial t} \qquad \textbf{2-48}$$

where the first two terms are the convective acceleration and the last term is the local acceleration. In keeping with the small amplitude assumption, the convective terms are considered negligible and the linear wave theory acceleration is determined by partial differentiation of the horizontal and vertical velocity components (Equations 2-46 and 2-47) with respect to time. The result is

$$a_x \cong \frac{\partial u}{\partial t} = \frac{2\pi^2 H}{T^2}\frac{\cosh k(d+z)}{\sinh kd}\sin(kx - \omega t) \qquad \textbf{2-49}$$

and

$$a_z \cong \frac{\partial w}{\partial t} = -\frac{2\pi^2 H}{T^2}\frac{\sinh k(d+z)}{\sinh kd}\cos(kx - \omega t) \qquad \textbf{2-50}$$

Notice that the particle acceleration has a 90° phase difference from the respective velocity components.

The size of the particle orbits depends on the particle displacements. Since the horizontal and vertical particle displacements (ξ, ε) are related to the respective velocity components ($u = \partial\xi/\partial t$ and $w = \partial\varepsilon/\partial t$), the displacements are

$$\xi = \int u\,dt = -\frac{H}{2}\frac{\cosh k(d+z)}{\sinh kd}\sin(kx-\omega t) \quad \textbf{2-51}$$

and

$$\varepsilon = \int w\,dt = \frac{H}{2}\frac{\sinh k(d+z)}{\sinh kd}\cos(kx-\omega t) \quad \textbf{2-52}$$

The pressure variation under waves is another important consideration. A linear wave theory expression for pressure is obtained by substituting the velocity potential (Equation 2-30) into the linearized Bernoulli equation (Equation 2-26) that yields

$$p = -\rho g z + \frac{\rho g H}{2}\frac{\cosh k(d+z)}{\cosh kd}\cos(kx-\omega t) \quad \textbf{2-53}$$

This shows the pressure is a combination of the normal hydrostatic pressure and a dynamic pressure due to the motion of the water particles. Under the crest the water particles are accelerating downward and the hydrostatic pressure is increased by the dynamic pressure, and the reverse is true under the trough. Wave height instruments are frequently placed on the sea floor to measure the pressure variation in order to evaluate the wave height and period conditions on the surface. However, the instrument must be at a depth of less than half the wave length in order to measure the dynamic pressure fluctuations due to the surface waves.

In the case of shallow water, the hyperbolic expressions in linear wave theory can be simplified as

$$\frac{\cosh k(d+z)}{\sinh kd} \approx \frac{1}{kd} \quad \text{and} \quad \frac{\sinh k(d+z)}{\sinh kd} \approx 1 + \frac{z}{d} \quad \textbf{2-54}$$

and for deep water the simplifications are

$$\frac{\cosh k(d+z)}{\sinh kd} \approx \frac{\sinh k(d+z)}{\sinh kd} \approx e^{kz} \quad \textbf{2-55}$$

A summary of linear wave theory expressions are tabulated in Table 2-2 for the relative water depths of shallow, intermediate, and deep water.

Additional water wave types include the rogue, internal, tsunami, storm surge, and ship waves. The rogue wave is a freak wave or unusually high wave caused by many waves traveling at different phase velocities and superimposing at a particular location. Waves traveling at the interface of water layers of slightly different density within the ocean are called internal waves. A tsunami is a large wave caused by ocean bottom seismic activity. The rise in water level due to tropical storms or hurricanes is known as the storm surge. The motion of a ship through the water causes ship waves. The most common wave height used in ocean engineering is called the significant wave height which is the average of the highest one third of the waves and is designated H_s or $H_{1/3}$.

Wave Energy and Power

The total energy in a wave is the sum of the kinetic and potential energy. Potential energy is the result of displacing the free surface, and kinetic energy is a consequence of the water particle movement throughout the fluid. Knowledge of the wave energy is needed to evaluate wave propagation, power necessary to produce waves, and the energy available for water wave energy extraction systems.

Table 2-2. Summary of linear wave theory relationships using θ as the phase angle (kx-ωt).

RELATIVE DEPTH	SHALLOW d/L <1/25 or 1/20	INTERMEDIATE 1/20 or 1/25 < d/L < 1/2	DEEP d/L >1/2
Elevation profile	$\eta = \frac{H}{2}\cos(kx - \omega t) = \frac{H}{2}\cos\theta$ 0 π/2 π 3π/2 2π phase angle (ϑ)		
Celerity	$C = \frac{L}{T} = \sqrt{gd}$	$C = \frac{gT}{2\pi}\tanh kd$	$C = C_o = \frac{L}{T} = \frac{gT}{2\pi}$
Wave Length	$L = T\sqrt{gd} = CT$	$L = \frac{gT^2}{2\pi}\tanh kd$	$L = L_o = \frac{gT^2}{2\pi} = C_o T$
Group Velocity	$C_g = C\sqrt{gd}$	$C_g = nC = \frac{1}{2}\left[1 + \frac{4\pi d/L}{\sinh(4\pi d/L)}\right]C$	$C_g = \frac{C}{2} = \frac{gT}{4\pi}$
Horizontal Particle Velocity	$u = \frac{H}{2}\sqrt{\frac{g}{d}}\cos\theta$	$u = \frac{\pi H}{T}\frac{\cosh k(d+z)}{\sinh kd}\cos\theta$	$u = \frac{\pi H}{T}e^{kz}\cos\theta$
Vertical Particle Velocity	$w = \frac{\pi H}{T}\left(1 + \frac{z}{d}\right)\sin\theta$	$w = \frac{\pi H}{T}\frac{\sinh k(d+z)}{\sinh kd}\sin\theta$	$w = \frac{\pi H}{T}e^{kz}\sin\theta$
Horizontal Particle Acceleration	$a_x = \frac{\pi H}{T}\left(\sqrt{\frac{g}{d}}\right)\sin\theta$	$a_x = \frac{2\pi^2 H}{T^2}\frac{\cosh k(d+z)}{\sinh kd}\sin\theta$	$a_x = 2H\left(\frac{\pi}{T}\right)^2 e^{kz}\sin\theta$
Vertical Particle Acceleration	$a_z = -\frac{2H\pi^2}{T^2}\left(1 + \frac{z}{d}\right)\cos\theta$	$a_z = -\frac{2\pi^2 H}{T^2}\frac{\sinh k(d+z)}{\sinh kd}\cos\theta$	$a_z = -2H\left(\frac{\pi}{T}\right)^2 e^{kz}\cos\theta$
Horizontal Particle Displacement	$\xi = -\frac{HT}{4\pi}\sqrt{\frac{g}{d}}\sin\theta$	$\xi = -\frac{H}{2}\frac{\cosh k(d+z)}{\sinh kd}\sin\theta$	$\xi = -\frac{H}{2}e^{kz}\sin\theta$
Vertical Particle Displacement	$\varepsilon = \frac{H}{2}\left(1 + \frac{z}{d}\right)\cos\theta$	$\varepsilon = \frac{H}{2}\frac{\sinh k(d+z)}{\sinh kd}\cos\theta$	$\varepsilon = \frac{H}{2}e^{kz}\cos\theta$
Pressure	$p = \rho g(h - z)$	$p = -\rho g z + \frac{\rho g H}{2}\frac{\cosh k(d+z)}{\cosh kd}\cos\theta$	$p = \rho g \eta e^{kz} - \rho g z$

Potential energy (E_p) is determined by evaluating a small column of fluid as illustrated in Figure 2-9 of mass (dm) relative to the bottom. The potential energy per unit width of wave crest over one wave length is expressed as

$$dE_p = g z_c \, dm \tag{2-56}$$

where z_c is the vertical distance to the centroid of the mass and g is the gravitational acceleration. Then for a progressive wave of wave height H, the average potential energy is evaluated by integrating over one wave length

$$E_p = \frac{1}{L}\int_0^L \rho g \frac{(d+\eta)^2}{2} dx \tag{2-57}$$

and after incorporating Equation 2-24, the result is

$$E_{p\,total} = \frac{\rho g d^2}{2} + \frac{\rho g H^2}{16} \tag{2-58}$$

By subtracting the potential energy without waves, the potential energy due to waves ($E_{p_{waves}}$) is obtained,

$$E_{p\,waves} = \frac{\rho g H^2}{16} \tag{2-59}$$

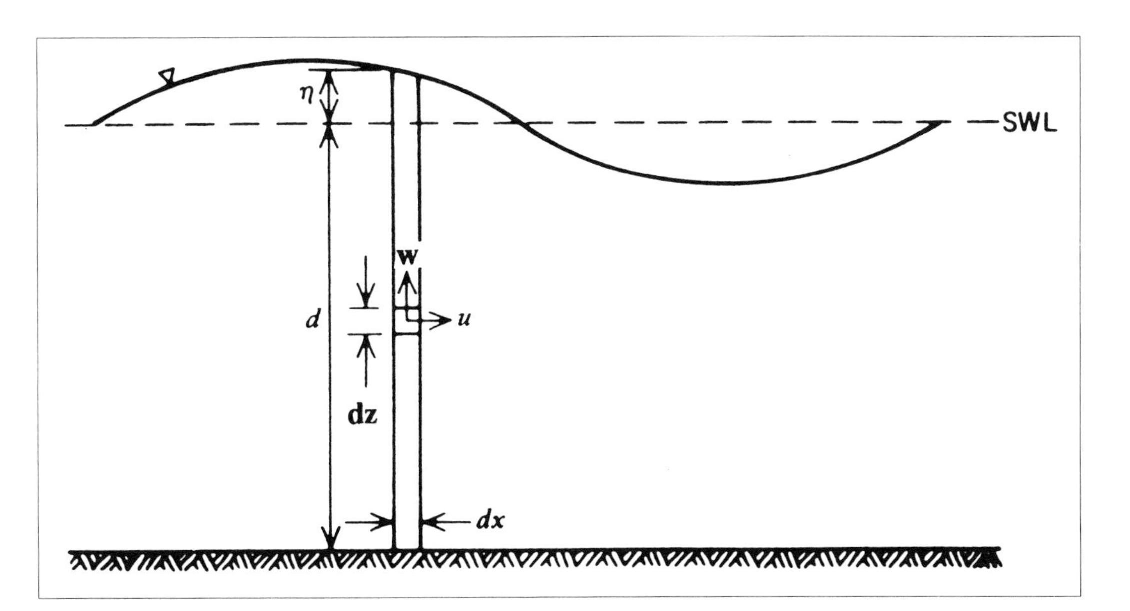

Figure 2-9. Schematic for evaluation of wave potential energy.

The movement of the water particles results in the kinetic energy of the waves. The average kinetic energy (E_k) is determined by integrating over the depth and averaging over a wave length,

$$E_k = \frac{1}{L}\int_0^L \int_{-d}^{\eta} \rho \frac{u^2 + w^2}{2} dz dx \quad \textbf{2-60}$$

Incorporation of the expressions for the horizontal and vertical orbital velocity from linear wave theory (Equations 2- 46 and 2- 47) gives

$$E_k = \frac{\rho g H^2}{16} \quad \textbf{2-61}$$

The total average energy per unit width of wave crest over the length of the wave is the sum of the kinetic and potential energy and is expressed as

$$E = E_p + E_k = \frac{\rho g H^2}{8} \quad \textbf{2-62}$$

and the total energy (E_L) in one wave length (L) per unit width of wave crest is

$$E_L = \frac{\rho g H^2 L}{8} \quad \textbf{2-63}$$

The energy per unit time or power (P) of the wave is expressed as

$$P = E_L nC = E_L C_g \quad \textbf{2-64}$$

where

$$n = \frac{1}{2}\left(1 + \frac{2kd}{\sinh 2kd}\right) \quad \textbf{2-65}$$

Wave Group Velocity

Waves are commonly formed as a small group of waves, and the group of waves travel at a group velocity that is normally less than the celerity of any of the individual waves. An example is waves generated by a storm. Prediction of the arrival time of the waves should be based on the group celerity, or group velocity, and not the celerity of any of the individual waves. The group celerity (C_g) is

$$C_g = nC \quad \textbf{2-66}$$

The expression for n is shown in Equation 2-65 above and varies from 0.5 for deep water to 1.0 for shallow water.

Wave Breaking

When the wave orbital velocity at the crest is equal to the wave celerity the wave becomes unstable and begins to break. Additionally, when a wave moves up a slope (shoals) the increasing crest velocity approaches the decreasing phase velocity which also results in breaking. The limiting condition for wave breaking in any water depth was developed by Miche (1944) and is

$$\left(\frac{H}{L}\right)_{max} = \frac{1}{7}\tanh kd = \frac{1}{7}\tanh\frac{2\pi d}{L} \quad \textbf{2-67}$$

For deep water waves, the wave breaks when the height is 1/7 of the wave length, and for the shallow water case, breaking occurs when

$$\left(\frac{H}{L}\right)_{max} = \frac{1}{7}\frac{2\pi d}{L} \; or \; \left(\frac{H}{d}\right)_{max} = 0.9 \quad \textbf{2-68}$$

Breaking waves are generally classified as spilling, plunging, collapsing, and surging as illustrated in Figure 2-10.

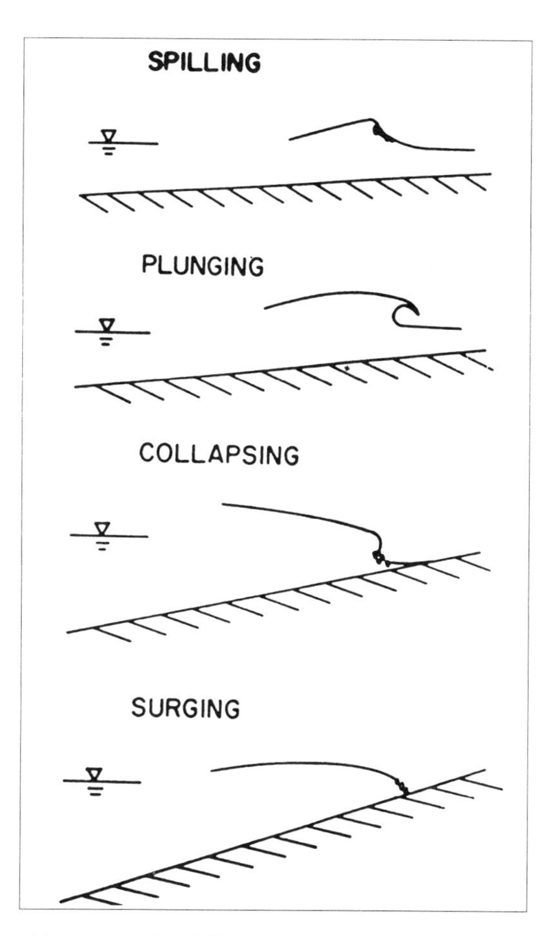

Figure 2-10. Illustration of breaking wave classifications. Reprinted with permission from Sarpkaya and Isaacson, 1981, *Mechanics of Wave Forces on Offshore Structures*, (Full citing in references.)

Advanced Wave Theories

Linear (Airy, small amplitude) wave theory has been described, but in some cases, the assumptions for linear wave theory are not satisfied and non-linearities are important. For these cases, advanced wave theories have been developed. More advanced texts on water wave theory such as Wiegel (1964), Kinsman (1965), Ippen (1966), Sarpkaya and Isaacson (1981), Dean and Dalrymple (1984), USACE (1984), Horikawa (1988) and Chakrabarti (1990) should be consulted. The most common advanced wave theories used in ocean engineering are Stokes 2nd - 5th order, Cnoidal, and Stream Function. The range of validity for the various periodic wave theories is illustrated in Figure 2-11. Linear random wave theory is commonly used for irregular waves, and a new hybrid wave theory has recently been developed (Zhang et al. 1993) which considers effects of the interactions of different wave components.

For higher order wave theories such as Stokes 2nd - 5th order, the expression for the wave elevation profile is

$$\eta = A\cos\theta + A^2 B_2 \cos(2\theta) + A^3 B_3 \cos(3\theta) + \ldots + A^n B_n \cos(n\theta) \qquad \textbf{2-69}$$

where A = H/2 for first and second orders but is less than H/2 for higher orders and B_n are special functions of the wave length and depth. Linear theory uses the first term only and the higher order Stokes theories use the respective higher order terms. For example, Stokes 2nd order theory uses the first two terms and Stokes 5th uses the first five terms. When use of these higher order theories is necessary, tables prepared by Skjelbreia (1959) and Skjelbreia and Hendrickson (1962) are useful in reducing error in the use of the higher order equations. Computer software are also available for linear wave theory, such as the ACES program distributed by the Corps of Engineers (USACE, 1992), and Stokes 3rd and 5th order theories.

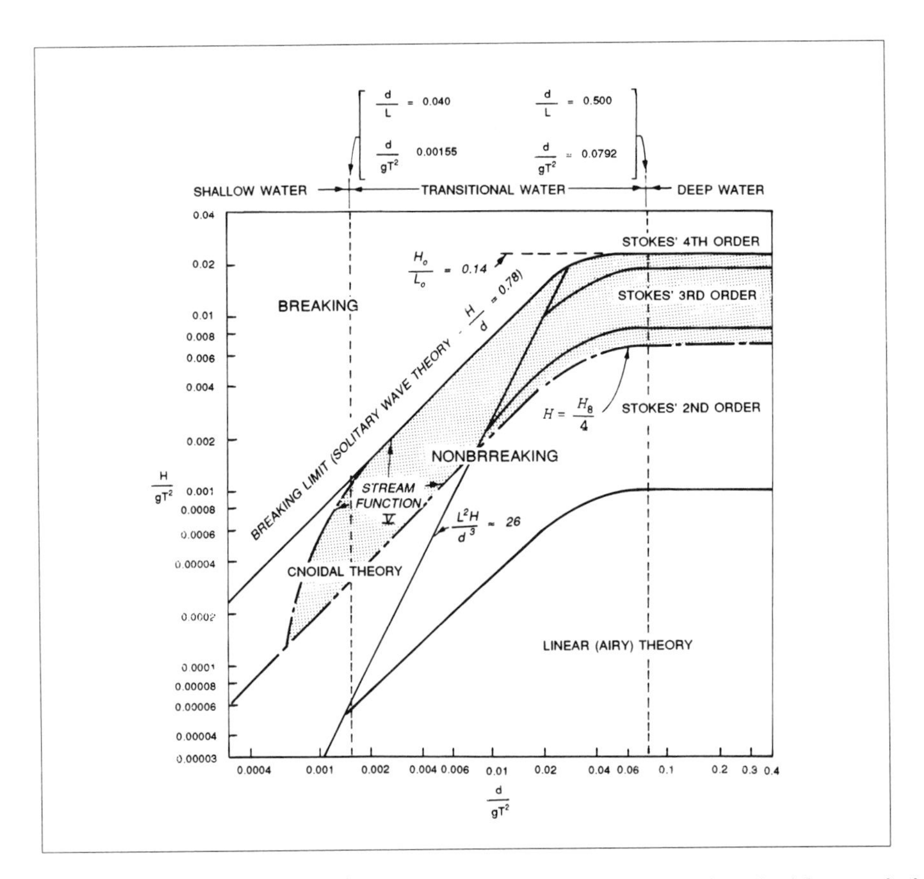

Figure 2-11. Suggested range of applicability of various wave theories. Reprinted with permission from Le Mehaute, 1969 "An Introduction to Hydrodynamics and Water Waves," Water Wave Theories, Vol. II, (Full citing in references).

A comparison of linear and Stokes 2nd order theory is illustrated in Figure 2-12 that shows elevation profile as a function of θ. The Stokes profile shows the crests are more peaked and higher, and the troughs are shallower and flatter than those in linear wave theory. Linear waves are symmetric about the SWL and Stokes waves are not. Water particles follow closed orbits under linear waves, and the orbits are open under Stokes waves that indicates a mass transport in the direction of wave propagation.

Water waves propagating in shallow water are often best described using Cnoidal wave theory originally developed by Korteweg and DeVries (1895). The theory is quite complicated and references such as Wiegel (1960, 1964) and Masch (1964) summarize the theory and present graphical results. The approximate valid range for Cnoidal wave theory is $d/L < 1/8$. Figure 2-13 shows the elevation profile for a Cnoidal wave that shows the crests are very peaked and the troughs are long and flat just below the SWL. It approaches a solitary wave when the wave period becomes very long, and as H/d becomes very small it approaches linear wave theory. A solitary wave is a wave of translation that is above the SWL and therefore has a crest but no trough. The ACES computer program (USACE 1992) computes characteristics of Cnoidal waves that have applications in the coastal zone.

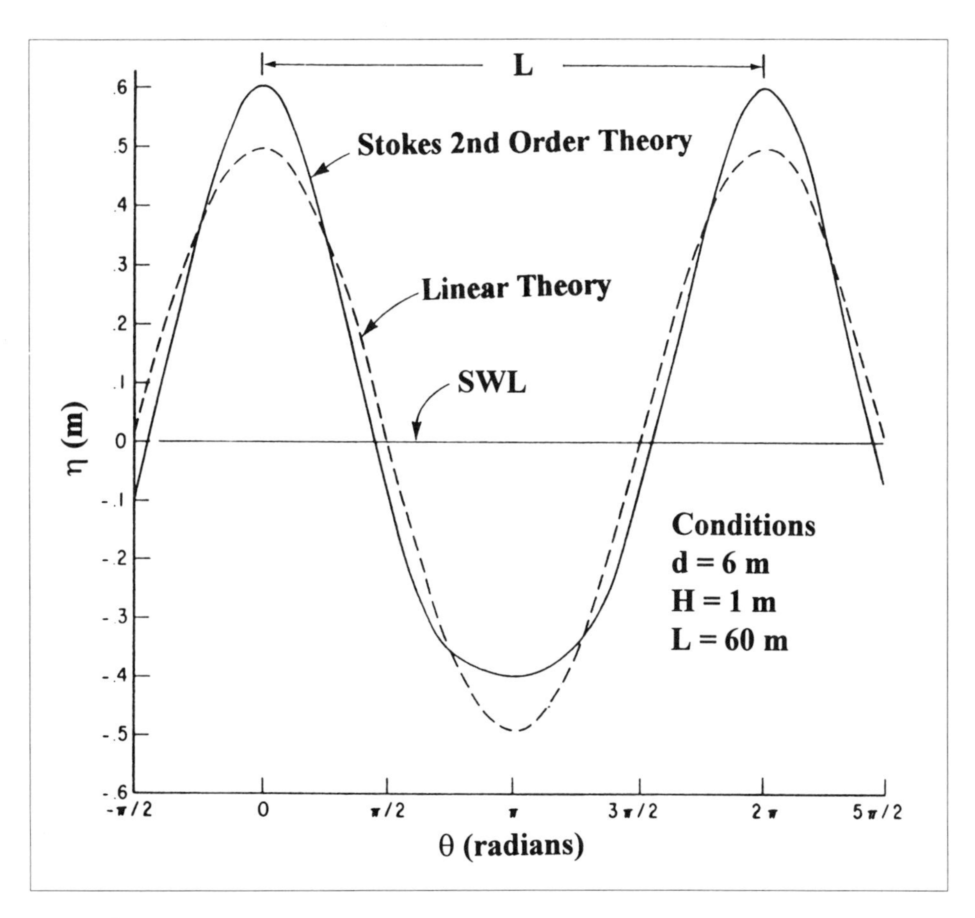

Figure 2-12. Comparison of linear and Stokes second order wave elevation profiles.

Stream function theory, developed by Dean (1965, 1967) and Monkmeyer (1970), is a nonlinear wave theory that uses sums of sine or cosine functions. The stream function results tend to more accurately predict wave phenomena observed in the laboratory than other theories. Dean (1974) presents tables and graphical representations for using stream function theory, and a computer program is also available to evaluate wave characteristics from stream function theory.

Ocean waves consist of many waves with different wave heights and periods which interact with one another. These component waves may also be traveling in different directions. Linear random wave theory has been developed to evaluate the wave characteristics in irregular sea conditions. Spectral analysis of the irregular wave elevation profile is used to decompose the wave profile into its wave components. Subsequently, linear wave theory is used to evaluate the wave characteristics (i. e. velocity, acceleration, pressure) for each component wave, and then each component wave is superposed linearly to determine the characteristics for the irregular wave. The interaction of the component waves is more complicated and in some cases the nonlinear interactions are important. A hybrid wave model has been developed (Zhang et al. 1993) that incorporates the important nonlinear interactions of irregular seas in determining the wave characteristics that must be used in the design process.

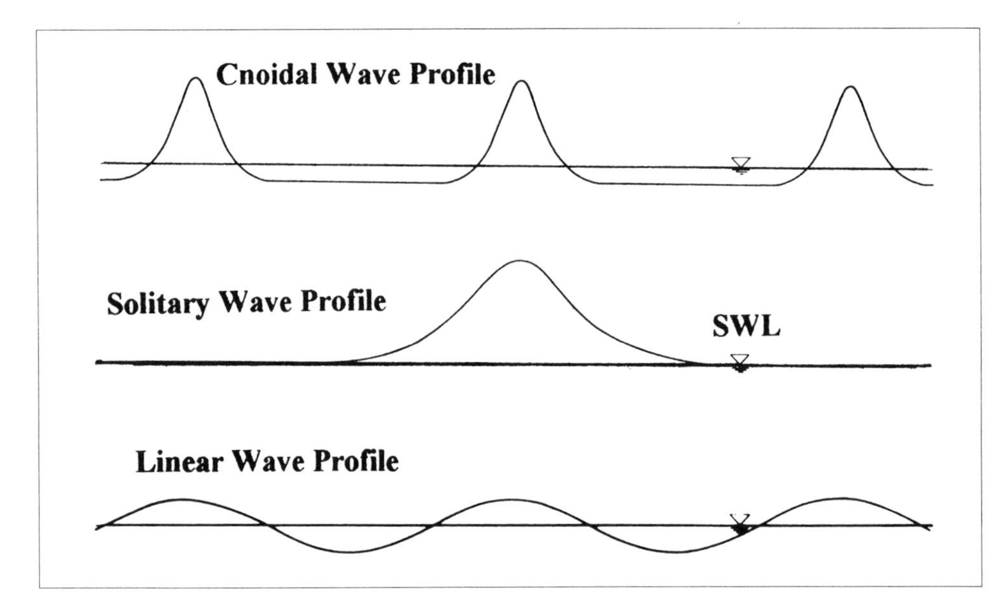

Figure 2-13. Example wave elevation profile for a Cnoidal wave and its limiting cases of solitary and linear waves. Reprinted with permission from Sarpkaya and Isaacson ,1981, *Mechanics of Wave Forces on Offshore Structures.* (Full citing in references.)

ICE

The consideration of the effects of ice on coastal and offshore facilities is important in temperate and arctic regions. Important ice parameters are thickness, concentration, persistence, and the variable mechanical properties. Forces resulting from ice interaction with structures include horizontal thrust, horizontal impact, vertical uplift, abrasion, and freeze-thaw damage. Information on ice and ice forces on structures can be found in various publications of the USACE Cold Regions Research and Engineering Laboratory (CRREL) and the International Association of Hydraulic Research (IAHR). The proceedings of the International Conferences on Port and Ocean Engineering under Arctic Conditions (POAC) is another source of current research related to ice effects and loads. Additional sources of information on ice and ice forces include API (1982), Gaythwaite (1990), Caldwell and Crissman (1983), Cammaert and Muggeridge (1988), Carstens, (1980), Chen and Leidersdorf (1988), Eranti and Lee (1986) Ingmanson and Wallace (1989), McClelland and Reifel (1986), Pickard and Emery (1990), Tsinker (1995), USACE (1982), and Wortley (1984).

Types of Ice

There are different types of ice such as fast ice, pack ice, smooth ice, ice islands, and icebergs that occur in the oceans. Ice in the sea forms differently than in freshwater lakes and rivers. Sea ice starts when crystals form a hard ice rind that is typically 5 cm (2 in) thick. Agitation of the sea surface results in disc shape cakes of ice called pancake ice that are approximately 0.4 to 1 m (16 to 39 in) in diameter. As freezing continues, the cakes form a continuous ice sheet that is also called an ice floe. These ice sheets may range in length from 10 m (32.8 ft) to 10 km (6 mi).

In the Arctic, the ice sheets in the nearshore area are called fast ice, and it forms quickly in the fall and winter and may possibly reach a thickness of 2 m (7 ft). Usually, the fast ice melts in the summer leaving the nearshore free of ice cover. Fast ice is attached to the shore and is stationary. Pack ice is seaward of the fast ice, and it is not attached to the shore. Pack ice is continually moving and consists of a varied ice coverage of large masses that interact with each other. Typically, pack ice is located in water depths of 15 m (50 ft) or greater. The thicker portion of pack ice often touches the bottom and prevents it from getting closer to shore. Pack ice includes ice that has been in existence for more than two years (multiyear ice), and this allows brine drainage that results in stronger ice than the fast ice. Pack ice may often move at the same speed as the local ocean current, and consequently, it is a significant threat to ships and structures in its path.

Fast ice may also be referred to as smooth ice as a result of its smooth appearance. However, adjacent smooth ice sheets may interact along their boundaries and create a pressure ridge. The ridge can extend a considerable amount above and below the water surface with the lower portion being the keel and the upper portion being the sail. The ridge consists of broken pieces of ice that are not to strongly bonded for a first year ridge. However, multiyear ridges can be strongly bonded and their movement toward and subsequent contact with offshore structures can create a large force on the structure. Evaluation of these ice forces is important and the references previously mentioned should be consulted.

Ice islands and icebergs are fresh water ice features that are glacial in origin and have calved or broken off from the ice shelves or glaciers. They subsequently drift into the ocean and may be incorporated into the moving pack ice. Ice islands usually draw more that 12 m (40 ft) of water so they can't drift into very shallow waters. The diameter of these ice islands can be as large as 100 m or several hundred feet. Further offshore and in deeper water, these ice islands can reach depths of near 45 m (150 ft) and breadths of 5 km or several miles. These ice islands or icebergs are a significant concern for ships and offshore structures that happen to be in their path. In the northern hemisphere, icebergs are very irregular in shape (Figure 2-14), and tabular icebergs are more regular in shape being long, flat, and table like. The similar ice features in the southern hemisphere are called ice islands.

Properties

Fresh water freezes at a temperature of 0 °C (32 °F), but its maximum density occurs at 4 °C. For sea water the temperature at which it freezes depends on the salinity, and the temperature at which the maximum density occurs is a function of the salinity, as illustrated in Table 2-3. Selected physical properties as a function of `salinity are tabulated in Table 2-4.

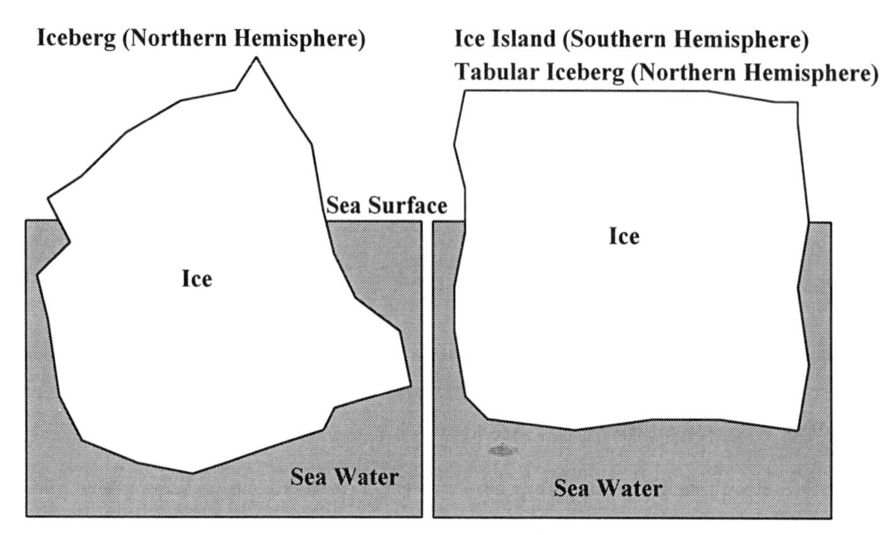

Figure 2-14. Schematic of an iceberg and ice island.

Table 2-3. Temperatures of the freezing point and maximum density for fresh and sea water (Pickard and Emery 1990).

Salinity, o/oo	0	10	20	24.7	30	35
Temperature of freezing, °C (°F)	0 (32)	-0.5 (31.1)	-1.08 30.1	-1.33 (29.6)	-1.63 (29.1)	-1.91 (28.6)
Temperature of maximum density, °C (°F)	3.98 (39.2)	1.83 (35.3)	-0.32 (31.4)	-1.33 (29.6)	-1.63 (29.1)	-1.91 (28.6)

Table 2-4. Properties of -20 °C fresh water ice and sea ice as a function of temperature (Patel 1989).

Property	Salinity (o/oo)			
	0	4	8	15
Density (kg/m^3) (slug/ft^3)	918 1.78	921 1.79	924 1.79	929 1.80
Tensile Strength MN/m^2) (psi)	0.1-0.3 14.5-43.5			
Compressive Strength (MN/m^2) (psi)	12 1741			
Young's modulus (GN/m^2) (psi)	9.75 1.41x10^6	8.78 1.27x10^6	7.33 1.06x10^6	6.12 0.88x10^6
Poisson's ratio	0.33			

REFERENCES

Airy, G. B. "Tides and Waves." *Encyclopedia Metropolitana* 192(1845).

American Petroleum Institute (API). "Bulletin on Planning, Designing and Constructing Fixed Offshore Structures in Ice Environments." Bull. 2N, Washington: American Petroleum Institute (API), 1982.

Bishop, J. M. *Applied Oceanography*. New York: John Wiley & Sons, Inc., 1984.

Caldwell, S. R., and R. D. Crissman (Editors). *Design for Ice Forces*. A State of Practice Report by Tech. Council on Cold Regions Engineering, New York: ASCE, 1983.

Cammaert, A. B., and D. B. Muggeridge. *Ice Interaction with Offshore Structures*. New York: Van Nostrand Reinhold, 1988.

Carstens, T., Editor. "Working Group on Ice Forces on Structures." SR-80-26, prepared by Int. Assoc. for Hyd. Res., US Army Cold Regions Research and Engineering Laboratory (CRREL), Hanover, 1980.

Chakrabarti, S. K. *Nonlinear Methods in Offshore Engineering*, Amsterdam: Elsevier, 1990.

Chen, A. T., and C. B. Leidersdorf (Editors). "Arctic Coastal Processes and Slope Protection Design," State of the Art Report by the Technical Council on Cold Regions Engineering of the ASCE, New York, 1988.

Dean, R. G. "Evaluation and Development of Water Wave Theories for Engineering Application." Vols. 1 and 2, Special Report 1, US Army, Coastal Engineering Research Center, Fort Belvoir, 1974.

Dean, R. G. "Relative Validities of Water Wave Theories." *Proceedings of the Conference on Civil Engineering in the Oceans I*, September 1967.

Dean, R. G. "Stream Function Representation of Nonlinear Ocean Waves." *J. Geophysical Research* 70.18(1965):4561-4572.

Dean, R. G., and R. A. Dalrymple. *Water Wave Mechanics for Engineers and Scientists*. Englewood Cliffs: Prentice-Hall, 1984.

Eckart, C., "The Propagation of Gravity Waves from Deep to Shallow Water." *Gravity Waves*, Circular No. 521, Washington: National Bureau of Standards, 1952.

Ekman, V. W. "On the Influence of the Earth's Rotation on Ocean Currents." *Royal Swedish Academy of Science, Arkivfor Matematik, Astronomi Och Fysik* 2.11(1905):1-53.

Eranti, E., and G. C. Lee. *Cold Regions Structural Engineering*. New York: McGraw-Hill, 1986.

Gaythwaite, J. W. *Design of Marine Facilities*. New York: Van Nostrand Reinhold, 1990.

Horikawa, K. *Nearshore Dynamics and Coastal Processes*. Tokyo: University of Tokyo Press, 1988.

Ingmanson, D. E. and W. J. Wallace. *Oceanography: An Introduction*. Belmont: Wadsworth Publishing Company, 1989.

Ippen, A. T. *Estuary and Coastline Hydrodynamics*. New York: McGraw-Hill, 1966.

Kinsman, B. *Wind Waves: Their Generation and Propagation on the Ocean Surface*. New York: Dover Publications, Inc., 1965. Figure 2-6 reprinted with permission: "Source: Kinsman, *Wind Waves: Their Generation and Propagation on the Ocean Surface*, Copyright Dover Publications, Inc., 1965."

Korteweg, D. J. and G. DeVries. "On the Change of Form of Long Waves Advancing in a Rectangular Channel, and on a New Type of Long Stationary Waves." *Philos. Mag.*, 5th Ser., 39(1895): 422-443.

Le Mehaute, B. "An Introduction to Hydrodynamics and Water Waves." *Water Wave Theories*, Vol. II, TR ERL 118-POL-3-2, U.S. Department of Commerce, ESSSA, Washington, 1969.

Masch, F. D. "Cnoidal Waves in Shallow Water." *Proceedings of the Ninth Conference of Coastal Engineering*, American Society of Civil Engineers 1(1964).

McClelland, B., and M.D. Reifel (Editors). *Planning & Design of Fixed Offshore Platforms*, New York: Van Nostrand Reinhold Company Inc., 1986.

McCormick, M. E. *Ocean Engineering Wave Mechanics.* New York: John Wiley & Sons, 1973.

Miche, R. "Movements ondulatoires des mers en profondeur constante ou decroissante." *Annales des Points et Chauses*, 1944.

Millero, F. J., and A. Poisson. "International One-atmosphere Equation of State of Seawater." *Deep-Sea Research* 28A(1982): 625-629.

Monkmeyer, P. L. "Higher Order Theory for Symmetrical Gravity Waves." *Proceedings of 12th Coastal Engineering Conference*, Washington. pp. 543-562, 1970.

National Ocean Survey (NOS). "Tide Tables, East Coast, West Coast North and South America." Department of Commerce, National Oceanic and Atmospheric Administration, 1996.

Pickard, G. L., and W. J. Emery. *Descriptive Physical Oceanography: An Introduction*, Fifth Enlarged Edition, New York: Pergamon Press. 1990.

Pond, S., and G. L. Pickard. *Introductory Dynamical Oceanography*. 2nd Edition, New York: Pergamon Press, 1983.

Prandtl, L. *Essentials of Fluid Mechanics*, New York: Hafner Publishing Company, 1952.

Sarpkaya, T. and M. Isaacson. *Mechanics of Wave Forces on Offshore Structures.* New York: Van Nostrand Reinhold, 1981. Figures 2-10 and 2-13 reprinted with permission: "Source: Sarpkaya and Isaacson, *Mechanics of Wave Forces on Offshore Structures*, Copyright Van Nostrand Reinhold, 1981."

Skjelbreia, L. "Gravity Waves. Stokes' Third Order Approximation. Tables of Functions." University of California, Council on Wave Research, The Engineering Foundation, Berkeley, 1959.

Skjelbreia, L., and J. A. Hendrickson. *Fifth Order Gravity Wave Theory and Tables of Functions*. National Engineering Science Co., Pasadena, 1962.

Sorensen, R. M. *Basic Coastal Engineering*. New York: John Wiley & Sons, Inc., 1978.

Tsinker, G. P. *Marine Structures Engineering: Specialized Applications*. New York: Chapman and Hall, 1995.

US Army Corps of Engineers (USACE). "Automated Coastal Engineering System, User's Guide." Coastal Engineering Research Center, USAE Waterways Experiment Station, Vicksburg, 1992.

US Army Corps of Engineers (USACE). *Ice Engineering*. EM 1110-2-1612, Dept. of the Army, Office of the Chief of Engineers, Washington: US Government Printing Office, 1982.

US Army Corps of Engineers (USACE). *Shore Protection Manual*, Vol. I and II, Coastal Engineering Research Center, US Army Engineer Waterways Experiment Station, Superintendent of Documents, Washington: US Government Printing Office, 1984.

US Navy (USN). *Tables for Sea Water Density*. US Naval Oceanographic Office, Washington, 615(1952): 265.

Wiegel, R. L. "A Presentation of Cnoidal Wave Theory for Practical Application." *Journal Fluid Mechanics,* 7.2(1960).

Wiegel, R. L. *Oceanographical Engineering*. Englewood Cliffs: Prentice-Hall, 1964.

Wortley, C. A. "Ice Engineering Manual for Design of Small Craft Harbors and Structures." University of Wisconsin Sea Grant Inst., SG-84-417, Madison, 1984.

Zhang, J., R. E. Randall, L. Chen, C. A. Spell, J. K. Longridge and M. Ye. "Nonlinear Decomposition of a 2-D Wave Field." *Proceedings of Waves'93*, ASCE, New Orleans, July 1993.

PROBLEMS

2-1. Compute the wave length and classify the following waves:

Period (s)	Water Depth (m)	Wave Length (m)	Classification
8	25		
8	100		
8	1000		

2-2. Compute the deepwater wavelength and wave celerity for waves with periods of 4, 6, 8, and 12 s in the SI system of units.

2-3. A wave tank is 120 ft long, 3 ft wide and 4 ft deep and is filled with fresh water to a depth of 3 ft. The wave maker generates a wave which has a wave height of 0.75 ft and wave period of 1.1 s. Assume the density of water is 1.94 slugs/ft^3 as found in Appendix Table A-2 and calculate the wave celerity, length, group celerity, energy in one wavelength (E_L), and power.

2-4. A 0.5 ft wave is generated with a wave period of 1 s in the wave tank describe in problem 4 above. Determine the water particle velocity and acceleration at a depth of 2 ft below the still water level directly under the wave crest.

2-5. Determine the maximum wave height (before breaking) for a 1 s wave in the wave tank in problem 2-3.

2-6. Compute the density and sigma-t for seawater at standard atmospheric pressure when the salinity and temperature are 30 o/oo and 28 ^{o}C, respectively.

2-7. A wave tank is 150 m long, 3 m wide and 4 m deep. The water depth in the tank is 3 m and a wave 0.4 m wave height and period 1.2 s is generated. Determine the wave celerity, length, group celerity, energy in one wave length (E_L), and power. Calculate the water particle velocity and pressure at a depth of 1 m below the still water level (SWL) and 0.2 m ahead of the wave crest.

2-8. A 8 s wave propagates normal to shore. Consider shoaling only and evaluate the wave length in water depths of 300 ft, 150 ft, 60 ft and 15 ft.

2-9. A pressure gauge measures an average maximum pressure of 9 N/cm^2 having an average period of 10 s. The gauge is mounted 1 m above the bottom where the water depth is 9 m. Evaluate the wave height and length.

2-10. Deep water waves have wavelengths of 50, 100, and 150 m. Determine the respective wavelengths in 20 m of water using Appendix Tables A-5 and A-6.

CHAPTER 3: OFFSHORE STRUCTURES

INTRODUCTION

The use of offshore structures for the exploration and production of offshore petroleum reserves is being conducted in most continental shelf areas of the world and its beginnings date back to the 1950's. The Gulf of Mexico led the way and was followed by the coastal waters off Mexico and Brazil. Installation of these structures in the extremely harsh environment of the North Sea began in the 1960's. Undersea pipelines transport oil and gas to the shore and are another type of offshore structure that are designed and installed to provide a means of transporting the energy resources to land.

Man's utilization of the oceans as a resource stimulates the need for other structures that must operate in the ocean. Wave and tidal energy systems and the ocean thermal energy conversion (OTEC) systems have been and continue to be pursued for extracting energy from waves, tides, and ocean water temperature differences. Ocean mining systems have been developed and may be used in the future to recover ocean minerals such as manganese nodules from the sea floor. The oceans may also be used in the future to support offshore fish processing plants, floating airports, and floating communities where land is scarce.

Drilling and Producing Oil and Gas

Currently, the major use of offshore structures is for the exploration and production of oil and gas. Therefore, a brief explanation of some terminology and procedures for drilling and producing oil and gas is given. Since the construction of offshore structures is very expensive, mobile exploratory drilling rigs are used to drill wells to determine the presence or absence (dry hole) of petroleum at the offshore site. If oil is present in sufficient quantity, then the well is plugged until a permanent production platform is installed.

Offshore wells are drilled by lowering a drill string through a conduit (riser) which extends from the drill rig to the sea floor. The drill string (Figure 3-1) consists of a drill bit, drill collar, and drill pipe. The drill pipe sections are typically 9.1 m (30 ft) long and made of steel weighing about 2669 N (600 lb). Additional drill pipe sections are connected at the surface as the well deepens. The drill string is lowered through the riser to the seafloor where it passes through a system of safety valves called the blowout preventer (BOP) stack. The BOP is there to contain pressures in the well and to prevent a blowout.

A rotary table at the surface turns the drill string and the drill bit teeth grind away at the seafloor sediment and rock formations. Drilling fluid (mud) is pumped into the drill pipe from a mud tank on the surface, and the mud flows through small holes in the drill bit. The drilling mud collects the cuttings of rock cut by the drill bit and brings them to the surface. The drill mud flows to the surface through the annulus between the well casing and the drill string below the sea bottom (mud line) and the riser and the drill string above the mud line. A strainer at the surface removes the cuttings from the drilling mud, and the mud is then recirculated through the mud tank and pumped to the drill string. Discharge of these fluids and cuttings into US ocean waters are governed by environmental regulations administered by the Environmental Protection Agency (EPA).

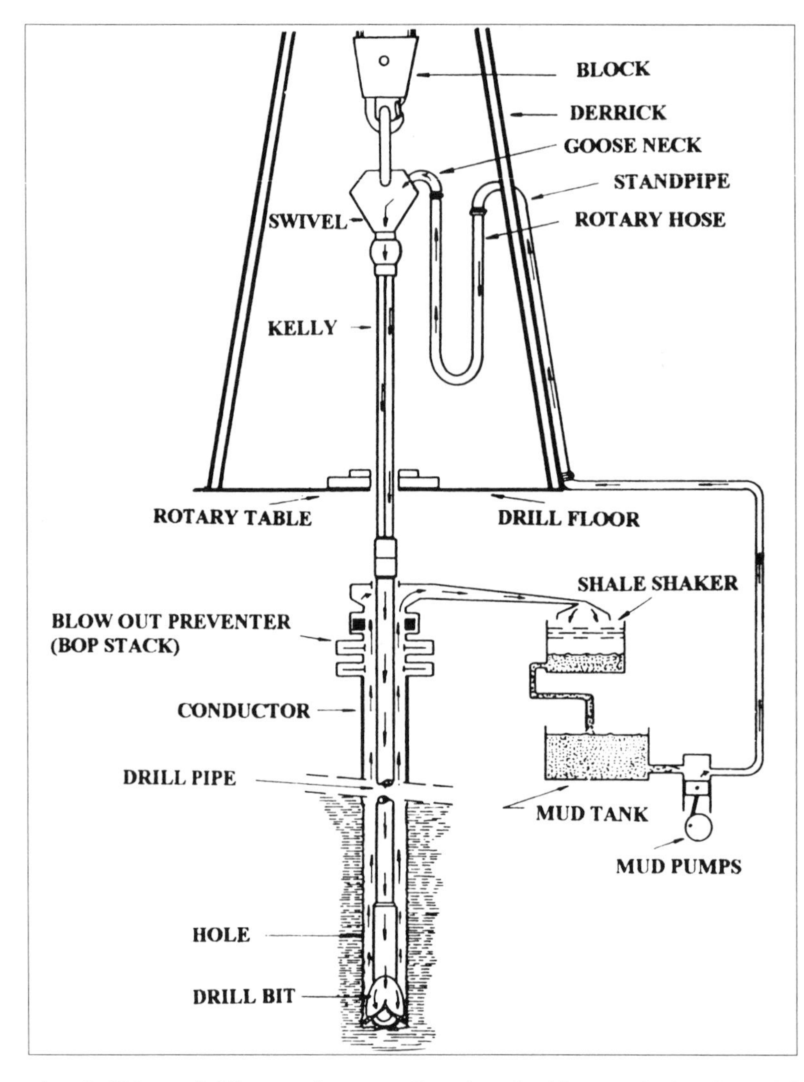

Figure 3-1. Schematic of offshore drilling equipment. Reprinted with permission from Mather, 1995, *Offshore Engineering: An Introduction*. (Full citing in references).

The weight of the mud exerts a pressure greater than that in the rock formations, and therefore it keeps the well under control. As the bit penetrates further into the rock formations, strings of steel pipe casing are run into the well and cemented into place in order to seal off the walls of the well and keep the hole from collapsing. It is possible to send instruments down a wireline and into the well to determine the existence of oil or gas. If oil and gas are found then steel production casing is set in place. This production casing is used as the conduit for bringing oil and gas safely to the surface.

Some platforms are capable of drilling and producing. The fluids from the well contain a mixture of oil, gas, and water that is processed by special equipment before sending it ashore through a pipeline or transporting it to shore by a tanker. The processing of the well fluid mixture is known as producing, and the equipment used is called the production equipment. Production platforms are designed to support the production equipment.

TYPES OF OFFSHORE STRUCTURES

Offshore Drilling Systems

Jack-up Drilling Rig

In order to drill in shallow offshore water, special drilling equipment has been developed. One of the most common shallow water exploratory drilling rigs is called the Jack-up Rig. It can drill in waters to depths of approximately 122 m (400 ft). The jack-up rig (Figure 3-2) is designed like a barge with movable elevator legs that can be extended to the sea floor. These rigs are typically towed like a barge to the drilling site with the legs (usually three legs) extended vertically above the barge deck. At the site, the legs are jacked down through the water column and into the sea floor. As the legs engage the sea floor, the drilling deck is raised out of the water and into the air. Deck space provides room for drilling equipment, supplies, and quarters for a crew. Helicopters and ocean supply boats ferry workers and equipment to the rig. The drill deck is well above the height of the highest expected waves. After the drilling is complete, the procedure is reversed and the drilling deck is lowered to the water and the legs are jacked up above the drill deck. A tow boat is then used to move the rig to another location.

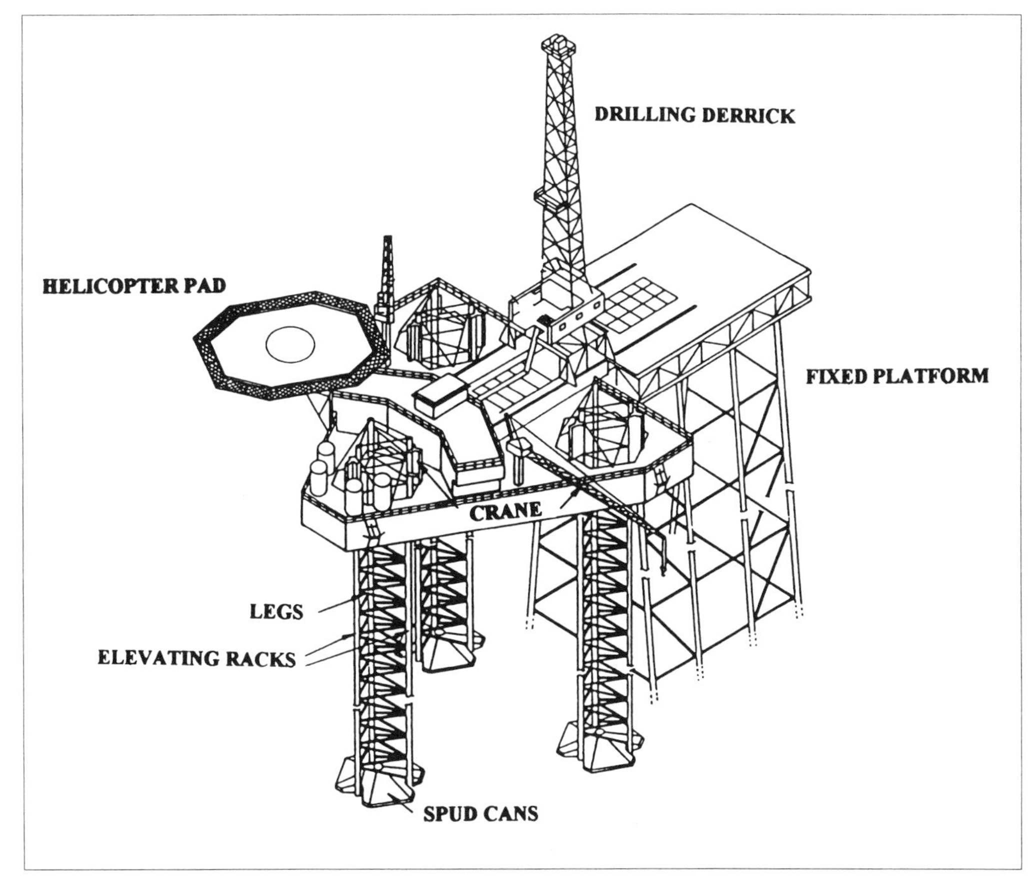

Figure 3-2. Example jack-up drill rig. Reprinted with permission from Patel, 1989, *Dynamics of Offshore Structures*. (Full citing in references).

Semisubmersible

The semisubmersible (Figure 3-3) is another type of offshore drilling rig. It is constructed of large vertical columns connected to very large pontoons at the bottom. This structure supports the drilling deck that accommodates the drilling derrick, equipment, supplies, and crew accommodations. Similar to the jack-up rig, supply boats and helicopters ferry equipment and personnel between the drill site and shore. The semisubmersible drilling rig is used in water depths ranging from 91.5 to 915 m (300 to 3000 ft), and it is typically towed to the site and moored to the bottom. It floats high in the water when it is being moved to a site, and then the pontoons are flooded to partially submerge the rig so that a majority of its structure is below the water, but the deck is well above the water surface. With only the columns exposed to the wave environment, the semisubmersible is a very stable platform for drilling operations.

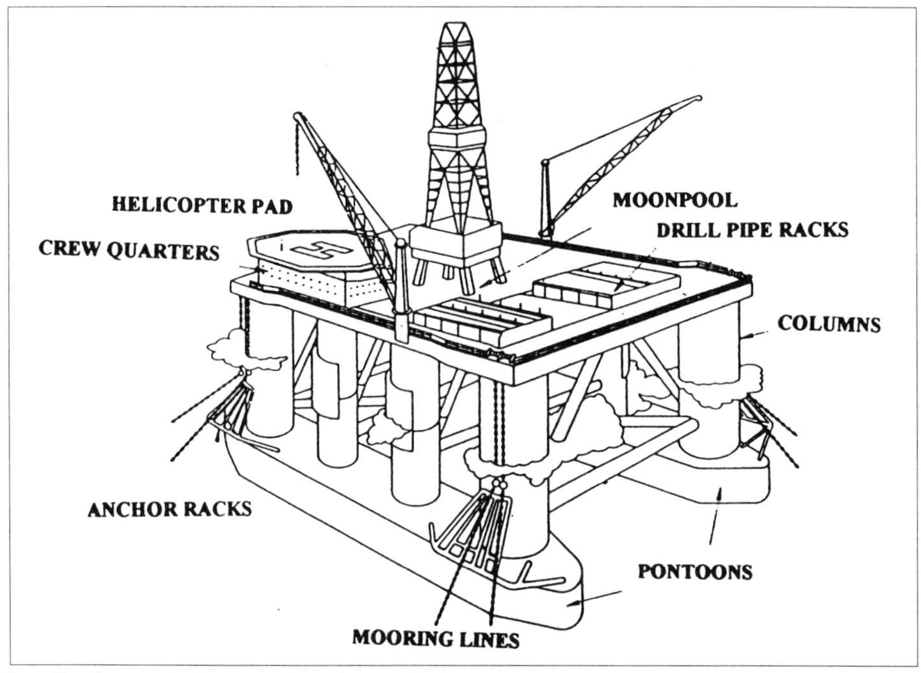

Figure 3-3. Sketch of a semisubmersible drilling rig. Reprinted with permission from Patel, 1989, Dynamics of Offshore Structures. (Full citing in references).

Drillship

A drillship is used to drill wells in water depths to 2439 m (8000 ft) or more. A ship shape hull is adapted to accommodate the drilling equipment as illustrated in Figure 3-4. The drilling derrick is usually positioned amidships, and a moon pool opening is located below the derrick for the drilling operation. Drillships have their own propulsion and consequently move under their own power. A dynamic positioning system is used to keep the drillship over the drilling location. Thrusters are added in the bow of the ship to assist in the positioning of the ship. A local acoustic positioning system or differential global positioning system (DGPS) is used to determine the position of the drillship relative to the well head on the seafloor. Environmental sensors measuring wind, wave, and current and acoustic sensors are used to automatically send signals to the ship's thrusters and propulsion system to keep it directly over the well. The conductor pipes, or drilling riser, are flexible to accommodate small inclination angles, and the conductor pipe can be disconnected should extreme weather approach.

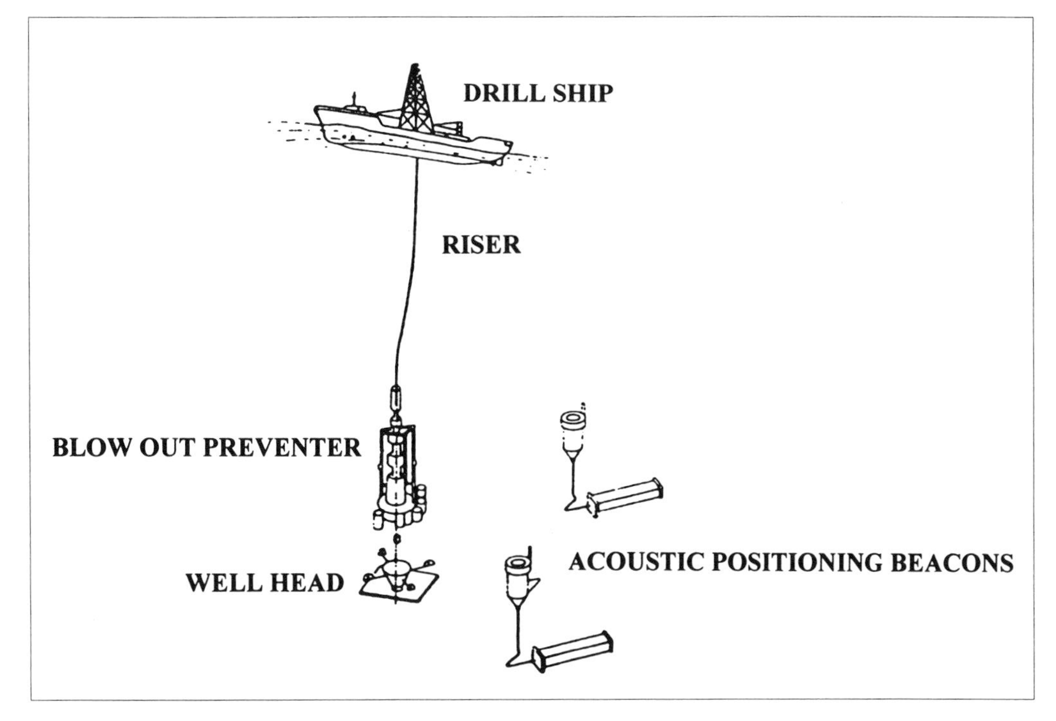

Figure 3-4. Example of dynamically positioned offshore drillship. Reprinted with permission from Patel, 1989, *Dynamics of Offshore Structures*. (Full citing in references).

Offshore Oil and Gas Platforms

Fixed Jacketed Structure

A fixed jacketed structure (Figure 3-5) consists of a steel framed tubular structure that is attached to the sea bottom by piles. These piles are driven into the seafloor through pile guides (sleeves) on the outer members of the jacket. The topside structure consists of drilling equipment, production equipment, crew quarters and eating facilities, gas flare stacks, revolving cranes, survival craft, and a helicopter landing pad. Drilling and production pipes are brought up to topside through conductor guides within the jacket framing, and the crude oil and gas travel from the reservoir through the production riser to topside for processing. The produced fluid is then pumped to shore through the export pipe line. The detailed design of the frame varies widely and depends on the requirements of strength, fatigue, and launch procedure. Structural members consist of X and K joints and X and K braced members. The platform phases include design, construction, load-out, launch, installation, piling, and hook-up before it begins producing. The design life of the structure is typically 10 to 25 years. This is followed by the requirement to remove and dispose of the platform once the reservoir is depleted.

Launching of these structures is usually accomplished with barges. Some structures are floated off the barge and righted using barge cranes, and others are designed with flotation to be self righting. Since these structures are made of steel, the effects of corrosion must be consider due to its exposure to the ocean environment. Anodic and cathodic protection systems are employed and maintained to protect against corrosion of the structure.

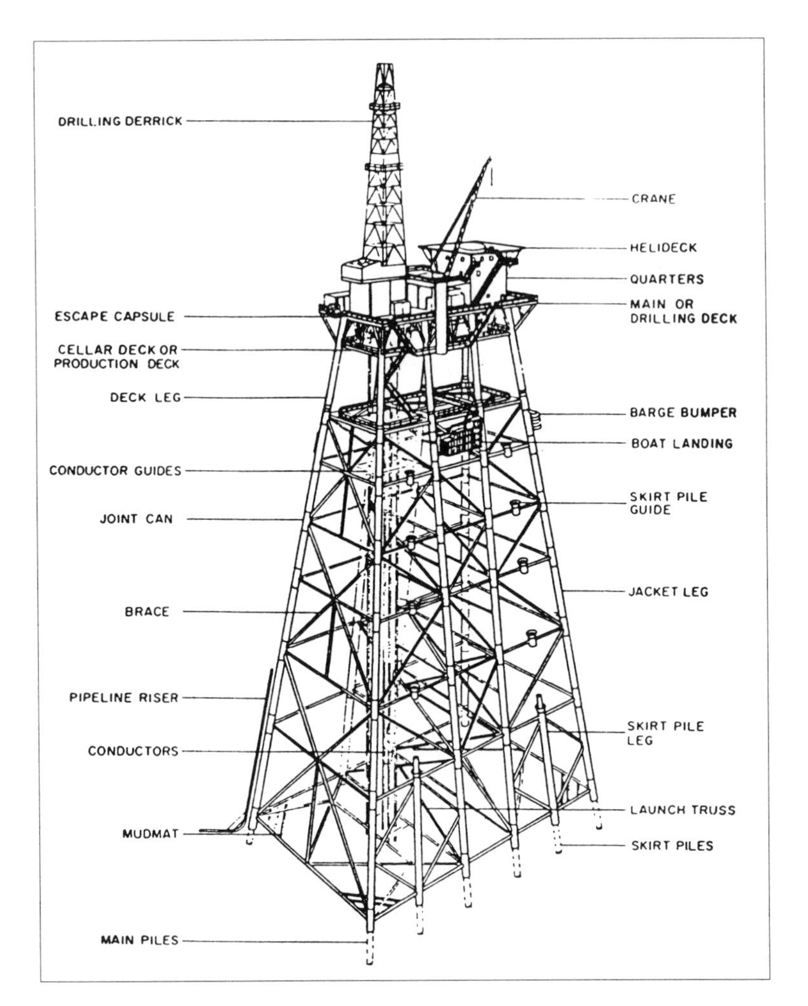

Figure 3-5. Steel jacketed platform. Reprinted with permission , McClelland and Reifel, 1986, *Planning and Design of Fixed Offshore Platforms*. (Full citing in references).

Gravity Structure

The concrete gravity structures were pioneered in the North Sea. The first structure was installed in the Ekofisk field in 1973, and over 17 platforms were installed by 1982. Advantages of these structures are that they have the ability to store oil and construction and testing can be completed before floating the structure and towing it to an offshore location. This type of structure is more tolerant to overloading and degradation due to exposure to sea water than steel platforms. Disadvantages include greater costs than for similar steel structure. More steel is sometimes required for the reinforcing members than required for an equivalent steel jacketed structure. Foundation settlement is expected over the life of the structure which reduces the clearance between the mean water level and the underside of the structure.

Figure 3-6 illustrates the Condeep design for the Beryl A concrete gravity structure that was installed in 1975. The concrete caisson at the base is used for oil storage and is 100 m (328 ft) wide with circular cellular oil storage tanks of 20 m (65.6 ft) diameter. Three water piercing support towers are 94.5 m (310 ft) high and are also used for drilling and oil production conductors. These structures use a construction technique known as slip-forming in which steel reinforcing members are set into the concrete and prestressed to obtain the required overall structural properties. Although concrete platforms are more expensive, they do offer advantages of lower maintenance and higher deck payload. However, construction of concrete platforms has slowed since the early 1980's compared to steel structures.

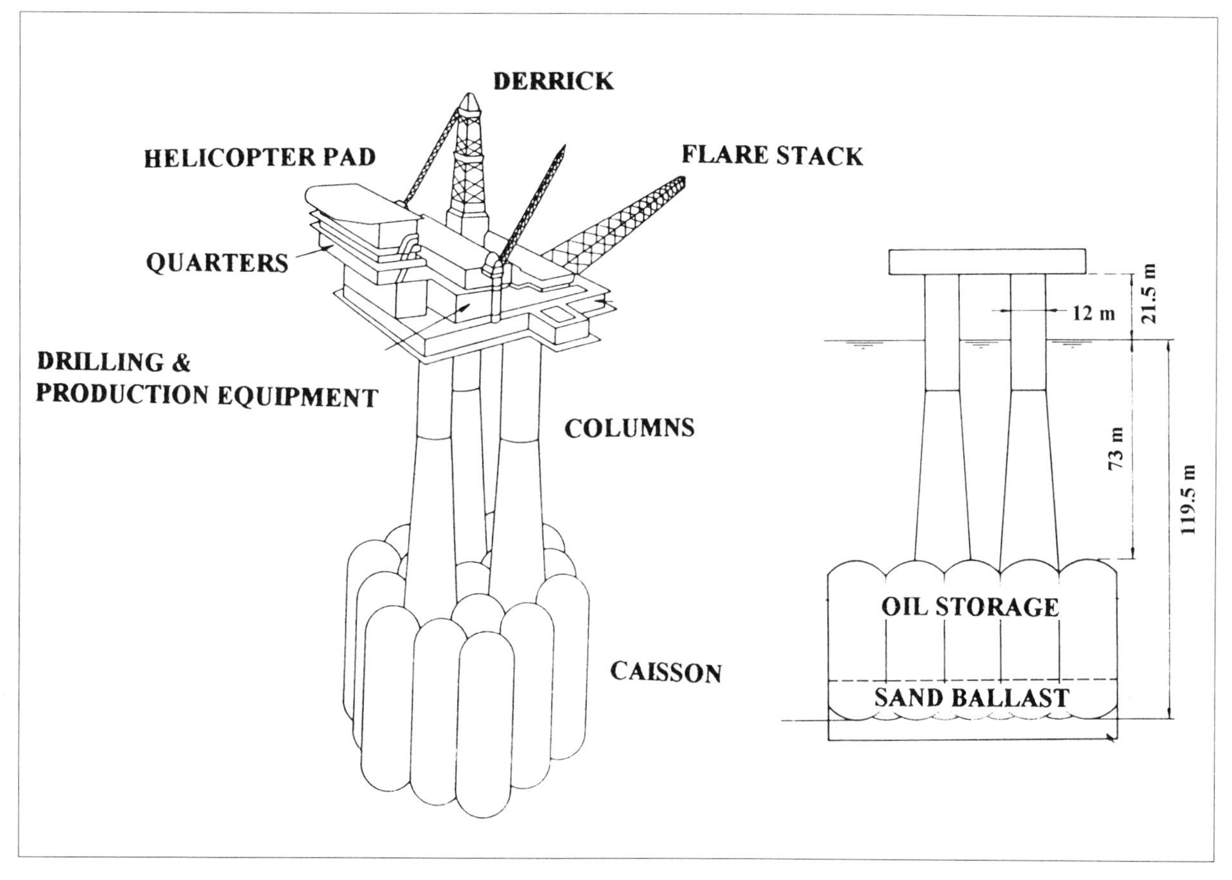

Figure 3-6. Example Condeep concrete gravity structure. Reprinted with permission, Furnes and Loset, 1980, "Shell structures in offshore platforms: design and application." (Full citing in references).

Compliant Structures

As offshore drilling and production proceeds into deeper waters, the weight and cost of fixed structures is increasing exponentially as illustrated by cost comparison shown in Figure 3-7. Compliant structures and floating production systems move with the applied environmental forces resulting from wind, current and waves. These structures are much lighter and cost considerably less. The more common compliant structures are the tension-leg platform (TLP), articulated tower, compliant tower, single anchor leg mooring system (SALM), and floating production systems (FPS).

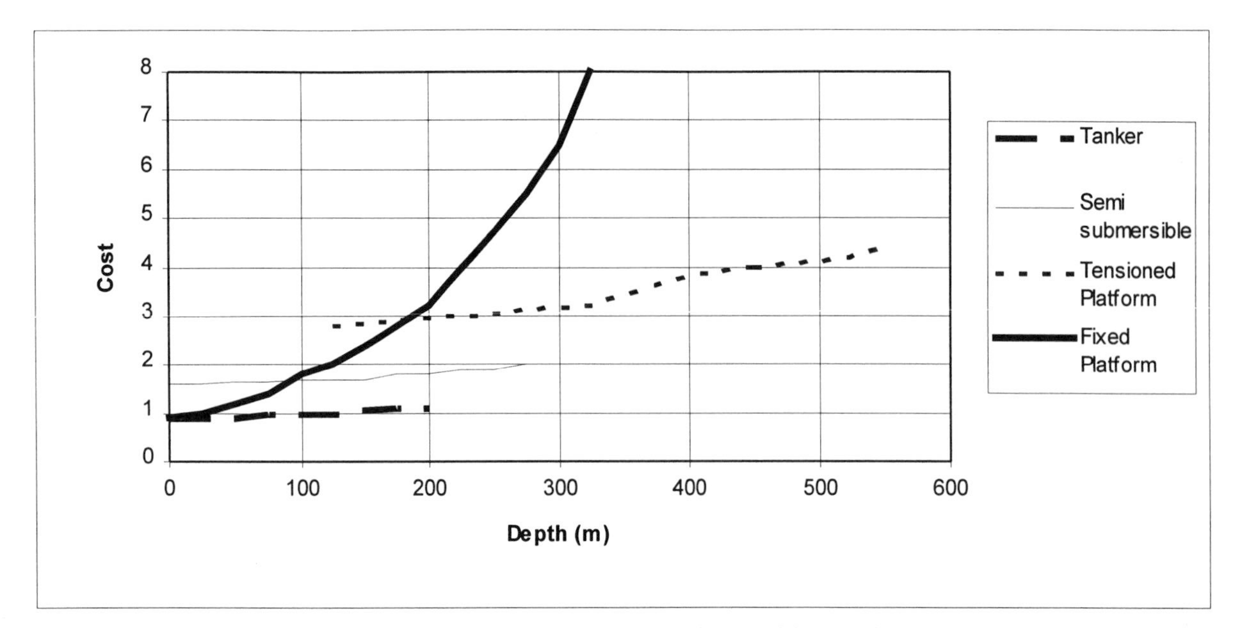

Figure 3-7. Comparison of cost for offshore structures. Reprinted with permission Hamilton and Perrett, 1986. "Deep water tension leg platform designs." (Full citing in references).

Tension Leg Platform

A popular compliant structure is the tension leg platform (TLP) as illustrated in Figure 3-8. The semisubmersible floating structure is tethered to the seafloor with vertical legs that are kept in tension by the excess buoyancy of the platform (approximately 15 to 25% of the platform displacement). The tension legs are in sufficient tension that heave, roll, and pitch motions due to waves are essentially eliminated. Sway, surge, and yaw motions are experienced, but the tether induced restoring forces are capable of keeping the vessel on station above the well heads. Marine risers carry the petroleum products from the wells to the processing equipment on the deck of the platform, and the processed oil and gas is pumped to shore through an export pipeline. The first TLP was installed in 148 m (485.4 ft) of water by CONOCO in the North Sea Hutton field in 1984. The second TLP, named the Auger, was installed by SHELL in the Gulf of Mexico in 1993. This type of compliant structure is a likely candidate for future deep water platforms for depths ranging from 120 to 1500 m (394 to 4920 ft).

Guyed Tower

Another compliant structure is the guyed tower as shown in Figure 3-9, which is a slender truss-steel structure supported on the sea floor by a spud-can foundation and held upright by multiple wire or chain guy lines. These guy lines connect to anchor piles and are equipped with heavy clump weights between the anchor and tower. The same guy wires also restrain the platform motion during typical operating weather conditions without lifting the clump weights off the bottom. During more extreme weather conditions the guy wires are designed to lift the clump weights off the bottom, and the clump weights create a larger restoring force to resist the larger wave forces. The cost of this type of structure is considerably less than a fixed steel jacketed structure. The Lena guyed tower was installed in 1983 in the Gulf of Mexico. It had a steel weight of 24,000 tons and deck weight of 19,640 tons. The tower was placed in 305 m

(1000 ft) water depth and utilized 20 galvanized spiral wound steel wire ropes as guy lines with 179 ton clump weights placed in a symmetrical pattern around the platform.

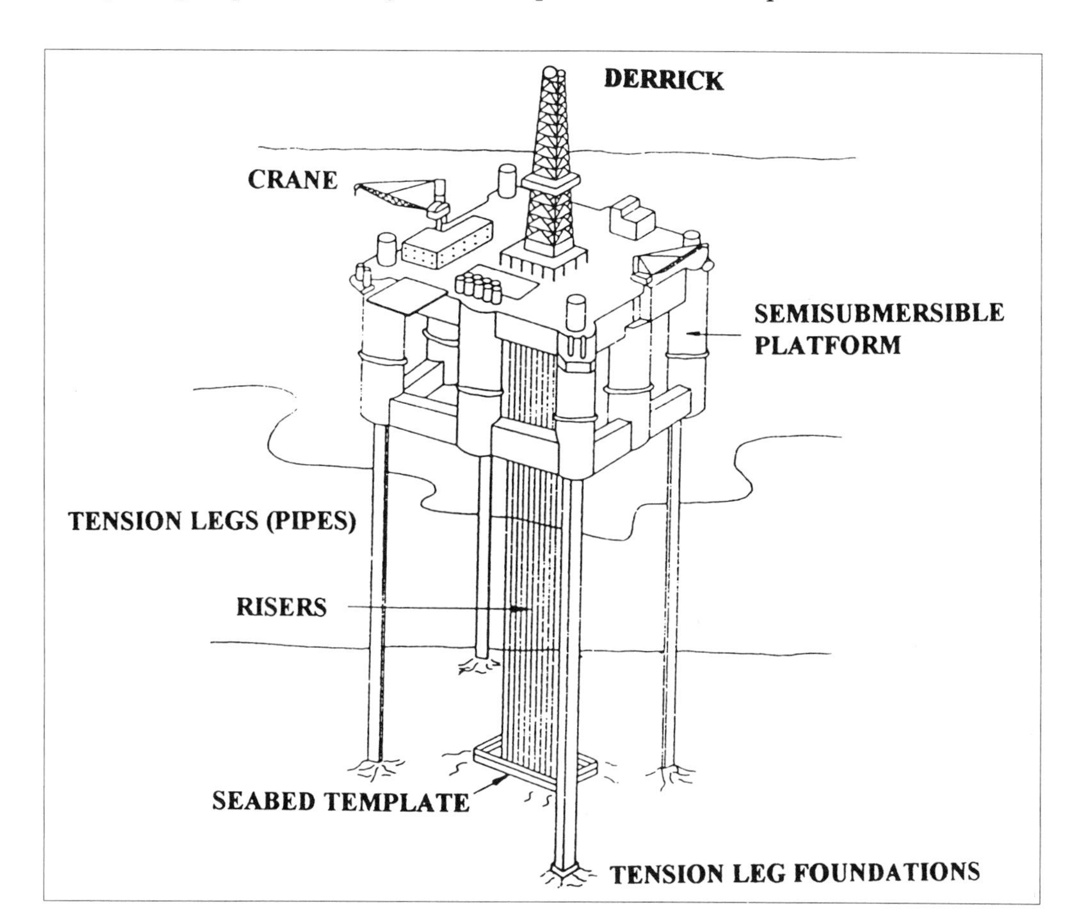

Figure 3-8. Example tension leg platform. Reprinted with permission from Patel, 1989, *Dynamics of Offshore Structures*. (Full citing in references).

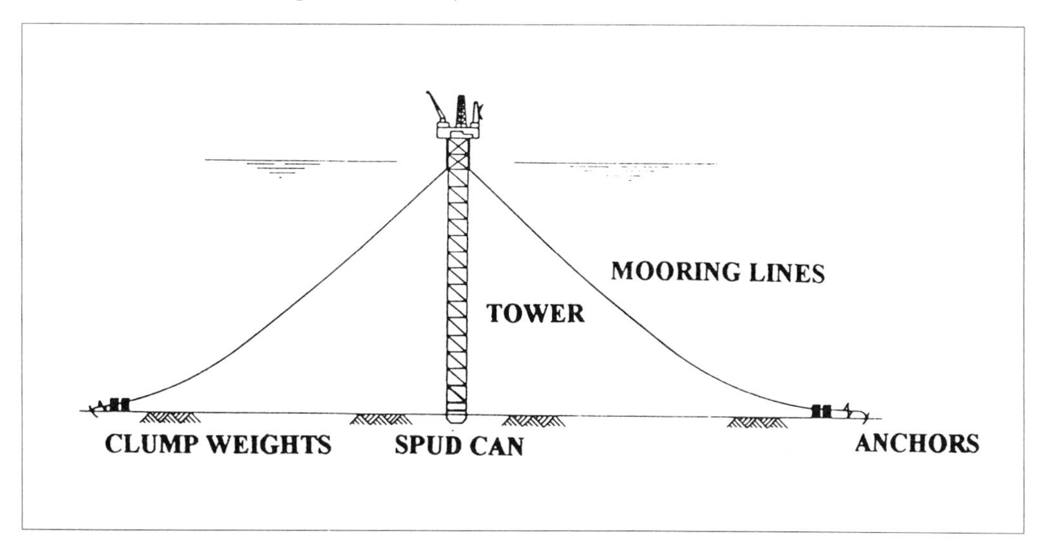

Figure 3-9. Example guyed-tower compliant offshore platform. Reprinted with permission from Patel, 1989, *Dynamics of Offshore Structures*. (Full citing in references).

Articulated Tower and Single Anchor Leg Moored Systems

For the development of small reservoirs in water depths up to about 200 m (656 ft), an articulated column (Figure 3-10), or a single anchor leg storage and tanker system (Figure 3-11), can be used in relatively calm weather areas. Crude oil is moved up the articulated tower and transferred to the tethered tanker for processing and storage. A shuttle tanker is brought alongside to receive the processed oil and transport it to shore. The single anchor leg mooring (SALM) system uses a yoke structure, buoyancy tank, and tensioned riser to moor a tanker. Processing and storage facilities are housed in the tanker, and oil export is accomplished with a shuttle tanker or an export oil pipeline.

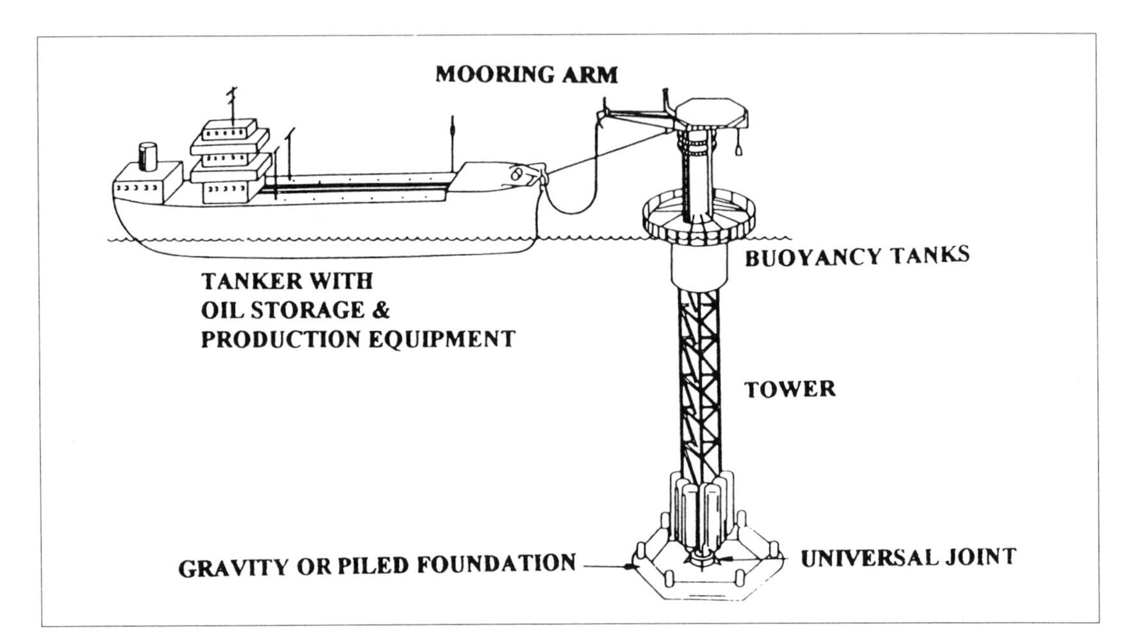

Figure 3-10. Example articulated tower. Reprinted with permission from Patel, 1989, *Dynamics of Offshore Structures*. (Full citing in references).

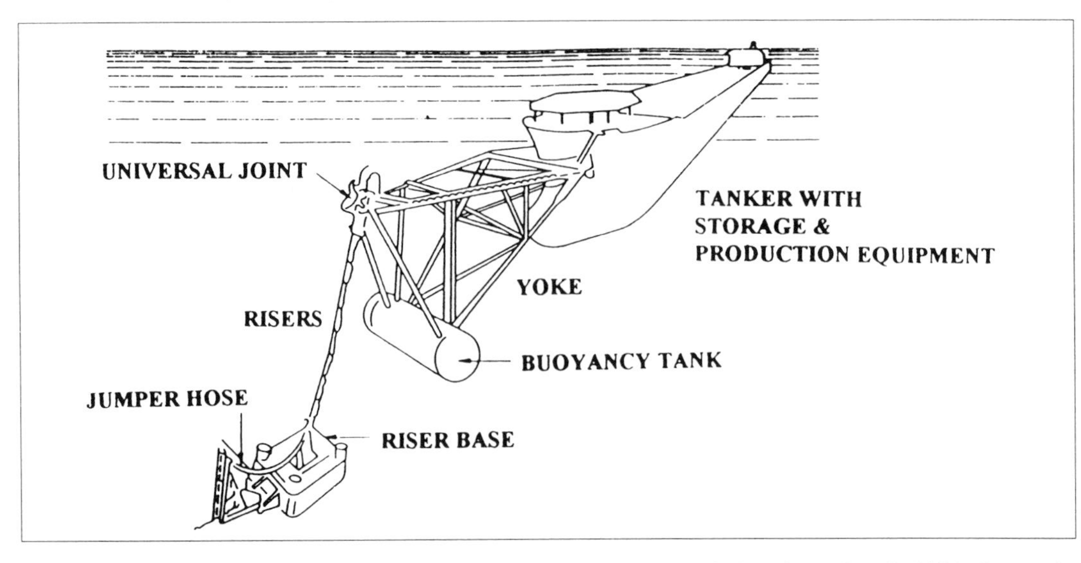

Figure 3-11. Single anchor leg mooring system. Reprinted with permission from Patel, 1989, *Dynamics of Offshore Structures*. (Full citing in references).

Floating Production Systems

An economical production system for small reservoirs is the floating production system (FPS) that consists of a converted or newly built semisubmersible that is moored to the seafloor using a catenary mooring system (Figure 3-12). Connection to the oil reservoir is through a rigid tensioned vertical multiple production risers or through flexible risers. This system is used in water depths of 70 to 250 m (229.6 to 820 ft) and supports only a relatively small deck load that limits the oil processing options compared to the capabilities of a fixed platform. The compliance of the system and the catenary mooring system create some risks such as damaged risers and moving to far off station in severe environmental conditions. The platform has no oil storage capabilities and vessel motions in severe weather conditions can limit or degrade the processing operations. Nonetheless, the floating production system provides an economic means for working small reservoirs that require only limited processing facilities.

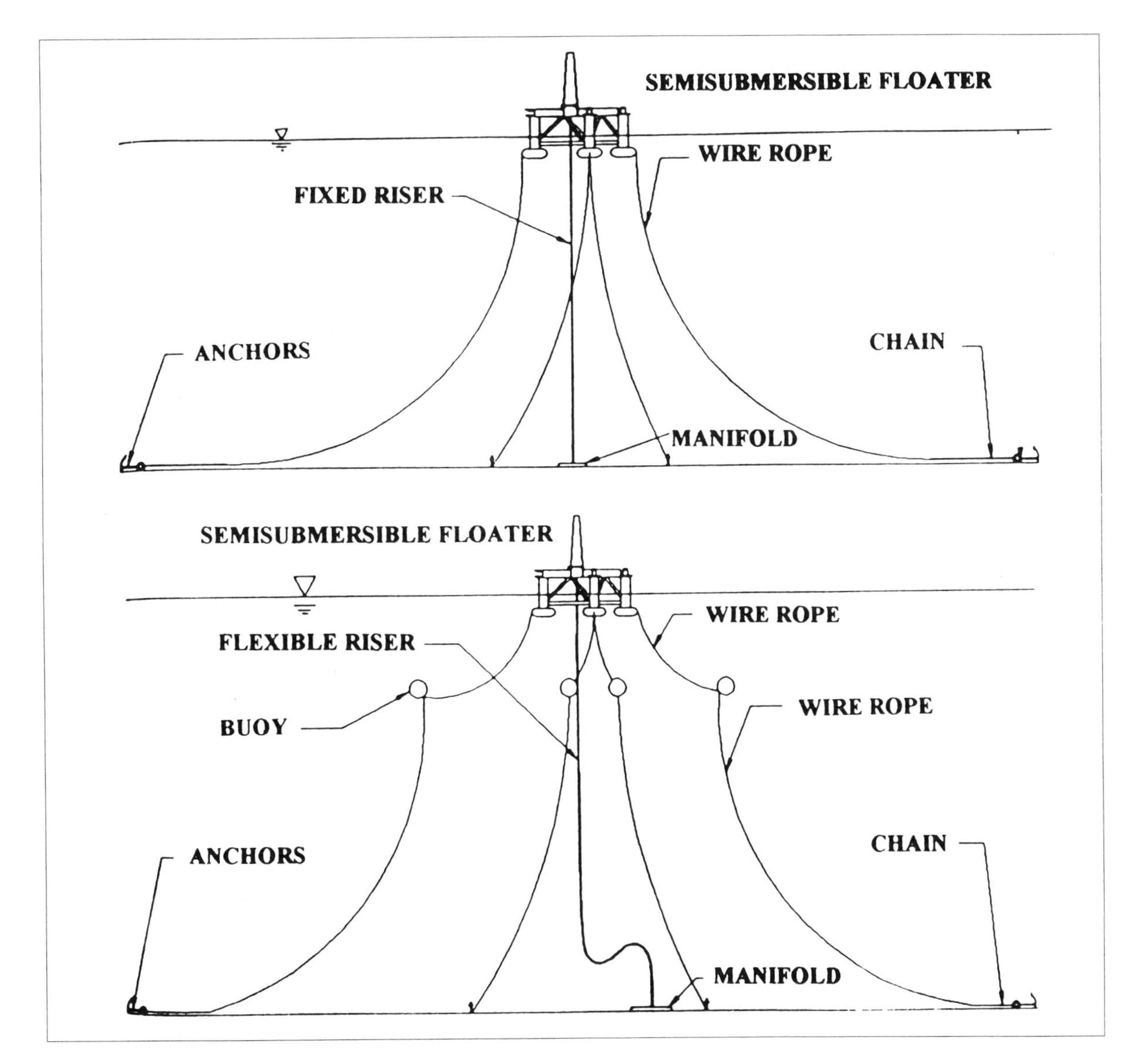

Figure 3-12. Example of a floating production system with fixed and flexible risers.

Spar Platform

A new compliant structure being designed and developed for deep water offshore installations is called a spar platform (Figure 3-13). Its response characteristics in deep water make it a viable possibility for deep water (> 915 m or 3000 ft) petroleum production.

Figure 3-13. Spar platform. (Reprinted with permission of Chevron and Deep Oil Technology).

WAVE FORCES ON OFFSHORE STRUCTURES

Background

A critical task for ocean engineers is the analysis and computation of forces on ocean structures caused by water waves. There are several good references such as Barltrop and Adams (1991), Chakrabarti (1987), Faltinsen (1990), Hsu (1984), Patel (1989), and Sarpkaya and Isaacson (1981) describing the complex analytical techniques. The American Petroleum Institute (API) publishes guidelines for evaluating wave forces as well as the complete design of fixed (API RP-2A 1993) and tension leg platforms (API RP-2T 1987), and Det Norske Veritas (DNV) also publishes rules for design, construction and inspection of offshore structures (DNV 1977). The major conferences addressing offshore structures are the Offshore Technology Conference (OTC), Offshore Mechanics and Arctic Engineering Conference (OMAE), International Society of Offshore and Polar Engineering Conference (ISOPE), and the ASCE Specialty Conferences (Civil Engineering in the Oceans I-V). A brief introduction to the fundamental concepts of analyzing wave and current forces on offshore structures is briefly described.

The methods for calculating wave forces can be divided into different approaches which are determined by the size of the structural member and the height and wavelength of the incident waves. Ratios of these parameters are often used to classify which force calculation procedure is to be used. The ratios used are diameter divided by wavelength (D/L) and wave height divided by diameter (H/D). The second ratio is better represented by the Keulegan-Carpenter number ($K = U_m T/D$) in which U_m is the peak water particle velocity and T is the wave period. When D/L is less than 0.2 the Morison Equation is used and when it is greater than 0.2, Diffraction Theory is used the calculate the wave forces. Table 3-1 describes the different wave load regimes.

Table 3-1. Guide for evaluating wave load calculation procedures.

K	D/L < 0.2	D/L > 0.2
$K > 25$	Drag dominated. Morison equation with C_m and C_d . $Re > 1.5 \times 10^6$; $C_m = 1.8$, $C_d = 0.62$ $10^5 < Re < 1.5 \times 10^6$; $C_m = 1.8$, C_d varies from 1.0 to 0.6	Morison equation should not be used for computing wave forces. Diffraction theory used.
$5 < K < 25$	Drag and inertia dominated range Morison equation applicable, but C_m and C_d values show large scatter. Flow behavior and load are complex and uncertain. $Re > 1.5 \times 10^6$; $C_m = 1.8$, $C_d= 0.62$.	
$K < 5$	Inertia dominated range. Morison equation or Diffraction theory is used. $C_m = 2.0$ Effect of drag is negligible	

Definitions: Keulegan-Carpenter Number, $K = U_m T/D$; Reynolds Number, $Re = U_m D/\nu$; C_m= inertia coefficient; C_d = drag coefficient; U_m = peak velocity; T= wave period; ν = kinematic viscosity; and D = diameter.

Design Wave Concept

The design wave concept is a well known and fairly simple concept for the design of offshore structures and is also used the API guidelines for design. It uses a wave of large height (H) and a corresponding wave period with a probability of occurrence such that it represents the largest wave that the structure is expected to encounter in a finite time interval called the return period. This is usually accomplished by defining the maximum height wave that is likely to occur over a large number of years (50 to 100 yr). In this way, the structure is designed to resist the worst case wave that has a very rare occurrence. This approach is realistic for designing against structural failure due to large waves, but it does not permit fatigue failure to be considered. Examples of design wave and wind parameters for 100 year return periods in several offshore areas are tabulated in Table 3-2.

The Morison Equation

Tubular members of offshore platforms and subsea pipelines are frequently exposed to ocean waves. In the design of these structures ocean engineers must determine the forces acting on the members. If the diameter to wavelength ratio (D/L) is less than 0.2, then the Morison

Equation (Morison et al., 1950) can be used to evaluate these forces. This equation assumes that the wave properties are unaffected by the presence of the structure. Therefore, the total wave load can be expressed as the sum of the inertia forces due to wave fluid acceleration and of the drag forces resulting from the wave fluid velocity.

Table 3-2. One hundred year return period design wave and wind parameters for selected United States offshore waters (API 1987).

Parameter	Offshore Area					
	Gulf of Mexico (TX/LA)	Gulf of Mexico (MS/AL/FL)	Southern California	Central California	Washington/ Oregon	Gulf of Alaska (Kodiak)
Maximum Wave Height m (ft)	22 (72)	21.3 (70)	13.7 (45)	18.3 (60)	25.9 (85)	27.4 (90)
Maximum Wave Period s	14.5	14.3	16.2	17.1	17.8	17.3
Maximum average one-hour wind at 10 m (33 ft) kph (mph)	157.7 (98)	157.7 (98)	93.3 (58)	111 (69)	111 (69)	111 (69)

The differential form of the Morison equation is written as

$$dF = C_m \rho \, dV \, \dot{u}_n + \frac{1}{2} C_d \rho \, dA \, |u_n| u_n \qquad \textbf{3-1}$$

where dF is the total wave force on the member element of volume (dV) and the projected area (dA), $\dot{u}_n$ and u_n are instantaneous wave fluid acceleration and velocities normal to the member axis, ρ is the fluid density and C_d and C_m are the drag and inertia coefficients. The modulus or absolute value sign in the drag force term is used to ensure the drag force is in the direction of the wave velocity. The Morison equation was initially developed for vertical cylinders, or piles, but it can also be used for pipes of arbitrary orientation, provided C_m and C_d coefficients are chosen for the applicable orientation.

Vertical Cylinder

As an example application, consider an element dz of a vertical pile that is a circular cylinder of radius "r" as shown in Figure 3-14. The incremental force dF acting on the element in the direction of wave propagation is the sum of the drag and inertia force components. Integration of dF from the sea bottom to the mean water level yields the total force (F) and the moment about the sea bed (M).

$$F = C_d \rho r \int_{-d}^{0} |u| u \, dz + C_m \rho \pi r^2 \int_{-d}^{0} \dot{u} \, dz \qquad \textbf{3-2}$$

and

$$M = C_d \rho r \int_{-d}^{0} (d+z)|u| u \, dz + C_m \rho \pi r^2 \int_{-d}^{0} (d+z) \dot{u} \, dz \qquad \textbf{3-3}$$

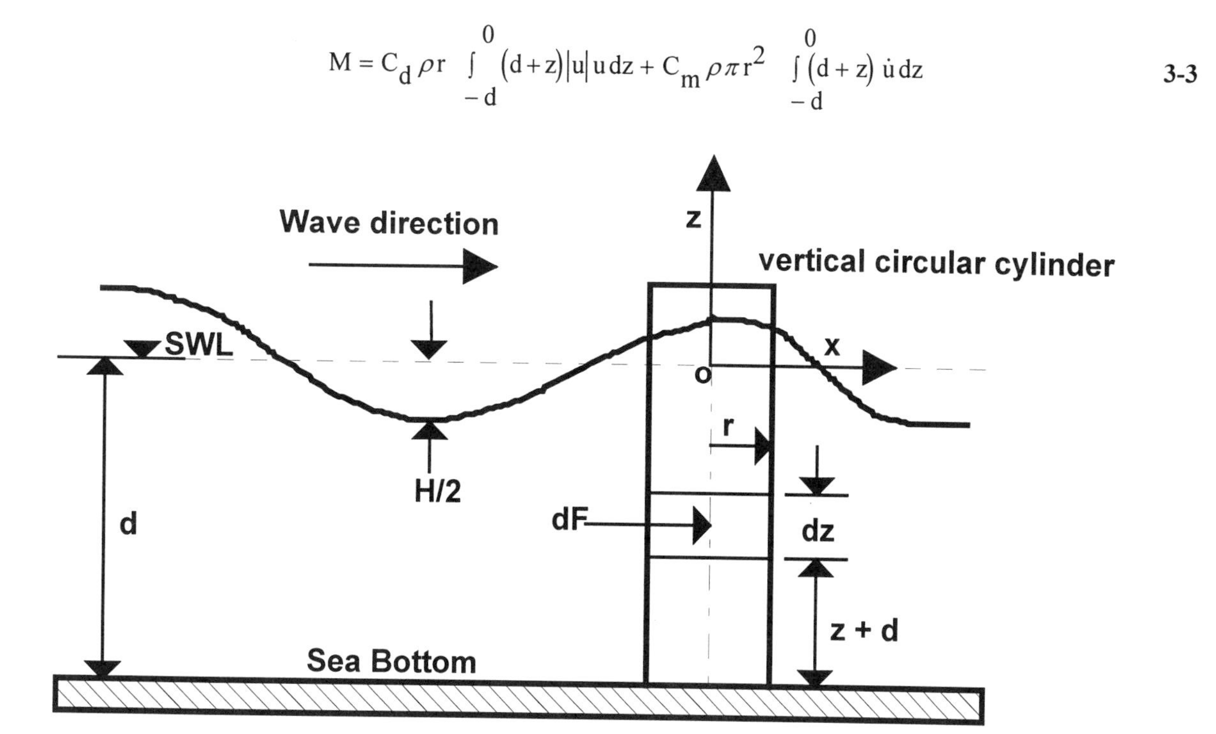

Figure 3-14. Schematic for wave force on vertical circular cylinder.

where u and $\dot{u}$ are the horizontal components of the wave orbital velocity and acceleration respectively. The above equations can be integrated after substituting linear wave theory expressions for u and $\dot{u}$ and setting x = 0 and $\theta = \omega t$. The result is

$$F = \frac{2\pi \rho r H^2 L}{T^2}\left[A_1 C_m \sin\theta + A_2 C_d |\cos\theta| \cos\theta\right] \qquad \textbf{3-4}$$

and

$$M = \frac{2\pi \rho r H^2 L^2}{T^2}\left[A_3 C_m \sin\theta + A_4 C_d |\cos\theta| \cos\theta\right] \qquad \textbf{3-5}$$

where

$$A_1 = \frac{\pi r}{2H}$$

$$A_2 = \frac{1}{16 \sinh^2 kd}\left[2kd + \sinh 2kd\right]$$

$$A_3 = \frac{\pi r}{4H \sinh kd}\left[1 + kd \sinh kd - \cosh kd\right]$$

$$A_4 = -\frac{1}{64 \sinh^2 kd}\left[2k^2 d^2 + 2kd \sinh 2kd + 1 - \cosh 2kd\right]$$

Figure 3-15 shows a typical time history of the inertia, drag, and total force on a vertical cylinder, and Figure 3-16 shows drag coefficients for a cylinder as a function of Reynolds number and relative roughness. Inertia and drag coefficients for a circular cylinder as a function of Reynolds and Keulegan-Carpenter number are shown in Figure 3-17 and Figure 3-18 respectively.

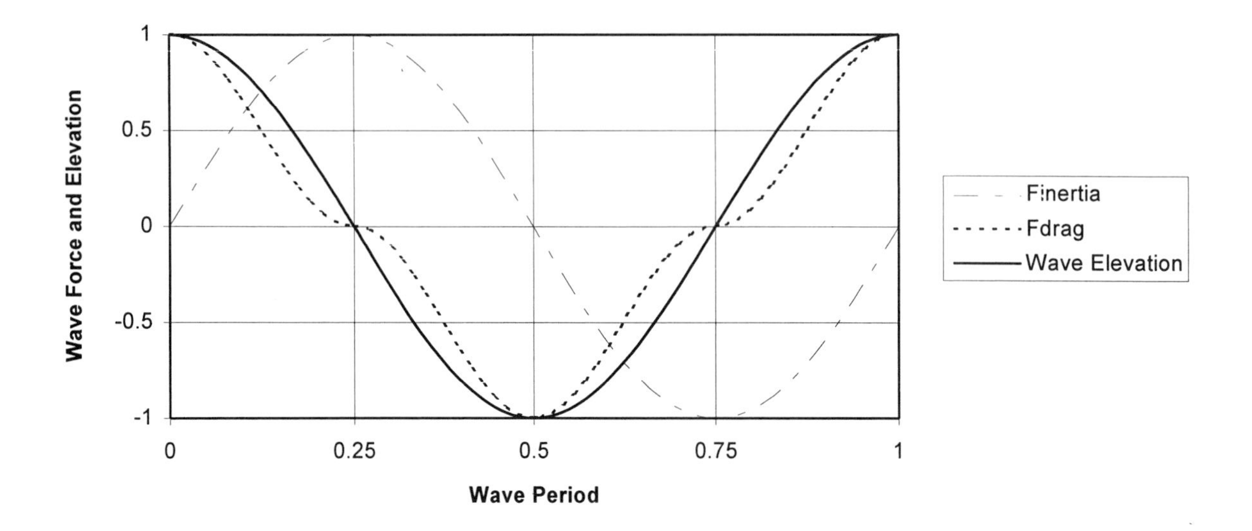

Figure 3-15. Typical time history over a single wave period of wave force on vertical circular cylinder.

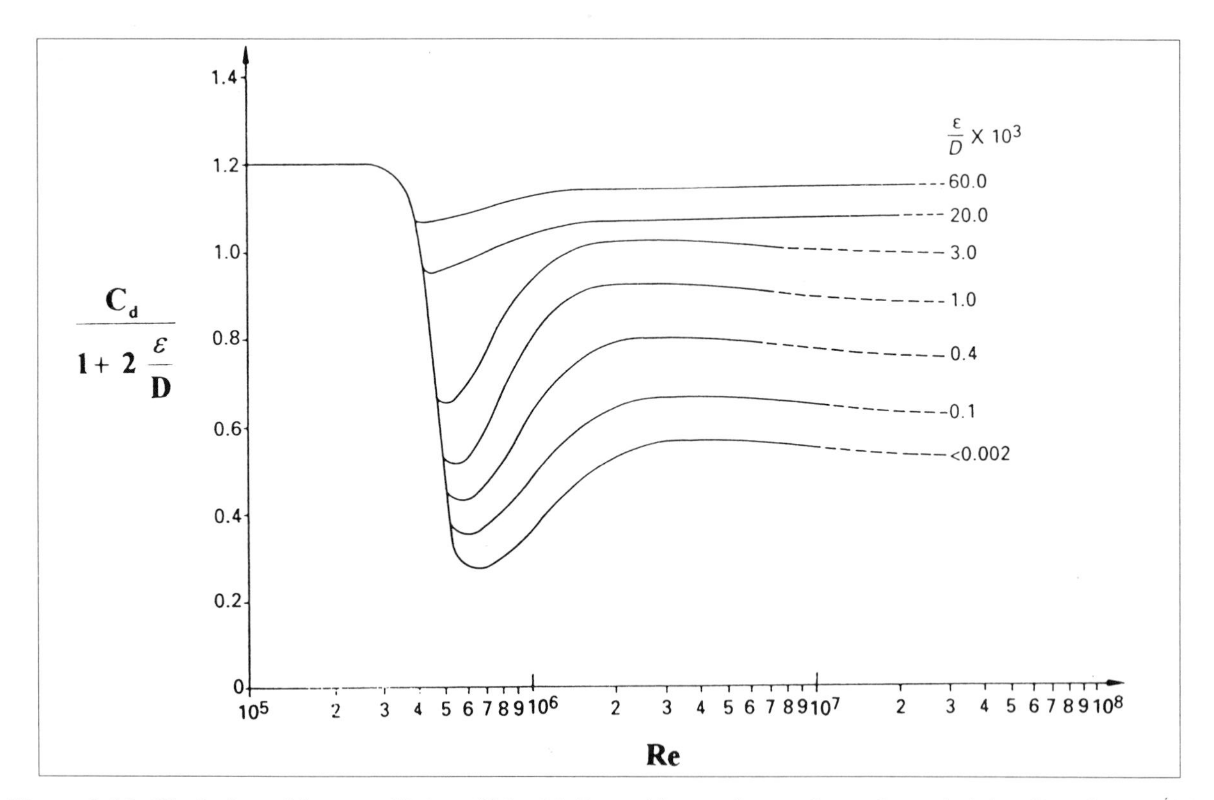

Figure 3-16. Variation of drag coefficient (C_d) with Reynolds number and roughness height. Reprinted with permission from Patel, 1989, *Dynamics of Offshore Structures*. (Full citing in references).

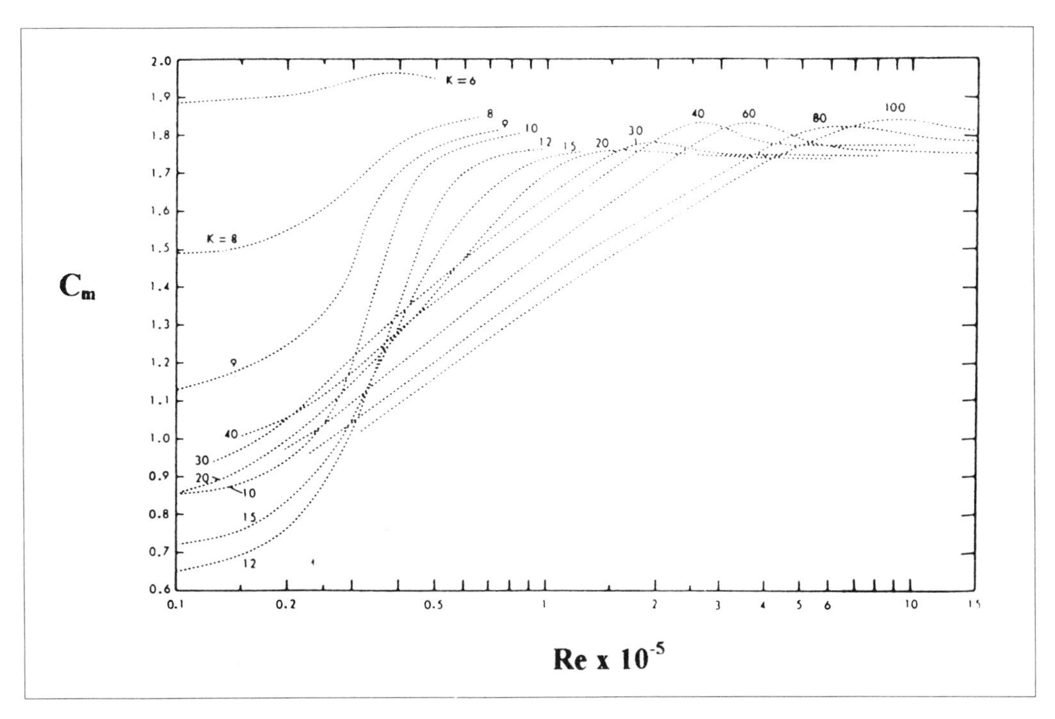

Figure 3-17. Inertia coefficient for cylinder as a function of Reynolds and Keulegan-Carpenter numbers. Reprinted with permission, Sarpkaya and Isaacson, 1981, *Mechanics of Wave forces on Offshore Structures*. (Full citing in references).

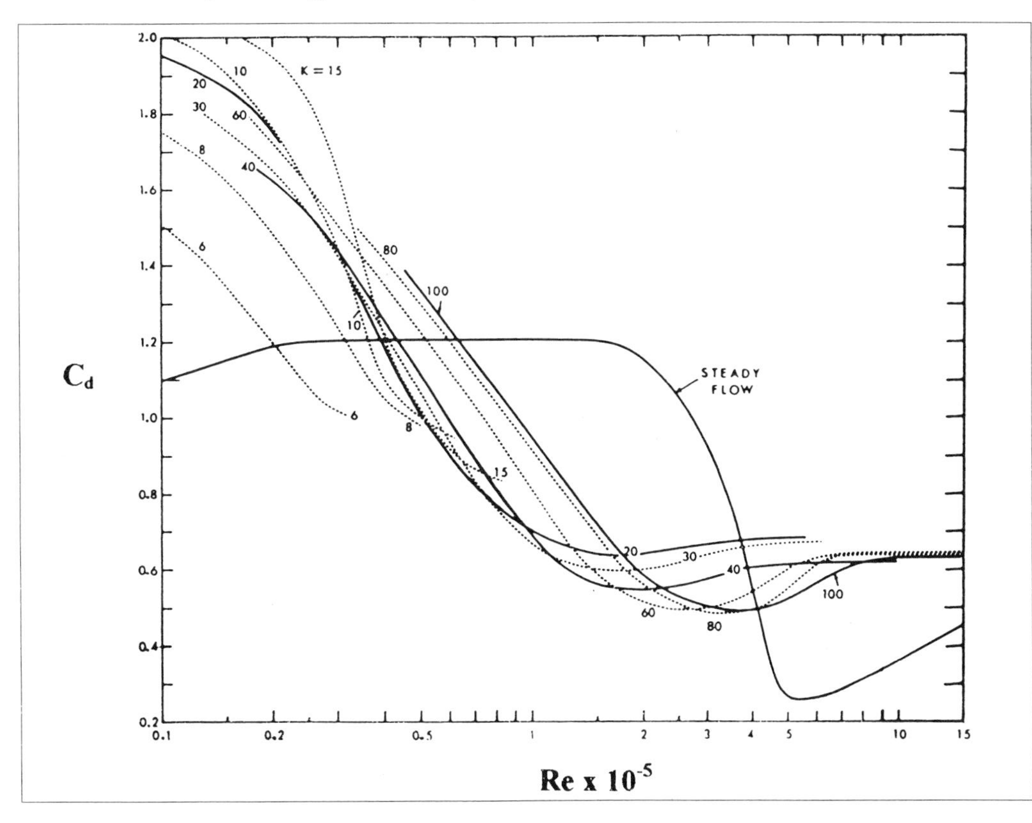

Figure 3-18. Drag coefficient for cylinder as a function of Reynolds and Keulegan-Carpenter numbers. Reprinted with permission, Sarpkaya and Isaacson, 1981, *Mechanics of Wave forces on Offshore Structures*. (Full citing in references).

Example Problem 3-1

A fixed jacketed structure is located in 30 m (98.4 ft) water depth and is subjected to a 4 m (13.2 ft) high, 11 s wave. The main legs of the structure are 1 m (3.3 ft) diameter vertical circular steel pipes. Calculate and plot the drag, inertia, and total force variation for one leg over one wave period. Assume linear wave theory is valid. The solution to this example problem is tabulated in Table 3-3, and the plot of the forces is shown in Figure 3-19.

Table 3-3. Tabulated results for example problem 3-1.

Given:	Wave Period (T) s	11
	Gravity (g) m/s^2	9.81
	Depth (d) m	30
	Wave Height (H) m	4
	Diameter (D) m	1
	Density (ρ) kg/m^3	1030
	Kinematic viscosity (ν)m^2/s	1.17E-06
Find:	$F_{inertia}$, F_{drag}, F_{total} on one leg over one wave period and plot	
Solution:	Deep water wave length (Lo) m Lo = 1.56*T^2	188.76
	Relative water depth d/Lo. This is an intermediate water depth.	0.158932
	Wave length for intermediate water depth L using Equation 2-40	164.7661
	Wave Number (k) k=2π/L	0.038134
	Maximum Horizontal Velocity (umax) m/s using Equation 2-47	1.400403
	Keulegan-Carpenter Number (K=umax*T/D)	15.40443
	Reynolds Number (R=umax*D/kinematic viscosity)	1196925
	Drag coefficient Cd, Table 3-1	0.62
	Inertia coefficient Cm Table 3-1	1.8
	Evaluate forces using Morison Equation (Equation 3-4)	
	A2=2kd +sinh 2kd/(16 sinh^2(kd))	0.225118
	A1=π(a/2H)	0.196349
	C1=2*π*ρ*a*H^2*L/T^2	70500
	Results from Morison Equation for theta varying between 0 and 2π (one wave period)	

Theta (rad)	$F_{inertia}$ (N)	F_{drag} (N)	F_{total} (N)	Theta (rad)	$F_{inertia}$ (N)	F_{drag} (N)	F_{total} (N)
0	0	9839.91	9839.91	3.2	-1454.49	-9806.38	-11260.9
0.2	4950.191	9451.534	14401.72	3.4	-6367.25	-9197.35	-15564.6
0.4	9703.034	8347.721	18050.75	3.6	-11026.2	-7913.02	-18939.2
0.6	14069.05	6702.739	20771.79	3.8	-15245.5	-6156.14	-21401.6
0.8	17874.17	4776.295	22650.47	4	-18857	-4204.1	-23061.1
1	20966.71	2872.531	23839.24	4.2	-21716.8	-2365.08	-24081.9
1.2	23223.37	1292.011	24515.38	4.4	-23710.8	-929.414	-24640.2
1.4	24554.19	284.2635	24838.45	4.6	-24759.5	-123.768	-24883.3
1.570796	24916.74	1.05E-09	24916.74	4.8	-24821.2	75.33506	-24745.8
1.8	24265.1	-507.944	23757.16	5	-23893.3	791.7608	-23101.5
2	22656.72	-1704.06	20952.67	5.2	-22012.8	2159.938	-19852.9
2.2	20145.09	-3407.89	16737.2	5.4	-19254.8	3963.861	-15290.9
2.4	16830.34	-5350.45	11479.89	5.6	-15729.1	5918.73	-9810.37
2.6	12844.61	-7225.04	5619.575	5.8	-11576.4	7715.914	-3860.46
2.8	8346.811	-8735.7	-388.893	6	-6962.12	9071.678	2109.556
3	3516.25	-9643.95	-6127.7	6.2	-2070.32	9771.977	7701.66
3.141593	-0.00863	-9839.91	-9839.92	6.283185	7.03E-05	9839.91	9839.91

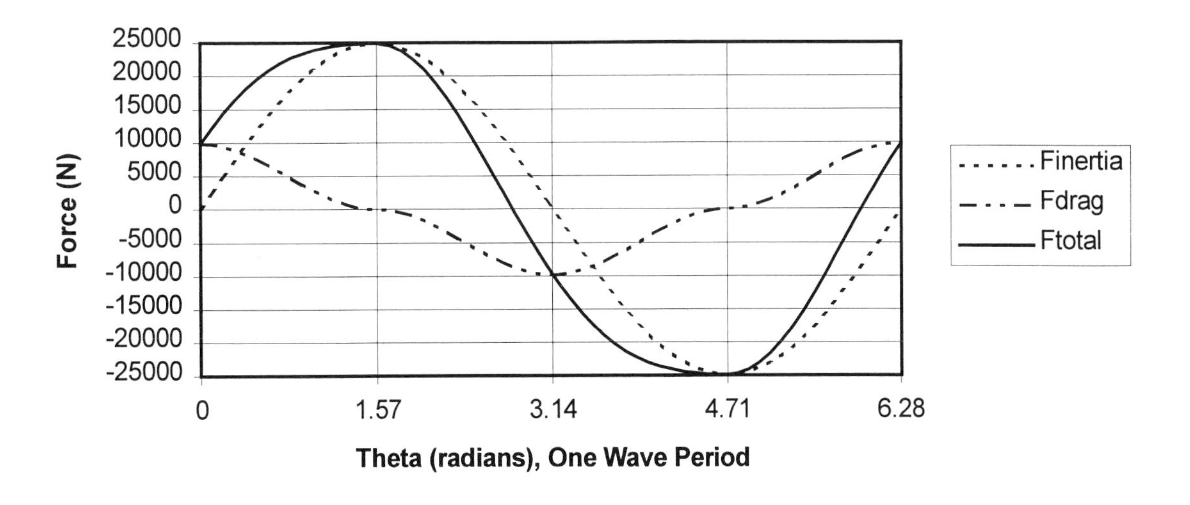

Figure 3-19. Inertia, drag, and total force for vertical circular cylinder in example problem.

Horizontal Cylinders

In the case of wave forces on horizontal submerged cylinders of length L with its longitudinal axis normal to the direction of wave propagation as shown in Figure 3-20, the wave forces are expressed as

$$F_h = C_m \rho \pi r^2 L \dot{u} + C_d \rho r L|u|u \qquad \text{3-6}$$

and

$$F_v = C_m \rho \pi r^2 L \dot{v} + C_d \rho r L|v|v \qquad \text{3-7}$$

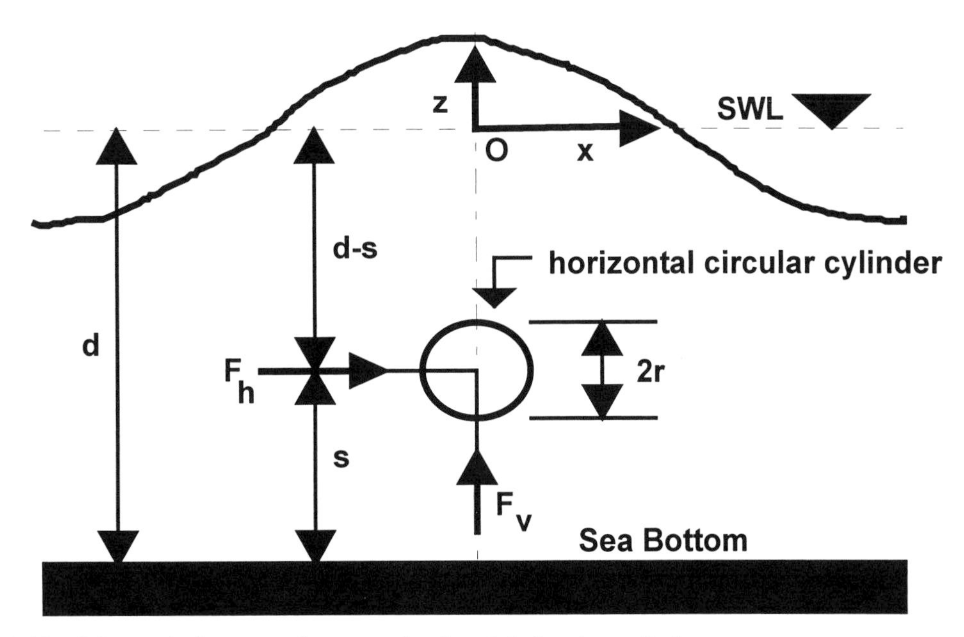

Figure 3-20. Schematic for wave forces on horizontal circular cylinder.

For the case of the horizontal submerged cylinder with its longitudinal axis inline with the direction of wave propagation, the expression for the horizontal force is no longer applicable, and the expression for the vertical force must be integrated over the length of the cylinder.

The Morison equation can also be applied to inclined circular cylinders. One approach is to resolve the wave velocity components into the total normal velocity component before computing the forces. The equation is written in vector form using the normal component approach with the wave force component normal to the cylinder axis on an element of length ds as

$$d\vec{F} = C_m \rho \pi r^2 \dot{\vec{q}}_n \, ds + C_d \rho r |q_n| q_n \, ds \qquad \textbf{3-8}$$

where $\vec{q}_n$ is a wave velocity vector normal to the cylinder axis at the length element. Questions still remain to be answered concerning wave forces on inclined cylinders and research continues on this subject.

WIND AND CURRENT FORCES

Winds and currents cause forces on submerged and exposed members of offshore and coastal structures (i.e. offshore platforms, semisubmersibles, fishing piers, marina docks and piers, etc.). The procedure for evaluating forces due to wind or current are similar with the major difference being the physical properties (i.e. density, viscosity) of the fluid medium. The general equation for evaluating the wind and current forces on slender structures is

$$F = \frac{1}{2} C_d \rho A U^2 \qquad \textbf{3-9}$$

where U is the fluid velocity, ρ is the fluid density, A is the frontal area facing the flow (i.e. length of cylinder times the diameter) and C_d is the drag coefficient that is determined from experience and experiment.

The value of C_d is sometimes very controversial, but in general it is known for typical shapes, Reynolds numbers, and relative roughness. Relative roughness (ε/D) is the ratio of the roughness distance (ε) to the diameter (D) or other characteristic length dimension. Figure 3-21 shows the relationship of the drag coefficient (C_d) for various body shapes as a function of Reynolds number (Re) for steady flow.

There are two types of wind speeds that are considered in the design of offshore and coastal structures. These are called sustained and gust wind speeds. The sustained wind speed is defined as the average wind speed over a time of one minute at an elevation of 10 m above the still water level (SWL). Similar to the evaluation of wave forces, it is common practice to define a sustained wind speed that occurs for a 50 to 100 year return period. The wind velocity above the still water level varies with height due to the boundary layer effect. The speed at a height (z) above SWL is related to the speed at the 10 m height by

$$U_z = U_{10}\left(\frac{z}{10}\right)^{0.113} \qquad \textbf{3-10}$$

where U_{10} is the 1-min mean sustained wind speed 10 m (32.8 ft) above the SWL. The average wind speed measured over a time of 3 s at an elevation of 10 m above SWL is the gust wind speed. The use of the 50 and 100 year return period are again used for design purposes.

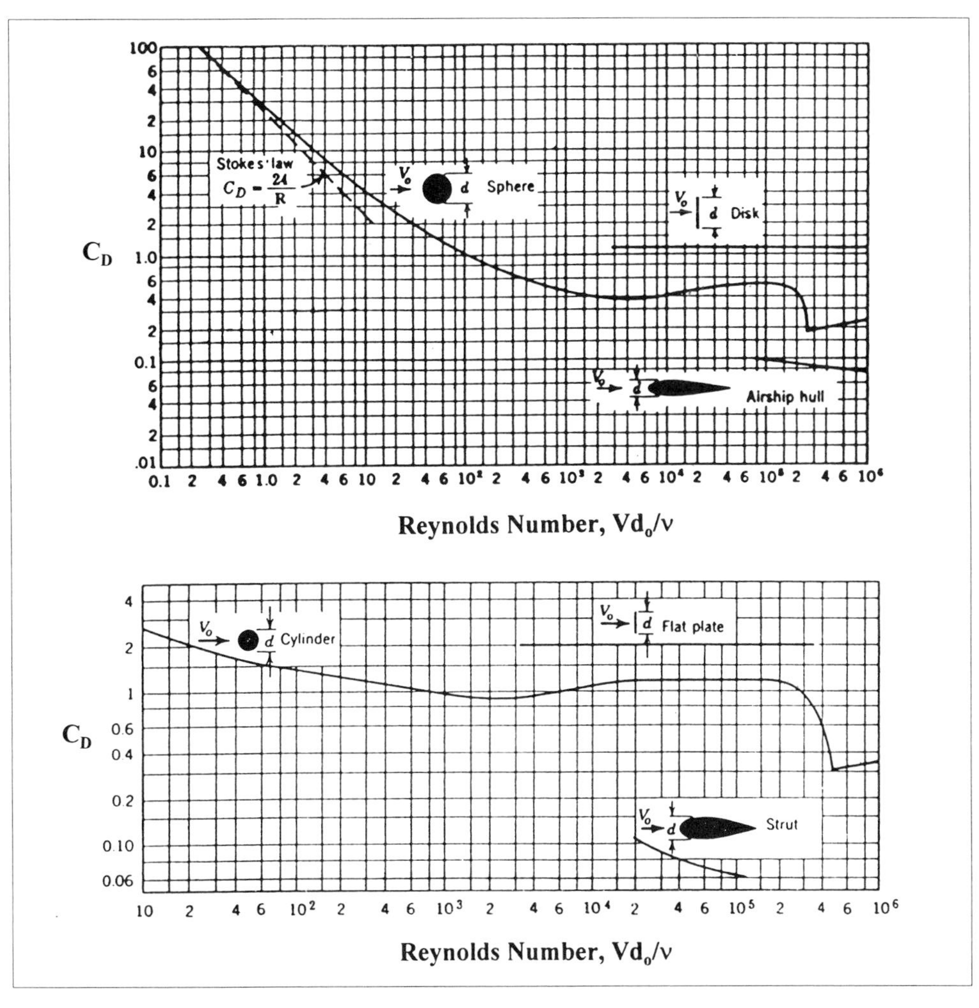

Figure 3-21. Drag coefficient as a function of Reynolds number. Reprinted with permission Vennard and Street, 1982, *Elementary Fluid Mechanics*. (Full citing in references).

Adjustments for elevation are determined by

$$\left(U_z\right)_{gust} = \left(U_{10}\right)_{gust}\left(\frac{z}{10}\right)^{0.1} \tag{3-11}$$

Values of C_d are typically around 1.1 to 1.3, but higher values can be used in design to incorporate a safety factor. For long slender members with length to diameter ratios greater than 5, C_d for sharp edged sections is generally between 1.5 to 2.0. For cylinders with diameters less than 0.3 m, C_d = 1.2 and for diameters greater than 0.3, C_d = 0.7. For shorter members the expression below is often used

$$C_{ds} = C_d\left(0.5 + \frac{0.1L}{D}\right) \tag{3-12}$$

where C_d and C_{ds} are drag coefficients for long and short members respectively, L is the member length, and D is diameter. Additional corrections may be made for members located behind each other in the direction of flow which is a shielding effect. Typically, members less than seven diameters away are considered to be shielded. Information regarding shielding is limited and most calculations assume no shielding, which provides an additional safety factor of the design.

OFFSHORE PIPELINES

Background

The transport of offshore gas and oil after production at offshore platforms often requires the transport of these fluids through horizontal pipelines that are placed on the seafloor or buried just beneath the seabed-seawater interface. These pipelines are classified as flowlines, gatherlines, trunk lines and loading/unloading lines according to Mousselli (1981). Flowlines are usually small diameter and may be bundled, and they connect the well to the platform or subsea manifold. The gatherlines are small to large diameter pipelines that connect between platforms, and they may also be bundled lines for oil, gas, condensate, or two phase flow. The combined flow from one or several platforms is accommodated through trunk lines that are large in diameter and transport products from different platforms that are often owned by different companies. Loading lines transport fluids between producing platforms or subsea manifolds to a loading facility through small to large diameter pipes.

Offshore pipelines are placed on the seafloor by large pipelay vessels (Figure 3-22) that weld lengths of pipe that are subsequently deployed over a device called a stinger which minimizes the bending stresses in the pipe. The pipe is laid in an "S" shaped fashion from the vessel to the seafloor. Typical vessels used to lay pipe are barges or semisubmersibles. Newer pipelay vessels are using a "J" configuration that welds the pipe in a vertical position and deploys the pipe vertically. In some cases the pipes are towed to location and deployed. These pipelines may also be placed in trenches and later covered with seabed sediments.

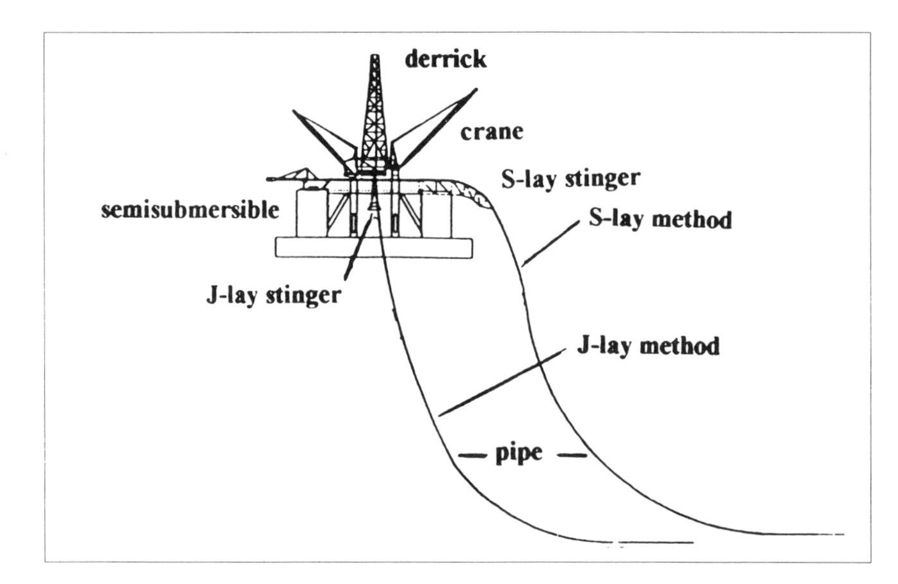

Figure 3-22. Example of a S-lay and J-lay pipeline vessel.

The design of offshore pipelines requires the consideration of many elements such as line sizing, hydrodynamic forces, geotechnical characteristics of the seafloor sediments and scour, structural analysis for bucking and internal pressure, and pipe lay analysis for effects of vessel motions. Permits are required from regulatory bodies before pipelines can be placed on the seafloor. The routes of pipelines must be predetermined to minimize effects of irregular sea bottom conditions resulting in unsupported lengths of pipe. In general, pipelines are designed so they do not float and so that they resist the corrosive effects of the ocean environment.

Forces on Pipelines Due to Waves and Currents

Offshore pipelines must resist forces caused by currents and waves while resting on the seafloor. These forces (Figure 3-23) include drag, lift, inertia, and frictional resistance between seabed and pipe.

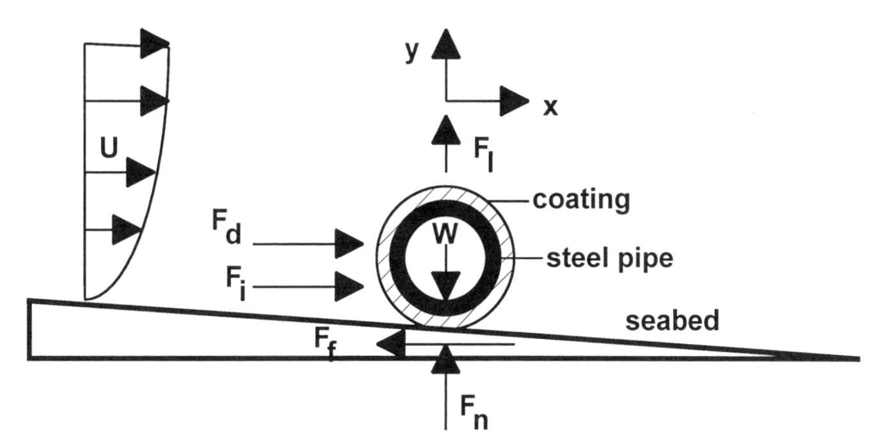

Figure 3-23. Schematic of forces acting on an offshore pipeline resting on the seabed.

For the pipeline to remain stable on the seabed, the forces acting on the pipeline must be in equilibrium. The static equilibrium equations for the pipeline horizontal and vertical forces are

$$F_d + F_i - F_f - F_w \sin\beta = 0 \qquad \textbf{3-13}$$

$$F_n + F_l - F_w \cos\beta = 0 \qquad \textbf{3-14}$$

where F_d is drag force, F_i is inertia force, F_f is the frictional resistance, F_w is the submerged unit weight of the pipe, F_n is the normal force, F_l is the lift force and β is the slope angle of the seabed. When a pipeline is resting on the seabed with only a small amount of embedment, the frictional resistance force F_f is related to the normal force F_n by

$$F_f = \mu F_n \qquad \textbf{3-15}$$

where μ is the coefficient of friction between the seabed and pipe. These three equations are combined to yield an equation for the minimum submerged pipe weight to remain on the seabed as

$$F_w = \frac{F_d + F_i + \mu F_l}{\mu \cos\beta + \sin\beta} \qquad \textbf{3-16}$$

The Morison equation is used to evaluate drag and inertia forces as discussed previously, but an effective velocity u_e is used due to the velocity profile near the seabed caused by the bottom boundary layer. The effective velocity may be determined using the 1/7th power law

$$\frac{U}{U_0} = \left(\frac{y}{y_0}\right)^{\frac{1}{7}} \tag{3-17}$$

where U_0 is the measured or theoretically determined horizontal particle velocity at a height y_0 above the seabed. The effective velocity is determined by integrating over the vertical distance equal to the pipe diameter.

$$u_e^2 = \frac{1}{D}\int_0^D u^2(y)\,dy \tag{3-18}$$

As a result, the effective velocity is

$$U_e^2 = 0.778\,U_0^2\left(\frac{D}{y_0}\right)^{0.286} \tag{3-19}$$

The lift force per unit length is evaluated using the expression

$$F_l = \frac{C_L}{2}\rho D|U_e|U_e \tag{3-20}$$

Evaluation of the drag, inertia and lift forces depends on the selection of the respective coefficients. The inertia and drag coefficients have been discussed previously, and the lift coefficient is illustrated in Figure 3-24 as a function of the Reynolds number.

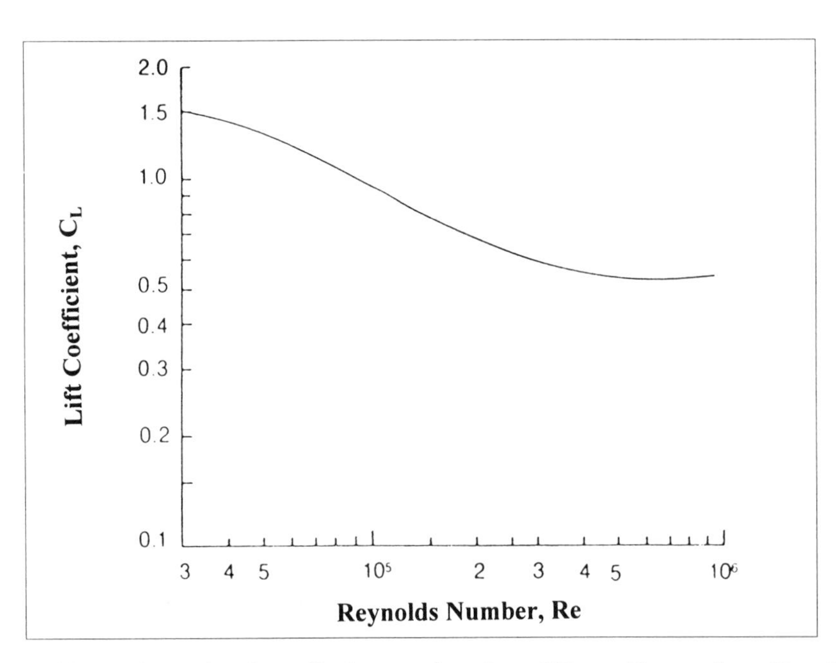

Figure 3-24. Lift coefficient for a circular cylinder as a function of Reynolds number. Reprinted with permission Mousselli, 1981, *Offshore Pipeline Design, Analysis, and Methods.* (Full citing in references).

The evaluation of forces on offshore pipelines depends heavily on the choice of several coefficients that have been determined by experimental measurements. The drag, lift and inertia coefficients for a cylinder in uniform (steady current) and oscillatory (under waves) flow have been discussed in this chapter, and the data are illustrated in several figures. These figures show the coefficients are a function of the Reynolds number and the Keulegan-Carpenter number. For practical pipeline design, Mousselli (1981) suggested values of these coefficients as shown in Table 3-4.

Table 3-4. Suggested hydrodynamic coefficients for practical pipeline design (Mousselli 1981).

Reynolds Number (Re)	Drag Coefficient (C_d)	Lift Coefficient (C_l)	Inertia Coefficient (C_m)
$Re < 5.0 \times 10^4$	1.3	1.5	2.0
$5.0 \times 10^4 < Re < 1 \times 10^5$	1.2	1.0	2.0
$1 \times 10^5 < Re < 2.5 \times 10^5$	$1.53 - \frac{Re}{3 \times 10^5}$	$1.2 - \frac{Re}{5 \times 10^5}$	2.0
$2.5 \times 10^5 < Re < 5.0 \times 10^5$	0.7	0.7	$2.5 - \frac{Re}{5 \times 10^5}$
$Re > 5.0 \times 10^5$	0.7	0.7	1.5
Note: Coefficients are for an exposed pipe with slight roughness.			

The coefficient of friction between the exposed pipe and the seabed is also needed for calculating the required weight of the pipe such that it remains on the bottom. The frictional resistance force resists the lateral movement of the pipe and depends also on the type of bottom sediment and type of pipe coating. A popular coating for offshore pipelines is concrete because it protects against corrosion and adds needed weight to keep the pipe on the seafloor. When a pipe lays on the seafloor, it tends to settle into the sediment, and the amount of settlement is called embedment. The amount of embedment also affects the frictional resistance. As expected the frictional resistance increases as the depth of embedment increases. Ranges of the coefficient of friction for concrete coated pipes with very small embedment and for different sediment types are tabulated in Table 3-5.

Table 3-5. Coefficients of friction for concrete coated pipes with only small embedment (Mousselli 1981).

Sediment	Coefficient of Friction (C_f)
Clay	0.3 - 0.6
Sand	0.5 - 0.7
Gravel	0.5

Example Problem 3-2

A concrete coated steel pipeline with a 15 in outside diameter is to be installed in 125 ft water depth where the seabed sediment is sand. The wave conditions at the location are a significant wave height and period of 12 ft and 10.5 s, respectively. An average current (U_o) measured at 3 ft above the bottom is 0.4 ft/s flowing normal to the pipeline route. The wave direction is assumed normal to the pipe and the bottom slope is 1 degree. Using linear wave theory, determine the required submerged unit weight for the pipeline. The results of this problem are outlined in Table 3-6.

Table 3-6. Results of example offshore pipeline problem 3-2.

Given:	Wave Period (T) s	10.50
	Gravity (g) ft/s^2	32.20
	Depth (d) ft	125.00
	Pi	3.14
	Wave Height (H) ft	12.00
	Diameter (D) ft	1.25
	Density slugs/ft3	2.00
	Kinematic viscosity ft^2/s	1.26E-5
	Bottom slope degrees	1.00
	z to top of pipe ft	-123.75
	Distance from seabed to top of pipe y_o ft	1.25
	Steady current ft/s	0.40
	Distance steady current measured above bottom ft	3.00
Find:	Submerged unit weight of offshore pipeline F_w	
Solution:	Deep water wave length (Lo) ft Lo = 5.12 T^2	564.48
	Relative water depth d/Lo. This is an intermediate water depth.	0.22
	Wave length for intermediate water depth L. Equation 2-40	530.98
	Wave Number (k) $k=2\pi/L$	0.01
	Horizontal Velocity at 1.25 ft (umax) ft/s	1.73
	Effective velocity due to wave (Ue_{wave}). Equation 3-19 ft/s	1.52
	Effective velocity due to current $Ue_{current}$. Equation 3-19 ft/s	0.10
	Total Effective velocity ($Ue_{current} + Ue_{wave}$) ft/s	1.62
	Keulegan-Carpenter Number (K=Ueffective*T/D)	13.60
	Reynolds Number (R=Ueffective*D/kinematic viscosity)	1.61E05
	Drag Coefficient C_d from Table 3-4	0.99
	Inertia coefficient C_m from Table 3-4	2.00
	Friction Coefficient C_f from Table 3-5	0.60
	Lift Coefficient C_L from Table 3-4	0.88
	Horizontal acceleration 0.625 ft above bottom due to waves. Equation 2-50	1.03
	F_{drag} maximum occurs at theta =0 lb/ft	3.26
	$F_{inertia}$ maximum occurs at theta = $\pi/2$ lb/ft	5.07
	F_{lift} maximum occurs when F_{drag} is maximum. lb/ft	2.88
	F_{weight} minimum occurs when F_{drag} and F_{lift} are max and $F_{inertia}$ is 0. lb/ft	4.28

Offshore pipelines require consideration of the bottom bathymetry and try to avoid routes that have ridges and valleys that cause problems in supporting the pipeline. When a pipeline spans valleys, the water currents tend to accelerate between the pipe and seabed causing increased scour. It can also result in vortex shedding that can result in flow induced vibrations of the pipeline. Pipelines passing over ridges experience increased stresses and bending at the top of the ridge that can lead to pipe failures without special design changes. Stability of the seafloor sediments, settlement and liquefaction are geotechnical considerations that must be addressed in the design and installation of offshore pipelines. For further information on these subjects the reader is referred to more advanced texts such as Mousselli (1981), Blevins (1990), and Herbich (1981).

DIFFRACTION THEORY

Wave loads on large bodies (D/L > 0.2) are usually determined using diffraction theory. For these large structures the presence of the structure in flow field can not be neglected. The governing equation is known as the Laplace equation with the sea bed and free surface boundary conditions as well as no flow through the body surface. This results in additional waves being formed called scattered or diffracted waves. The wave force is then due to both the incident and scattered waves. For this case, drag forces are negligible and the potential (frictionless) flow solution of the wave diffraction problem gives realistic solutions. Further explanations of diffraction theory are found in more advanced texts such as Chakrabarti (1987), Sarpkaya and Isaacson (1981) and Patel (1989) to mention a few.

MARINE FOUNDATIONS

Marine foundations support ocean structures such as the offshore platforms and moorings that have just been discussed. Coastal structures such as breakwaters, piers, groins, jetties must also be supported by foundations. Piled foundations are frequently used when sediments are relatively soft as found in the Gulf of Mexico and Persian Gulf. In hard bottom areas such as the North Sea, the sea floor can support gravity foundations. Ocean engineers sometimes receive an introduction to marine foundations in a basic geotechnical engineering course. Without adequate attention to the foundation design, the offshore or coastal structure may experience severe settlement and as a result, fail to perform its designed purpose. More advanced geotechnical texts and literature and, more likely, geotechnical engineers must be consulted to design these critical marine foundations.

Pile Foundations

Pile foundations are commonly used to support offshore structures especially in the Gulf of Mexico and other continental shelf regions where the sediments are relatively soft. Driven piles are one type of pile that is open ended and is driven into the seafloor with large impact hammers using steam, diesel or hydraulic power. The walls of the piles must be thick enough to withstand the stresses resulting from the pile driving operation. These stresses may be predicted using analytical techniques (API 1993). Drilled and grouted piles can be used when the sediments support an open hole. In this case, an oversized hole is drilled and a pile is lowered into the hole. Once the pile is placed in the hole, grout is placed in the annulus between the pile and the soil. In some cases, two piles are placed in the holes and subsequently grouted together. The third type of pile is the belled pile that is constructed with a flared bell at one end of the pile to provide increased bearing and uplift capacity.

The design of pile foundations must consider pile diameter, penetration, type of tip, wall thickness, number of piles, spacing, geometry, location, material strength, method of installation, restraining conditions at the mudline, and other appropriate parameters. Design is normally based upon allowable stress with safety factors of 1.5 to 2.0. The ultimate bearing capacity for axially loaded piles can be determined from

$$Q_d = Q_f + Q_p = f A_s + q A_p \qquad \textbf{3-21}$$

where Q_f is skin friction resistance (lb), Q_p is the total end bearing (lb), f is the unit skin friction capacity (lb/ft^2), A_s is the side surface area of the pile (ft^2), q is the unit end bearing capacity (lb/ft^2) and A_p is the total end area of the pile (ft^2). The unit skin friction (f) and end bearing capacity for cohesive sediments (silts and clays) can be determined by

$$f = \alpha C \qquad \textbf{3-22}$$

where C is the undrained shear strength of the soil and α is a dimensionless factor that is computed from

$$\begin{aligned} \alpha &= 0.5\,\psi^{-0.5} \quad & \psi \le 1.0 \\ \alpha &= 0.5\,\psi^{-0.25} \quad & \psi \rangle\ 1.0 \end{aligned} \qquad \textbf{3-23}$$

The term ψ is the ratio of the undrained shear strength (C) to the effective overburden pressure (p_o'). Determination of C and p_o' are discussed in API (1993). The unit end bearing capacity for cohesive soils is determined from

$$q = 9C \qquad \textbf{3-24}$$

The friction acts on both the inside and outside of the pile, and the total resistance is the sum of the inside and outside friction and the end bearing on the pipe wall annulus.

For cohesionless sediments (sands), the unit skin friction capacity is determined from

$$f = K p_o \tan\delta \qquad \textbf{3-25}$$

where K is the coefficient of lateral earth pressure, p_o is the effective overburden pressure, and δ is the friction angle between the soil and the pile wall. The value of K is normally 0.8 for open ended piles driven unplugged, and it is 1.0 for plugged or closed end piles. The unit end bearing capacity is found by

$$q = p_o N_q \qquad \textbf{3-26}$$

where p_o is the effective overburden pressure at the pile tip and N_q is the dimensionless bearing capacity factor. Some suggested design values for cohesionless soils are tabulated in Table 3-7 that should be interpreted as guidelines only.

The above discussion of marine foundations is just a beginning and is discussed in other texts, conference proceedings, and journal papers. The annual Offshore Technology Conference Proceedings (1968 - present) are good sources of technical and design considerations for marine foundations that support offshore structures. Similar foundations are used to support anchoring points for tension leg platforms. Shallow foundations are also important for coastal applications. Geotechnical textbooks and literature should be consulted for additional information in this technical area.

Table 3-7. Suggested design parameters for cohesionless sediments (API 1993).

Density	Sediment Description	Sediment-pile Friction Angle (δ) degrees	Unit Skin Friction (f) kips/ft^2	N_q	Unit End Bearing Capacity (q) kips/ft^2
Very loose Loose Medium	Sand Sand-silt Silt	15	1.0	8	40
Loose Medium Dense	Sand Sand-silt Silt	20	1.4	12	60
Medium Dense	Sand Sand-silt	25	1.7	20	100
Dense Very dense	Sand Sand-silt	30	2.0	40	200
Dense Very dense	Gravel Sand	35	2.4	50	250

REFERENCES

American Petroleum Institute (API). "Recommended practice for planning, designing and constructing fixed offshore platforms, Working stress design." API RP WSD 2A, 20th Edition, Washington, July 1993.

American Petroleum Institute (API). "Recommended practice for planning, designing and constructing tension leg platforms." API RP 2T, Washington, April 1987.

Barltrop, N. D. P., and A. J. Adams. *Dynamics of Fixed Marine Structures*, Third Edition. Oxford: Butterworth-Heinemann Ltd., 1991.

Blevins, R. D. *Flow-Induced Vibration*, 2nd Edition. New York: Van Nostrand Reinhold, 1990.

Chakrabarti, S. K. *Hydrodynamics of Offshore Structures.* Boston: Computational Mechanics Publications, 1987.

Det Norske Veritas (DNV). "Rules for the design, construction and inspection of offshore structures.," Oslo, 1977.

Faltinsen, O. M. *Sea Loads on Ships and Offshore Structures.* Cambridge: Ocean Technology Series, 1990.

Furnes, O., and O. Loset. "Shell structures in offshore platforms: design and application." *Engineering Structures*, 3(1980):140-152. Figure 3-6 reprinted with permission: "Source: Furnes and Loset. "Shell structures in offshore platforms: design and application." Copyright *Engineering Structures*, 3(1980):140-152."

Hamilton, J., and G. R. Perrett. "Deep water tension leg platform designs." *Proceedings of the Royal Institution of Naval Architects International Symposium on Developments in Deeper Waters*, October 6-7, 1986, p. 10. Figure 3-7 reprinted with permission: "Source: Hamilton and Perrett, "Deep water tension leg platform designs." Copyright *Proceedings of the Royal Institution of Naval Architects International Symposium on Developments in Deeper Waters*, October 6-7, 1986."

Herbich, J. B. *Offshore Pipeline Design Elements.* New York: Marcel Dekker Inc., 1981.

Hsu, T. H. *Applied Offshore Structural Engineering.* Houston: Gulf Publishing Co., 1984.

Mather, A. *Offshore Engineering: An Introduction.* London: Witherby & Co. Ltd., 1995. Figure 3-1 reprinted with permission: "Source: Mather, *Offshore Engineering: An Introduction,* Copy right Witherby & Co. Ltd., 1995."

McClelland, B. and M. D. Reifel, Editors. *Planning and Design of Fixed Offshore Platforms.* New York: Van Nostrand Reinhold Company, 1986. Figure 3-5 reprinted with permission: "Source: McClelland and Reifel, *Planning and Design of Fixed Offshore Platforms*. Copyright Van Nostrand Reinhold Company, 1986."

Morison, J. R., M. P. O'Brien, J. W. Johnson and S. A. Schaaf. "The force exerted by surface waves on piles." *Petroleum Transactions*, AIME 189(1950):149-157.

Mousselli, A. H. *Offshore Pipeline Design, Analysis, and Methods.* Tulsa: PennWell Publishing Co., 1981. Figure 3-24 reprinted with permission: "Source: Mousselli, *Offshore Pipeline Design, Analysis, and Methods*. Copyright PennWell Publishing Co., 1981."

Patel, M. H. *Dynamics of Offshore Structures.* London: Butterworths & Co., 1989. Figures 3-2, 3-3, 3-4, 3-8, 3-9, 3-10, 3-11, and 3-16 reprinted with permission: "Source: Patel, *Dynamics of Offshore Structures,* Copyright Butterworths & Co., 1989."

Sarpkaya, T., and M. Isaacson. *Mechanics of Wave Forces on Offshore Structures.* New York: Van Nostrand Reinhold Co., 1981. Figures 3-17 and 3-18 reprinted with permission: "Source: Sarpkaya and Isaacson, *Mechanics of Wave Forces on Offshore Structures.* Copyright Van Nostrand Reinhold Co., 1981.

Vennard, J. K., and R. L. Street. *Elementary Fluid Mechanics*, Sixth Edition. New York: John Wiley & Sons, 1982. Figure 3-21 reprinted with permission: "Source: Vennard and Street, *Elementary Fluid Mechanics*, Sixth Edition. Copyright John Wiley & Sons, 1982."

PROBLEMS

3-1. A smooth vertical stainless steel pipe is totally submerged in sea water where the water depth is 50 ft. The pipe is 40 ft long with an outside diameter of 6 in and fixed at the sea floor. For a uniform current of 3.4 ft/s, evaluate the total force and moment acting at the sea floor (mud line). The density of sea water is 1.99 slugs/ft^3 and kinematic viscosity is 1.26 x 10^{-5} ft^2/s.

3-2. A 20 m cylindrical pole extends 20 m vertically above the water surface and is supporting a small anemometer which measures the wind speed. The pole has an outside diameter of 10 cm. The average one minute mean sustained wind speed measured by the anemometer is 120 km/hr. Evaluate the total wind force on the pole using the one minute mean sustained wind speed at 10 m above the water as the uniform wind speed over the pole.

3-3. A fixed jacketed structure is located in 60 m water depth and is subject to a 5 m high, 11 s wave. The main legs of the structure are 0.9 m diameter vertical circular steel pipes. Calculate and plot the drag, inertia and total force variation for one leg over one wave period. Assume linear wave theory is valid.

3-4. A vertical cylindrical pile is located in 50 ft of water and has a diameter of 12 in and length of 70 ft. A 3 ft wave with a 5 s period impacts the pile. Evaluate the maximum drag and inertia force on the pile.

3-5. A fishing pier extends 1500 ft into the ocean and is supported by 9 in cylindrical timber piles. The pier platform is 15 ft above the mean high tide level. The depth of water at high tide is 50 ft at end of pier. The maximum vertically averaged longshore current is 3.5 ft/s and the average wind speed is 100 mph. Evaluate the maximum force and moment at the sea floor for a single pile at the seaward end of the pier.

3-6. A concrete coated steel gas pipeline is to be laid between two offshore platforms in 100 m water depth where the maximum environmental conditions include waves of 20 m wave height and 14 s period. The pipeline outside diameter is 46 cm, and the clay bottom slope is 1 on 100. Determine the submerged unit weight of the pipe. Assume linear wave theory is valid and that the bottom current is negligible.

3-7. A steel pipeline for a sand by-passing project is installed across the bottom of a ship channel. The tidal current in the channel reaches a maximum of 2.5 kts. The channel depth is 45 ft with no slope. The slurry (sand-water mixture) has an average specific gravity of 1.2 in the 9 in outside diameter pipeline. Determine the submerged unit weight of the pipe.

3-8. A 2 ft outside diameter pipeline laying on a sand seabed moved during a hurricane event. Determine the wave height for a 15 s period necessary to just move the concrete coated steel pipe located in 180 ft of water. The current normal to the pipeline is assumed to be 0.8 kts. Use linear wave theory. Crude oil with a specific gravity of 0.86 is assumed to be in the pipe line. The thickness of the steel pipe is 0.25 in and the thickness of the concrete coating is 1 in.

3-9. A 9 in diameter horizontal cross-member of a steel jacketed structure is located near middepth where the maximum uniform current is 4.5 ft/s. Determine the force on the 300 ft long cross-member.

3-10. A vertical cylinder is located in 120 m water depth. It extends from the seafloor through the water surface and supports an instrument monitoring system. The cylinder has an outside diameter of 0.8 m and is designed for a 10 m high and 15 s wave. Calculate the maximum inertia wave force on the cylinder assuming linear theory with C_m equal to 1.8 and C_d equal to 0.6.

3-11. A horizontal pipe cross-member of a steel jacketed structure is located near middepth where the maximum uniform current is 5.1 ft/s. The diameter of the member is 1.5 ft. Determine the maximum force per unit length of the pipe.

CHAPTER 4: COASTAL PROCESSES AND STRUCTURES

INTRODUCTION

The ocean engineer is also involved with engineering in the coastal zone where the oceans meet the land. This coastal zone and its circulation are important for understanding coastal process and solutions for coastal system design. Waves approach the coast and the changing bathymetry causes them to shoal and refract, and structures such as jetties, breakwaters, artificial islands, and port entrances cause waves to diffract. Wave forecasting and hindcasting are important for the design of coastal structures and determining effects of extreme meteorological events such as hurricanes, storm surges, tsunamis, cyclones, and typhoons on the coastal environment. Sediment transport occurs along the coastlines and affects the erosion and accretion of the beaches, and scour is experienced in areas of high localized currents. Dredging is an industry that is important to the construction and maintenance of navigable waterways, commercial shipping ports, and recreational marinas as well as for the maintenance and restoration of beaches.

Coastal engineering application areas include:

- Nearshore wave, current, wind, and water level design conditions.
- Design of breakwaters, jetties, groins, seawalls, revetments, piers, towers, and pipelines.
- Control of beach erosion.
- Stabilization of tidal entrances.
- Prediction of inlet and estuary currents and water levels and their effect on water quality, sediment movement, and navigation.
- Wave forecasting and hindcasting.
- Oil spill control and clean-up.
- Protection of coastal areas from storm surges and tsunamis.
- Design of harbors, marinas, ports, marine pipeline outfalls, and offshore islands.
- Wave refraction, diffraction, and reflection analysis.
- Dredging and the disposal of dredged material.

Some of the major conferences and publications related to coastal engineering are:

- International Conference of Coastal Engineering (ICCE), every two years.
- Specialty Conferences, sponsored by American Society of Civil Engineers (ASCE). Examples: Dredging '94, Coastal Practices, Sediment Transport, Coastal Structures, and Floating Breakwaters.
- World Dredging Conference (WODCON), World Dredging Association (WODA)
- Permanent International Association of Navigation Congresses (PIANC)
- British Hydromechanics Research Association (BHRA)

Journals and periodicals that discuss basic and applied coastal engineering research include:

- Journal of Coastal Research, The Coastal Education and Research Foundation (CERF).
- Journal of Hydraulic Research, International Association for Hydraulic Research, IAHR.
- Journal of Waterway, Port, Coastal and Ocean Engineering, ASCE.
- Coastal Engineering, Elsevier.
- Shore and Beach, American Shore and Beach Association.
- World Dredging and Marine Construction, Symcon Publishing Co.
- Ocean Engineering, Pergamon Press.
- The Dock and Harbor Authority, Foxlow Publishing Co.
- Terra et Aqua, International Association of Dredging Companies (IADC).

Some U. S. government publications related to coastal engineering are:

- US Army Waterways Experiment Station; Coastal Engineering Research Center, Hydraulics Laboratory, Environmental Laboratory; Technical Reports, Technical Notes, Miscellaneous Reports; Vicksburg, MS.
- US Army Corps of Engineers Districts; Chicago, Galveston, Jacksonville, Los Angeles, Mobile, New England, New Orleans, Portland, San Francisco, Wilmington, and other coastal districts; Project Reports.
- US Naval Civil Engineering Laboratory; Technical Reports, Technical Notes, Miscellaneous Reports; Port Hueneme, CA.

In many areas of the world the ocean impinges on land over a sandy beach, and this demarcation between land and ocean is called the shore or shoreline. The sloping sandy beach offers protection from waves, currents, and storms. A schematic of the coastal zone is illustrated in Figure 4-1.

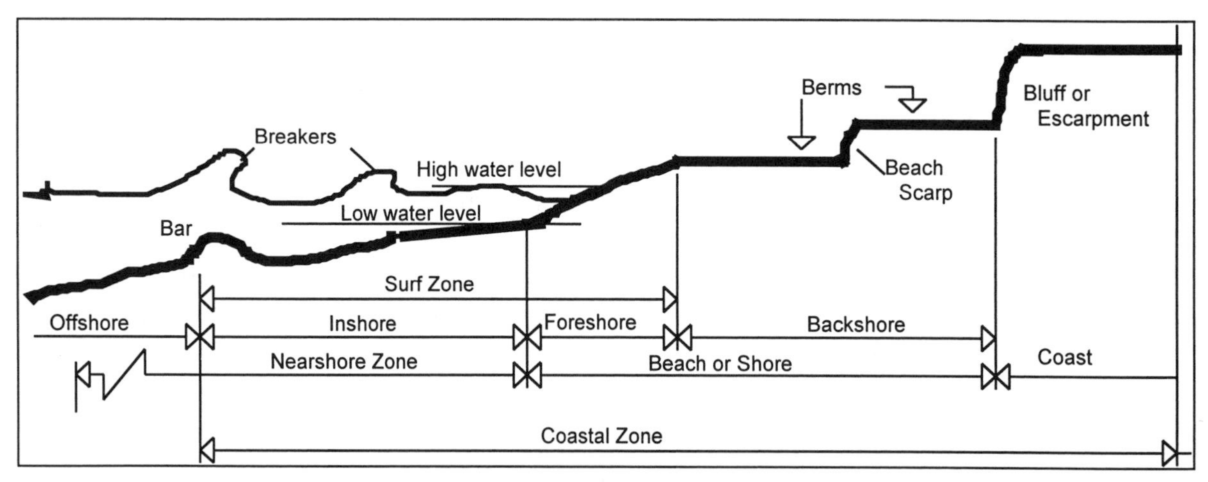

Figure 4-1. Schematic of typical beach profile and coastal zone.

The beach and nearshore zone are the location where ocean forces affect the land. This interaction is very complex and difficult to fully understand. These forces are due to wind waves, tides, currents, storm surges, hurricanes, and tsunamis. Most beaches have sediments ranging from fine sand to cobbles. The size and character of the sediments and the slope of the beach are related to the forces acting on the beach and the type of material available at the coast. Much of the beach material originates from land sources and is transported to the coast by rivers and streams. This material reaches the beach as sand and is transported by the longshore current. In some locations the material may be marine shell particles, volcanic ash, or coral reef particles. Silt and clay don't usually exist on the beach because the turbulence in the water is sufficient to keep it in suspension until it reaches more quiescent waters of lagoons, estuaries, or deeper offshore waters. However, there are some silt and clay (consolidated) beaches around the world.

In areas of flat sloped beaches, barrier islands often line the coasts and help protect the land from severe waves resulting from storms and hurricanes or cyclones. Protection of these barrier islands is important so that the mainland is not exposed to the most severe wave climates. Lagoons are shallow bodies of water usually found between the barrier islands and the mainland, and are a rich habitat for marine life and serve as safe harbors and navigable waterways. The narrow opening between the ocean and a lagoon or estuary is called an inlet. These inlets occur at fairly regular intervals along the coast, and changing environmental conditions result in these inlets opening and closing.

Beaches dissipate wave energy and are constantly adjusting to the wave environment. In normal conditions the beach can easily dissipate the wave energy, but in storm conditions, the beaches and dunes must sacrifice large amounts of sand and then redistribute the material in time after the storm has past. Natural and man-made underwater bars also protect beaches during storm events. An example of the effect of storms on a beach is illustrated in Figure 4-2.

Littoral transport is another dynamic response occurring along the beach and in the nearshore zone. It is defined as the transport of sediment in the nearshore zone by currents and waves. There are two types of littoral transport that are called longshore transport (parallel to the shore movement) and onshore-offshore transport (perpendicular to the shore). The material that is transported in this way is called littoral drift. Onshore-offshore transport is affected mostly by wave steepness, particle size, and beach slope. Steep waves move material offshore, and low (long period) waves move material onshore. Longshore transport is affected by breaking waves and their angle of approach to the beach or shoreline.

The attack of waves and currents on the beach can cause erosion. There are natural causes of beach erosion (e.g. waves and currents) and man-made causes due to efforts to protect the beaches and to a lack of understanding the physical processes. The various causes of beach erosion are tabulated in Table 4-1.

Table 4-1. Common natural and man-made causes of beach erosion.

Natural	Man-made
Sea level rise	Land subsidence
Variability of sediment supply	Interruption of sediment supply
Wave and surge overwash	Concentration of wave energy
Wind removal of beach sediment	Deepening and widening inlets
Longshore sediment transport	Changing natural coastal protection
Sorting of beach sediment	Removal of beach and dune material
Storm waves	Reduce sediment supply (river dams)

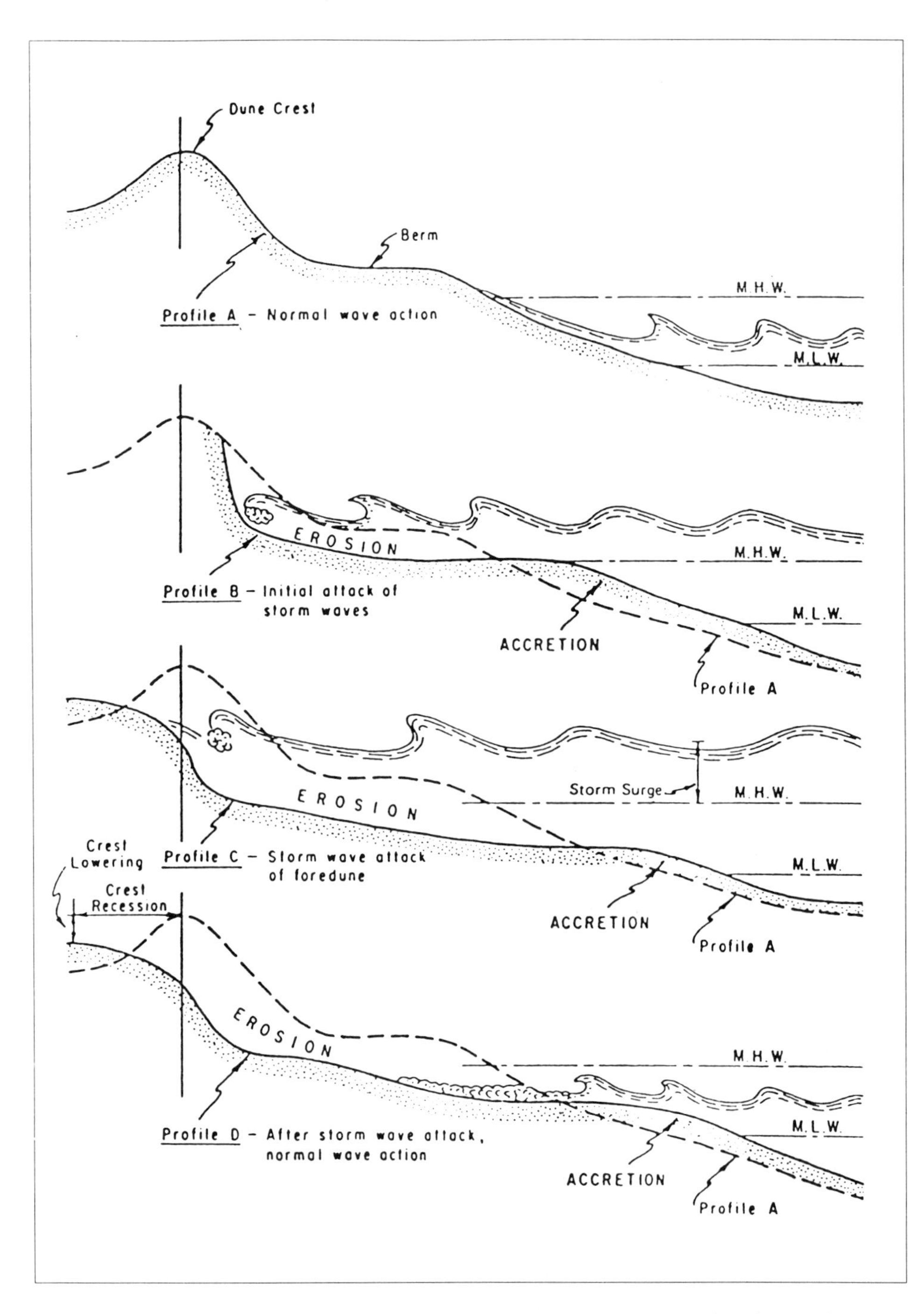

Figure 4-2. Effect of storm waves on beach and dune. Reprinted with permission from US Army Corps of Engineers (USACE),1984, *Shore Protection Manual.* (Full citing in references).

COASTAL STRUCTURES

There are many types of protective measures and modifications that are commonly constructed or put in place to attempt to counteract the erosion of beaches and coastlines. A summary of these protective methods is shown in Table 4-2.

Table 4-2. Summary of coastal protection methods.

Method	Brief Description
Beach nourishment or dune rebuilding	Sand placed on eroding beach or used to replace dune structure
Breakwaters	Large barriers to protect land or water behind them from wave attack.
Seawalls	Large wave resistant wall along shore.
Bulkheads	Smaller wave resistant wall along beach.
Revetments	Slope protection along waterways.
Groins	Structures perpendicular to shore to interrupt littoral transport or slow down sand losses.
Sand Bypassing	Movement of sand from accreting updrift side to eroding downdrift side of harbor or inlet.
Jetties	Structures used at inlets to stabilize entrance location.

Breakwaters

These structures protect a shore area, harbor, anchorage, or basin from waves and may be connected to the shore or placed offshore. Several types of breakwaters are in existence such as: rubble mound, cellular-steel sheet-pile, sheet-pile, stone-asphalt, and concrete caisson. An example of a cellular-steel sheet-pile and sheet-pile breakwater is illustrated in Figure 4-3.

Figure 4-3. Combined cellular-steel sheet-pile and sheet-pile breakwater at Port Sanilac, Michigan. Reprinted with permission from US Army Corps of Engineers (USACE),1984, *Shore Protection Manual.* (Full citing in references).

Jetties

These structures are constructed at mouths of rivers or tidal inlets to stabilize an entrance channel and prevent shoaling by littoral material. Rubble mound jetties are constructed with graded stone and capped with a top layer of large quarry stones or man-made armor units. Sheet-pile jetties are constructed with timber, steel, or concrete sheet piles which form an enclosure that is filled with stone or sand. The Humboldt Bay, California rubble mound jetty is shown in Figure 4-4.

Figure 4-4. Humbolt Bay, California rubble mound jetty. Reprinted with permission from US Army Corps of Engineers (USACE),1984, *Shore Protection Manual.* (Full citing in references).

Groins

This coastal structure is built normal to the beach and its purpose is to trap sediment on the beach and to retard erosion. Groin types include timber, steel, concrete, rubble mound, and asphalt. The design of these structures is described in US Army Corps of Engineers' Manuals. A rubble mound groin field is shown Figure 4-5.

Seawalls, Bulkheads, and Revetments

These structures separate land and water areas to prevent erosion and other damage due primarily to wave action. Seawalls are typically much larger structures designed to withstand the full force of storm waves. Bulkheads are much smaller in size and designed to retain shore material under less severe wave conditions. Revetments are designed to protect shorelines and waterways against erosion by currents and small waves. An example seawall is the Galveston, Texas curved face seawall in Figure 4-6.

Figure 4-5. Rubble mound groin field at Westhampton Beach, NY. Reprinted with permission from US Army Corps of Engineers (USACE), 1984, *Shore Protection Manual.* (Full citing in references).

Figure 4-6. Concrete curved face seawall at Galveston, Texas. Reprinted with permission from US Army Corps of Engineers (USACE), 1984, *Shore Protection Manual.* (Full citing in references).

Bulkheads are typically anchored vertical pile walls or gravity cellular steel structures. The structures are generally built with steel, timber, or concrete, and an example timber bulkhead is illustrated in Figure 4-7.

Figure 4-7. Timber sheet-pile bulkhead in Avalon, New Jersey. Reprinted with permission from US Army Corps of Engineers (USACE), 1984, *Shore Protection Manual.* (Full citing in references).

Revetments are either rigid, cast in place concrete, or flexible (articulated) armor units. The flexible units can tolerate minor consolidation or settlement. An interlocking concrete-block revetment is illustrated in Figure 4-8.

Figure 4-8. An interlocking concrete-block revetment at Jupiter Inlet, Florida . Reprinted with permission from US Army Corps of Engineers (USACE), 1984, *Shore Protection Manual.* (Full citing in references).

Beach Nourishment and Restoration

The protection and restoration of beaches is usually accomplished by land hauling or by direct pump-out from a dredge. There are many beaches along the US coastline that have been restored (e.g. Corpus Christi, TX; Miami Beach, Fl; Carolina Beach, NC; Rockway Beach, NY; Newport Beach, Ca; Sand Hill Cove Beach, RI; Virginia Beach, VA; Atlantic City, NJ to name a few). A before and after illustration of the Corpus Christi, TX beach is illustrated in Figure 4-9.

Figure 4-9. Restoration of beach in Corpus Christi, Texas. Reprinted with permission from US Army Corps of Engineers (USACE), 1984, *Shore Protection Manual.* (Full citing in references).

Sand Bypassing

Construction of jetties and breakwaters to protect harbors and tidal inlets along sandy coasts usually results in the interruption of the natural littoral transport along the coastline. As a result, the downdrift beach can be starved of sediment, and serious erosion takes place unless the sand is transferred or bypassed from the updrift to the downdrift side. Examples of barriers and sand bypassing schemes are shown in Figure 4-10.

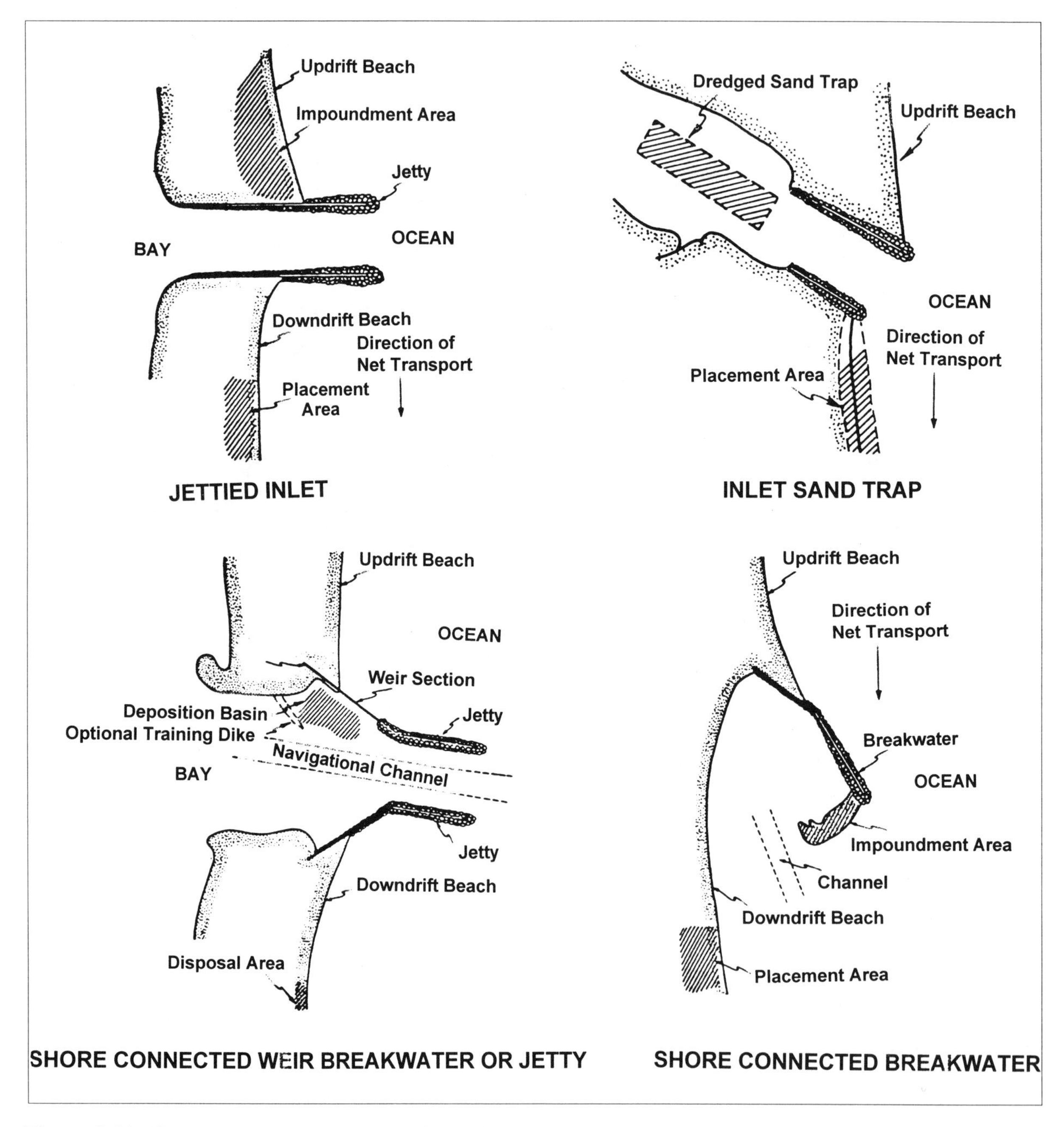

Figure 4-10. Examples of coastal barriers for which sand bypassing schemes have been used. Reprinted with permission from US Army Corps of Engineers (USACE), 1984, *Shore Protection Manual.* (Full citing in references).

Ports, Harbors, and Marinas

Marine facilities are typically specialized as to their operation which focuses on the type of cargo being handled. The major facility types include general, container, trailer, ferry, passenger/cruise ship, barge carrier, dry bulk and liquid bulk. Permanent berthing facilities and shipyards are not as active and such facilities include military and Coast Guard bases, shipyards, fishing and commercial small craft, research vessels and workboats and marinas (yachts and recreational boats). An example of a bulk cargo terminal and passenger/ferry facility are shown in Figure 4-11.

Figure 4-11. Example port and harbor faciltites. Reprinted with permission from Gaythwaite, 1990, *Design of Marine Facilities*. (Full citing in references).

WAVE REFRACTION, DIFFRACTION, AND REFLECTION

Refraction

The celerity of a wave approaching from deep water to the coast depends on the water depth in which it is propagating. As the celerity decreases with depth, the wavelength also decreases proportionally. The wave celerity varies along the crest of a wave moving at an angle to the underwater bathymetry, and this results in the wave crest bending to align itself with the bathymetry. This bending of the wave crests is called refraction. A similar refraction phenomena also occurs in the transmission of sound and light waves.

Refraction coupled with shoaling determines the wave height in a particular water depth for a wave originating in deep water. The change in wave direction causes convergence and divergence of wave energy that affects forces applied to coastal structures and beaches. It also has an effect on bottom topography due to the erosion and deposition of bottom sediments. Waves may also be refracted by currents that cause one section of a wave to travel faster than that of another. This effect has practical importance at tidal inlets where strong tidal currents can result in refraction. Examples of refraction are illustrated in Figure 4-12.

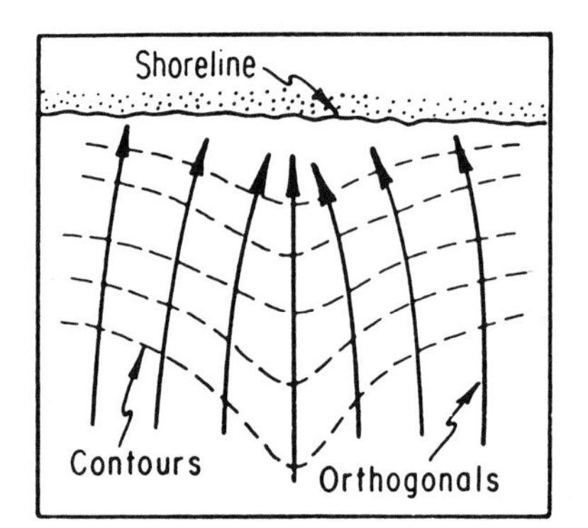

Submarine Ridge

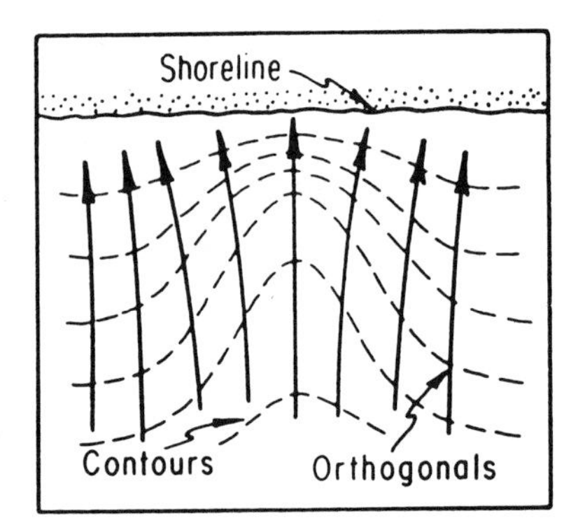

Submarine Canyon

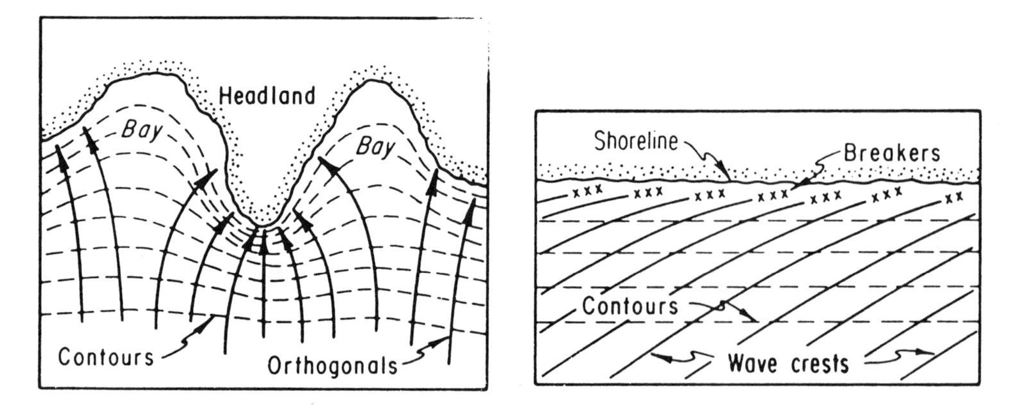

Irregular Shoreline **Straight Beach with Parallel Bottom Contours**

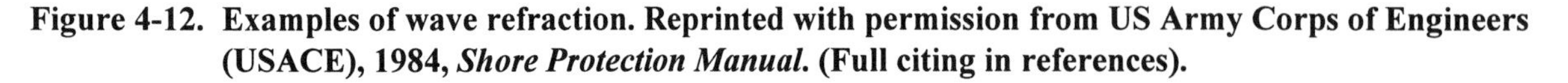

Figure 4-12. Examples of wave refraction. Reprinted with permission from US Army Corps of Engineers (USACE), 1984, *Shore Protection Manual.* (Full citing in references).

General Refraction Analysis

Refraction analysis assumes (1) constant wave energy between wave rays or orthogonals, (2) direction of wave advance is perpendicular to wave crest, (3) speed of wave of given period at a particular location depends only on depth at the location, (4) bathymetric changes are gradual, (5) waves are long-crested, constant period, small amplitude, and monochromatic, and (6) effects of currents, winds, and reflections are negligible. Recalling the expression for wave celerity

$$C^2 = \frac{gL}{2\pi}\tanh\left(\frac{2\pi d}{L}\right) \qquad \textbf{4-1}$$

and that in deepwater, it reduces to

$$C_o^2 = \frac{gL}{2\pi} \qquad \textbf{4-2}$$

In this equation the wave celerity is independent of depth, and consequently, refraction is not significant in deep water regions. However, in transitional and shallow water (d/L less than 0.5), wave celerity is dependent on the depth and refraction is significant. The total energy in a wave per unit crest width is given by

$$E = \frac{\rho g H^2 L}{8} \qquad \textbf{4-3}$$

In deep water only one half of the energy is transmitted forward with the wave. The amount of energy transmitted remains relatively constant as the wave moves toward the shore if the dissipative effects of bottom friction, percolation, and reflection are negligible. Therefore, the average deep water wave energy flux $(\overline{P}_o)$ transmitted across a plane between two adjacent orthogonals is

$$\overline{P}_o = \frac{1}{2} b_o \overline{E}_o C_o \qquad \textbf{4-4}$$

where b_o is the distance between orthogonals in deep water. In shallower water with the spacing (b), the average wave energy flux $(\overline{P})$ is

$$\overline{P} = nb\overline{E}C \qquad \textbf{4-5}$$

Equating the two equations 4-4 and 4-5 yields

$$\frac{\overline{E}}{\overline{E}_o} = \frac{1}{2}\left(\frac{1}{n}\right)\left(\frac{b_o}{b}\right)\left(\frac{C_o}{C}\right) \qquad \textbf{4-6}$$

Since the wave energy (E) is related to the wave height (H) as

$$\frac{\overline{H}}{\overline{H}_o} = \sqrt{\frac{\overline{E}}{\overline{E}_o}} \qquad \textbf{4-7}$$

then combining the two equations yields

$$\frac{H}{H_o} = \left[\left(\frac{1}{2}\right)\left(\frac{1}{n}\right)\left(\frac{C_o}{C}\right)\left(\frac{b_o}{b}\right)\right]^{\frac{1}{2}} \quad \textbf{4-8}$$

which can be written as

$$\frac{H}{H_o} = K_S K_R \quad \textbf{4-9}$$

where K_S is the shoaling coefficient and K_R is the refraction coefficient. These coefficients are defined as

$$K_R = \left[\frac{b_o}{b}\right]^{\frac{1}{2}} \quad \textbf{4-10}$$

$$K_S = \left[\left(\frac{1}{2}\right)\left(\frac{1}{n}\right)\left(\frac{C_o}{C}\right)\right]^{\frac{1}{2}} \quad \textbf{4-11}$$

The shoaling coefficient is useful in determining the wave height of a deep water wave as it propagates to shore without refracting. These shoaling coefficients are tabulated as a function of d/L in USACE (1984), Wiegel (1964), and in Appendix A. Bathymetric contours need to be parallel to the shore for no refraction. Refraction of waves can be determined either graphically or numerically. A graphical procedure is described in USACE (1984). A numerical technique or personal computer (PC) software program (Leenknecht et al. 1992) is contained in the Automated Coastal Engineering System (ACES) that is distributed by the Coastal Engineering Research Center at the US Army Engineer Waterways Experiment Station in Vicksburg, MS.

Diffraction

When wave energy is transferred laterally along a wave crest, the phenomenon is called wave diffraction. It is easily observed when a train of regular waves is interrupted by a physical barrier such as a breakwater or small island. If the lateral transfer of energy did not occur, then there would be perfectly calm water behind the barrier. The diffraction phenomena is shown in an aerial photograph (Figure 4-13) that shows actual diffraction of waves around a breakwater.

Wave diffraction in harbors and bays is an important consideration in the design of man-made structures that are used to protect against incident waves. Knowledge of wave diffraction is essential in the planning of such facilities. Proper design and location of harbor entrances to minimize silting and harbor resonance require the knowledge of wave diffraction.

Graphical wave diffraction analysis is described in USACE (1984) and Wiegel (1964). Numerical techniques are also available (USACE 1992). Most diffraction analyses assume (1) the fluid is inviscid and incompressible, (2) waves are of small amplitude (linear theory), (3) flow is irrotational and satisfies Laplace equation, and (4) the depth shoreward of the barrier is constant. When this last assumption is not satisfied, then refraction and diffraction occur simultaneously.

Figure 4-13. Aerial photograph of wave diffraction at Channel Islands Harbor breakwater in California. Reprinted with permission from US Army Corps of Engineers (USACE), 1984, *Shore Protection Manual.* (Full citing in references).

Reflection

Wave reflection occurs at coastal structures, sloping beaches and walls of wave tanks. The ratio of the reflected wave height, H_r, to the incident wave height, H_i, is called the reflection coefficient, C_r, and is defined as

$$C_r = \frac{H_r}{H_i} \qquad \textbf{4-12}$$

For waves approaching normal to the vertical impermeable walls, the reflection coefficient is unity. As the wall is inclined and has some degree of permeability, the reflection coefficient decreases. In harbor or ports, the waves approach obliquely to walls and very complicated wave patterns are formed.

In the case of beaches and man-made structures, the reflection of waves depends on the slope, roughness, and permeability. The surf similarity parameter (ξ) developed by Battjes

(1974) is commonly used to determine the reflection coefficient for a wave approaching normal to the coastal structure such as a plane sloping beach, rubble mound breakwater, and natural beach. The expression for the surf similarity parameter is

$$\xi = \frac{\tan \beta}{\sqrt{\frac{H_i}{L_o}}} \qquad \textbf{4-13}$$

where tan β is the bottom slope, H_i is the incident wave height and L_0 is the deepwater wavelength. Experimental data for the relationship between the surf similarity parameter and the reflection coefficient are shown in Figure 4-14.

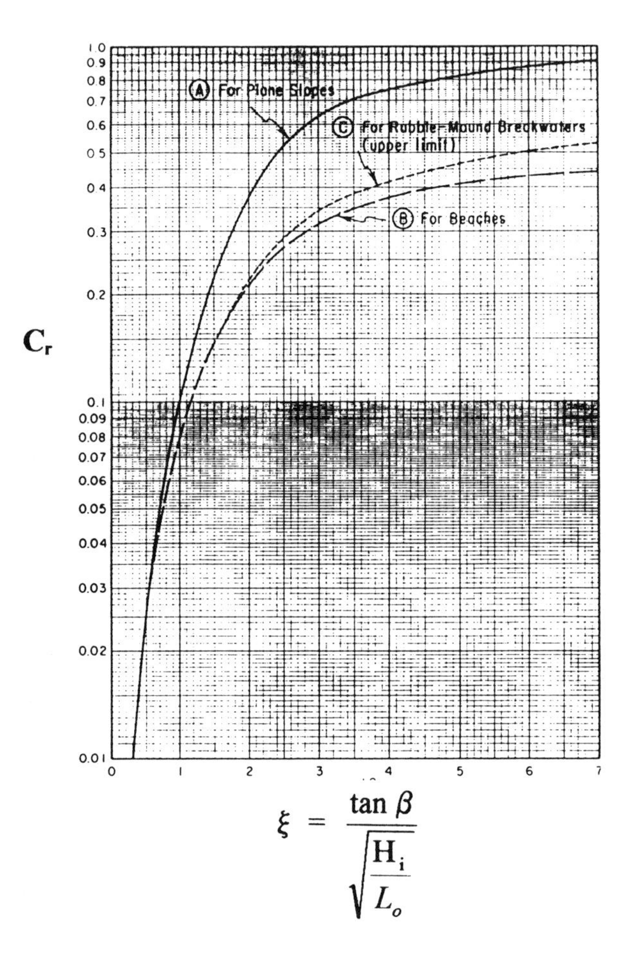

Figure 4-14. Coefficient of wave reflection as a function of the surf similarity parameter for plane slopes, beaches and rubble mound breakwaters. Reprinted with permission from US Army Corps of Engineers (USACE), 1984, *Shore Protection Manual.* (Full citing in references).

Additional information regarding wave reflection in enclosed basins such as harbors, lakes, and estuaries is addressed in Wiegel (1964), Ippen (1966), USACE (1984), and Horikawa (1988). The reflection of waves entering these enclosed bodies of water result in oscillations known as seiches. The period (T_i) of oscillation is approximated by Meriam's equation

$$T_i = \frac{2L_b}{i}\frac{1}{\sqrt{gd}} \qquad i=1,2,.... \qquad \text{4-14}$$

where L_b is the length of the basin, d is the depth, and "i" is the mode of oscillation. The fundamental period is obtained when "i" is one. As an example, Lake Erie is assumed to have an average depth of 18.6 m (61 ft) and a length of 354 km (220 mi). Using Equation 4-14, the fundamental period of oscillation is determined to be 14.6 hr which is in reasonable agreement with the observed value of 14.4 hr. This close agreement is not always expected due to variability of lake cross sections.

WAVE RUNUP

Wave runup occurs on beaches and coastal protection structures such as seawalls and breakwaters. A design concern is the determination of the distance the approaching water from waves advances up a beach, or structure, and possibly causes overtopping. The concept of wave runup is illustrated in Figure 4-15. Miche (1951) determined that the runup height (R) is a function of the beach slope and is expressed as

$$\frac{R}{H_o} = \sqrt{\frac{\pi}{2\beta}} \qquad \text{4-15}$$

where H_o is the deep water wave height and β is the beach slope angle.

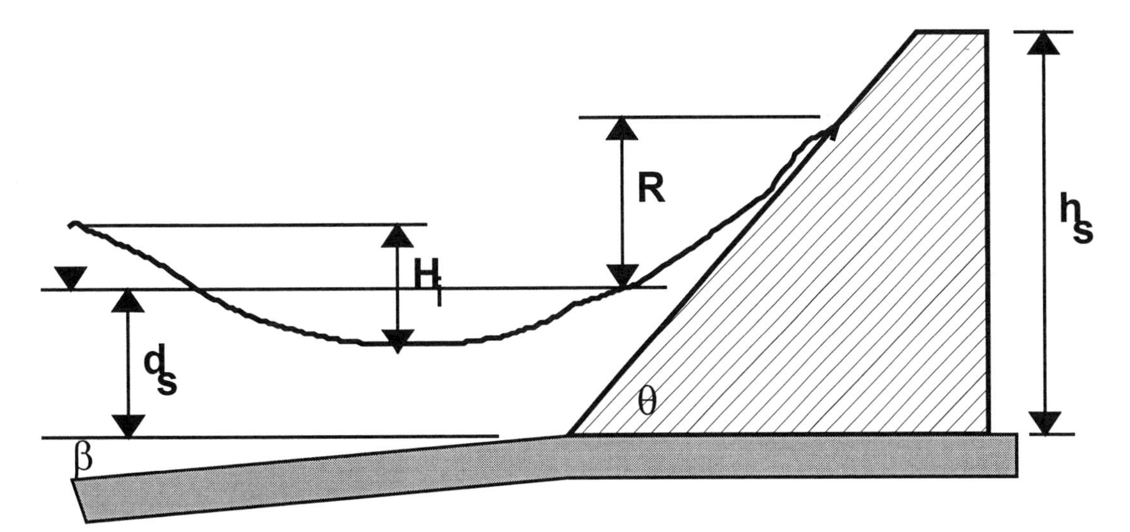

Figure 4-15. Schematic of runup and overtopping.

Later, Hunt (1959) used empirical data to relate the wave runup to the surf similarity parameter (Iribarren number), and the result is

$$\frac{R}{H} = \frac{\tan\beta}{\sqrt{\frac{H}{L_o}}} \quad \text{for } \tan\beta > 0.1 \qquad \text{4-16}$$

where H is the incident wave height and L_o is the deep water wavelength. In the case of the coastal structures as shown in Figure 4-15, the slope angle of the beach is replaced by the slope angle of the structure (θ). The slope roughness affects the runup, and empirical results are contained in the Shore Protection Manual (USACE 1984). Numerical prediction of wave runup is the discussed in the Automated Coastal Engineering System (ACES) manual (USACE 1992). Additional information on wave runup are discussed in Horikawa (1988) and USACE (1984).

WAVE FORECASTING AND HINDCASTING

The prediction of waves based upon past meteorological data is called hindcasting, and it is called forecasting when based upon predicted data. The same procedures are used for forecasting and hindcasting with the source of the meteorological data the only difference. There are many advanced numerical techniques for wave prediction that require advanced computers and access to weather data. The USACE distributes the ACES software for PC computers (USACE 1992) that also has the capability for wave prediction. For the purpose of getting a feel for the wave prediction process, a simplified method is briefly described to provide the opportunity for students to conduct some wave prediction calculations.

Ocean Wave Characteristics

Ocean waves are extremely complex and some idealization is required. In ocean engineering, ocean waves are commonly assumed to be distributed according to the Rayleigh distribution. Also, the significant wave height (H_s or $H_{1/3}$) and period (T_s, T_{ave}, or $T_{1/3}$) are commonly used and have been defined as the average of the highest one third of the waves over a given period of time. Other statistical wave height parameters include maximum wave height (H_{max}), root mean square wave height (H_{rms}), one tenth wave height (H_{10}), and average wave height (H_{ave}). Another wave height parameter is based on energy concepts (H_{mo}). Relationships between the various parameters are:

$$H_{mo} = 4\sigma \qquad \textbf{4-17}$$

where σ is the standard deviation of the wave record,

$$H_{rms} = \sqrt{\frac{1}{N}\sum_{i=1}^{N} H_i^2} \qquad \textbf{4-18}$$

$$H_{ave} = 0.886 H_{rms} \qquad \textbf{4-19}$$

$$H_s = 1.416 H_{rms} \qquad \textbf{4-20}$$

$$H_{10} = 1.27 H_s \qquad \textbf{4-21}$$

In deep water, H_s and H_{mo} are approximately equal, but in shallow water H_s can be as much as 30 % greater. It can also be shown that the wave height (H) in a monochromatic wave train with the same energy as an irregular wave train with significant height H_{mo} is equal to 0.71 H_{mo}.

Simplified Method for Wave Prediction

Although numerical methods are more accurate for wave prediction, this simplified approach allows the ocean engineer the opportunity to predict wave height (H_{mo}) and spectral peak period (T_m) for a given wind speed and fetch or duration. Again the H_{mo} is equal to the significant wave height in deep water and becomes less than H_s in shallow water. The significant wave period (T_s) and spectral peak period (T_m) are nearly equal. In this method the fetches are assumed to be short (80.5 - 120.7 km or 50 - 75 miles) and wind is assumed to be constant and uniform over the fetch. These conditions are rarely met exactly so the determined wave heights and period should not be expected to be more accurate than the accuracy of the input wind conditions.

Fetch is defined as the region over which the wind speed and direction are considered constant. For practical situations, wind speeds varying within ± 5 kts of the mean are considered constant, and wind directions varying ± 15° from the mean wind direction are considered constant. It has been demonstrated that the spectrum of an actively growing wind sea can be represented by a family of spectral shapes such as the JONSWAP (Hasselmann et al. 1973), or PM (Pierson-Moskowitz 1964). The condition at which the wind waves are in full equilibrium with the wind is called a fully arisen sea.

A combined empirical-analytical technique was developed by Sverdrup and Munk (1947), and it was revised by Bretschneider (1952, 1958) using empirical data that became known as the Sverdrup-Munk-Bretschneider (SMB) method. The field data from Mitsuyasu (1968) and Hasselmann (1973) have resulted in some additional revisions. This simplified wave prediction technique is convenient when limited data and time are all that are available. The geometry of the waterbody must be simple, and the wave conditions are considered fetch- or duration-limited. Fetch-limited means the winds have blown at a constant rate long enough for wave heights to reach equilibrium at the end of the fetch. Duration-limited conditions mean the wave heights are limited by the length of time the wind has been blowing. The equations developed for this simplified technique are summarized in USACE (1984). A wave forecasting nomograph for predicting deepwater significant wave height as a function of wind stress factor, fetch length, and wind duration are shown in Figure 4-16 and Figure 4-17. It is important to check fetch-limited wave calculations to see if they are duration-limited and vice versa.

The wave prediction equations are expressed in terms of the wind stress factor U_A or an adjusted wind speed at 10 m elevation. The average wind speed at 10 m is found from

$$U_{10} = U_z \left(\frac{10}{z}\right)^{\frac{1}{7}} \qquad \textbf{4-22}$$

where U_{10} and U_z are in m/s and z is in meters, and this equation is valid for elevations less than 20 m. The wind stress factor is computed as

$$U_A = 0.71 U_{10}^{1.23} \ (\mathrm{m/s}) \quad \text{or} \quad U_A = 0.589\, U_{10}^{1.23} \ (\mathrm{mph}) \qquad \textbf{4-23}$$

The wind stress factor is used to account for the nonlinear relationship between wind stress and wind speed. There are other factors, or adjustments, that may be necessary due to the

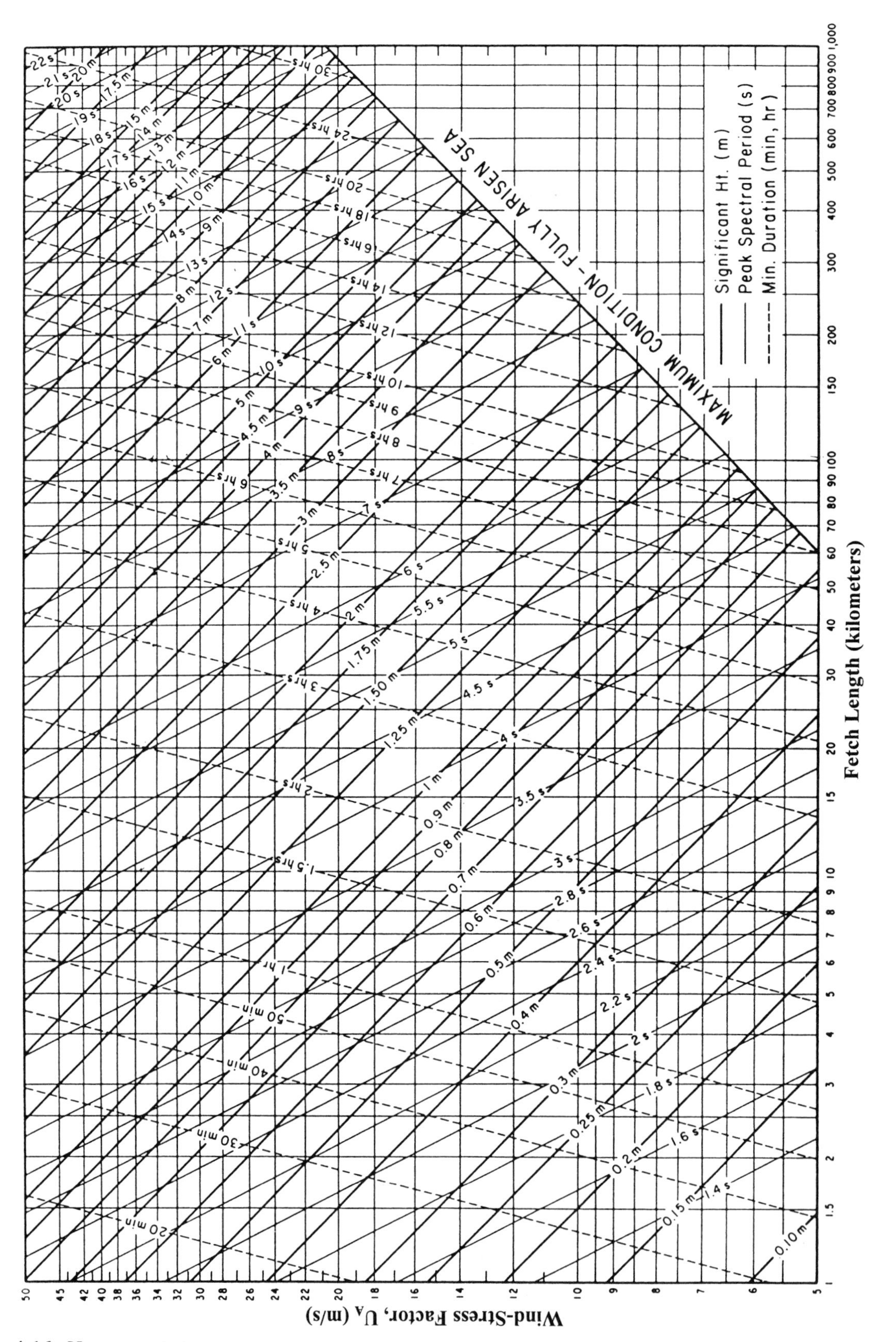

Figure 4-16. Nomograph for deepwater wave prediction curves for constant water depth in SI units. Reprinted with permission from US Army Corps of Engineers (USACE), 1984, *Shore Protection Manual*. (Full citing in references).

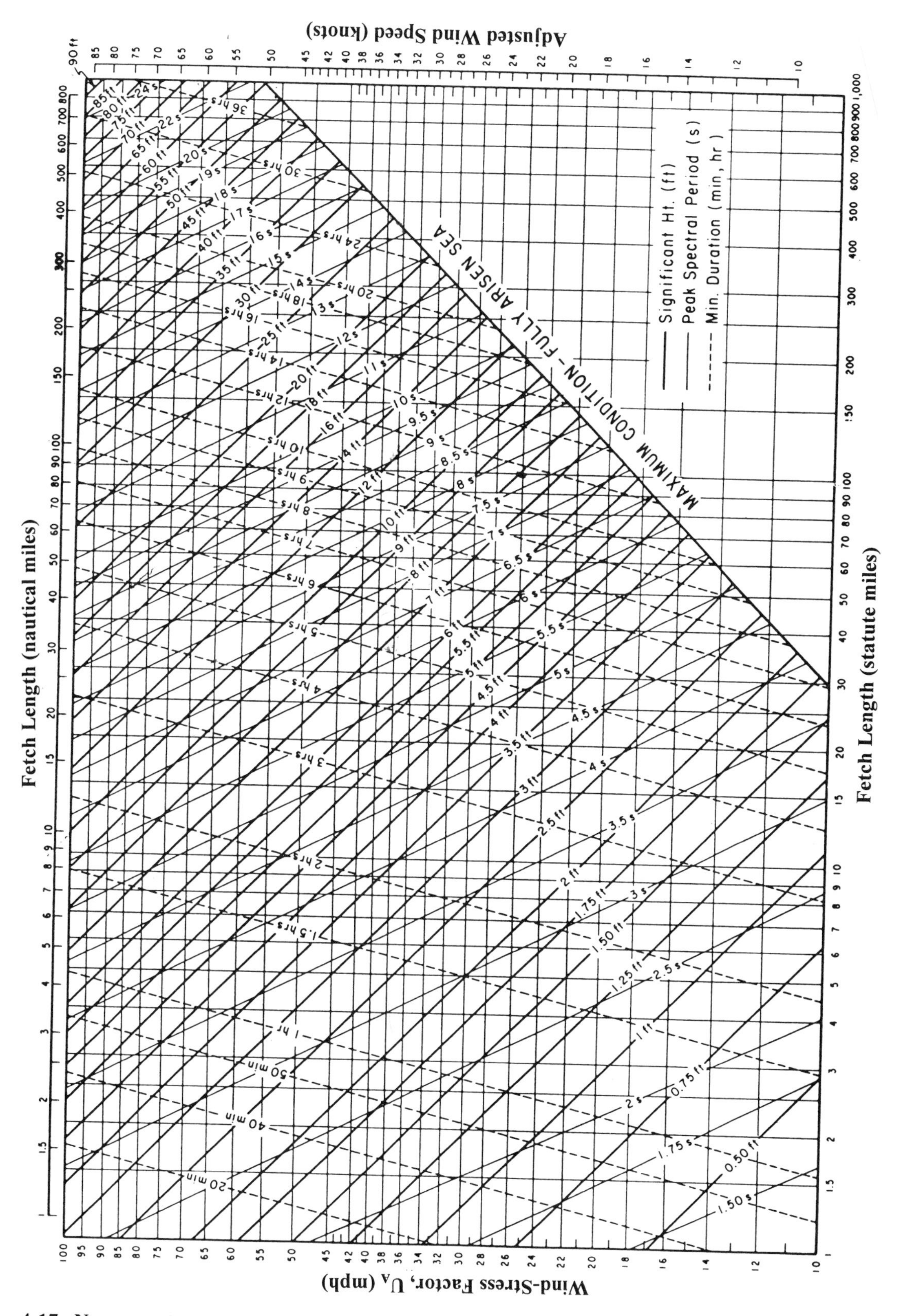

Figure 4-17. Nomograph for deepwater wave prediction curves for constant water depth in English units. Reprinted with permission from US Army Corps of Engineers (USACE), 1984, *Shore Protection Manual.* (Full citing in references).

measurement of wind over water or land, and wind values may be determined from weather maps of pressure fields. Simplified and numerical techniques for shallow water and hurricane wave forecasting techniques are also available, and these procedures are discussed in the Shore Protection Manual (USACE 1984).

As an example, consider an offshore platform located in 365.9 m (1200 ft) of water where the average wind speed has been measured as 59 kph (36.7 mph) at an elevation of 10 m (33 ft). The wind duration is 8 hr and the fetch length is 96.6 km (60 mi). It is desired to estimate the significant wave height and period at the platform. Also, if the wind duration is 4 hr, what is the estimated wave height and period? What is the fetch length and duration for a fully developed sea? Using Equation 4-23 the wind stress factor is computed as 79.7 kph (49.5 mph). Figure 4-17 is used to find H_s and T_s equal to 3.51 m (11.5 ft) and 8.0 s respectively for the 8 hr duration and H_s and T_s are 8.0 ft and 6.3 s for the 4 hr duration. The intersection of the 49.5 wind stress factor line with the fully arisen sea line yields a fetch of 1126.5 km (700 mi) and duration of 35 hr.

SEDIMENT TRANSPORT AND SCOUR

The transport of sediment in the littoral zone is the result of the interaction of the winds, waves, currents, tides, storms, and sediments. Beaches and shores tend to erode, accrete, or be stable. Erosion often adversely affects coastlines and undermines foundations for coastal structures. Therefore, it is important for the ocean engineer to understand and be able to predict erosion and accretion effects and rates.

Sediment Characteristics

Sediments are solid particles usually classified as clay, silt, sand, gravel, cobble, or boulder. The characteristics of the sediment are important inputs to coastal engineering design projects, and the median grain size (d_{50}) is probably the most commonly used descriptive characteristic. The grain size classification is illustrated in Table 4-3. A well sorted size distribution means the grain sizes are close to one size, and well graded means the sizes are evenly distributed over a wide range of sizes. The median diameter is the typical size of a sediment sample and is designated as d_{50}. This is the particle diameter size that divides the sample such that half the sample, by weight, contains particles coarser than the d_{50} size. The phi (Φ) size is related to the grain diameter by

$$\Phi = -\log_2 d \qquad \textbf{4-24}$$

where d is the diameter of the particle in mm.

A important sediment property is its vertical terminal or fall velocity. The ratio of the sediment fall velocity to the characteristic fluid velocity is often used as a measure of sediment mobility or transport. Sediments are frequently characterized as spherical in shape, and consequently, the fall velocity of a sphere is used to approximate the sediment fall velocity. The fall velocity for a sphere can be expressed as a single curve by relating the Reynolds number ($V_f d_s/\nu$) to the buoyancy index (B)

$$B = \left[\frac{\gamma_S}{\gamma} - 1\right]\frac{g d_s^3}{\nu^2} \qquad \textbf{4-25}$$

where γ_s is the solid specific weight, γ is the fluid specific weight, g is gravitational acceleration, d_s is sphere diameter, and ν is fluid kinematic viscosity. Figure 4-18 illustrates empirical results for sphere fall velocities.

Table 4-3. Sediment classification.

Unified Soils Classification	ASTM Sieve Size (s)	Grain Diameter Size (d) mm	Phi (Φ) Value
Cobble		d > 76	Φ < -6.25
Gravel	s < 4	4.76 < d < 76	-6.25 < Φ < -2.25
Coarse Sand	4 < s < 10	2.0 < d < 4.76	-2.25 < Φ < -1.0
Medium Sand	10 < s < 40	0.42 < d < 2.0	-1.0 < Φ < 1.25
Fine Sand	40 < s < 200	0.074 < d < 0.42	-1.25 < Φ < 3.75
Silt	s > 200	0.0039 < d < 0.074	3.75 < Φ < 8.0
Clay	s > 200	d < 0.0039	Φ > 8.0

Empirical results for the fall velocity of natural grain sediment such as sand are also illustrated on Figure 4-18 where the sphere diameter is replaced by the median grain diameter (d_{50}). Curve fit equations for evaluating the fall velocity are

$$V_f = \left(\frac{\gamma_S}{\gamma} - 1\right)\left(\frac{g d_{50}^2}{18\nu}\right) \qquad \text{for } B \langle 39$$

$$V_f = \left[\left(\frac{\gamma_S}{\gamma} - 1\right) g\right]^{0.7}\left(\frac{d_{50}^{1.1}}{6\,\nu^{0.4}}\right) \qquad \text{for } 39 \langle B \langle 10^4 \qquad \textbf{4-26}$$

$$V_f = \left[\left(\frac{\gamma_S}{\gamma} - 1\right)\left(\frac{g\, d_{50}}{0.91}\right)\right]^{0.5} \qquad \text{for } 10^4 \langle B$$

These equations for fall velocity are for ideal situations. Factors such as turbulence in the fluid, sediment concentration, and flocculation also affect the fall velocity.

Experimental tests by Richards (1908) using quartz particles under laminar, transitional, and turbulent conditions led to Equation 4-27 for terminal velocity where d_g is the grain diameter in mm and V_f is the terminal or fall velocity in mm/s. The range of application is for particles with grain diameter between 0.15 and 1.5 mm and $10 < N_R < 1000$. Results of many velocity measurements are summarized in Figure 4-19.

$$V_f = \frac{8.925}{d_g}\left[\sqrt{1 + 95(2.65 - 1)d_g^3} - 1\right] \qquad \textbf{4-27}$$

Sand is the most important littoral material, and it has a specific gravity of 2.65 that corresponds to a specific weight (γ_s) of 165.4 lb/ft^3. However, in tropical climates calcium carbonate is often more dominant littoral material. In temperate climates, quartz and feldspar grains are more abundant. Gravel is often found on rocky coasts. Cohesive material (clay, silt and peat) is common in littoral sediments in relative calm wave environments. In the US, the New England coast is commonly a rock headland separated by short beaches of sand and gravel. On the Atlantic coast from New York to Florida the beach material is typically sand (0.2 to 0.6

mm), but below Palm Beach, Florida the beaches become mostly calcium carbonate. Along the Gulf of Mexico coast from Florida to Mississippi the beaches are mostly fine white sands, and along the Louisiana-Texas the beaches are mostly fine sands that are darker in color. The Pacific Coast consists of sands (0.1-0.6 mm) off California, but more coarse material is predominant off Oregon and Washington. The Alaskan coast is long, and gravel is the predominant beach material. Beaches in Hawaii are predominantly white sands from calcium carbonate. The US Great Lakes have a combination of rocky and sandy shores. The characteristics of beach sediments are determined by visual comparison with a standard sieve analysis and settling velocity tube analysis.

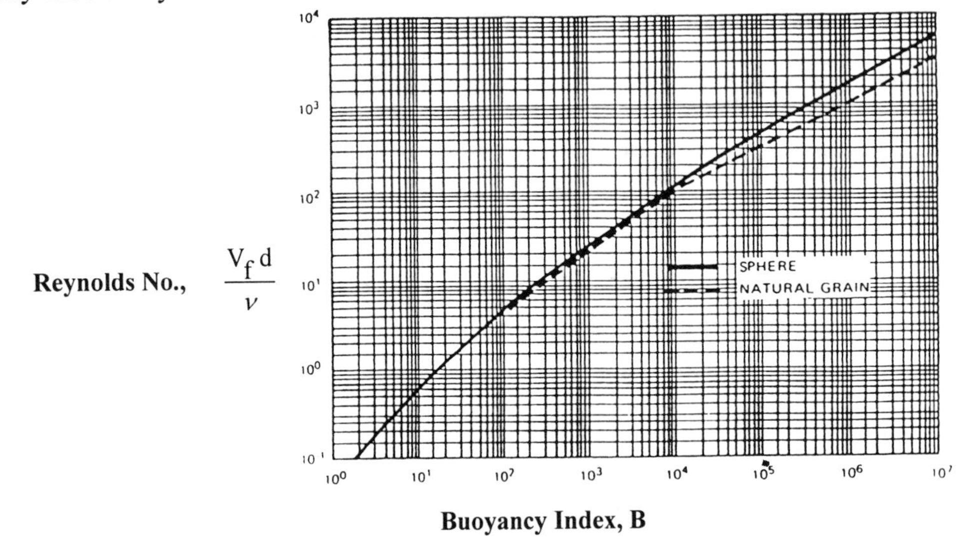

Figure 4-18. Fall velocity of spheres. Reprinted with permission from US Army Corps of Engineers (USACE), 1984, *Shore Protection Manual*. (Full citing in references).

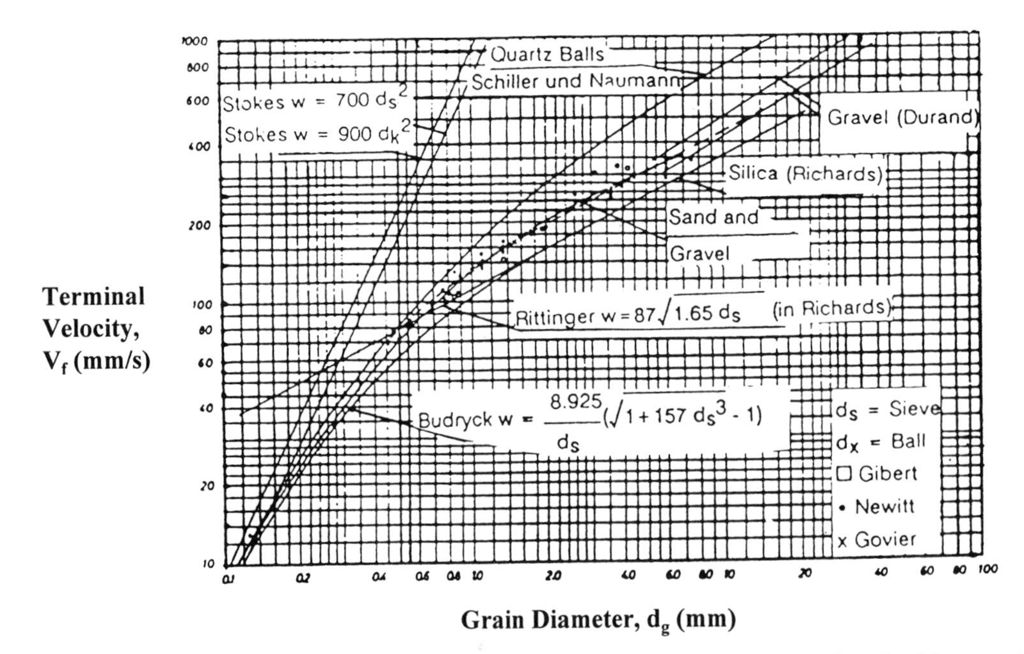

Figure 4-19. Terminal velocity experimental data and empirical equations. Reprinted with permission from Herbich, 1992, *Handbook of Dredging Engineering*. (Full citing in references).

Littoral Transport

The transport of sediment in the littoral zone is primarily the result of the waves arriving at the shore. Larger waves break farther offshore and set more sand in motion. Changes in wave height and period affect the movement of sand onshore and offshore. The angle between the breaking waves and shore determine the direction of the longshore current and longshore transport. Thus, knowledge of the wave climate is needed to understand the littoral processes. The mean wave conditions for the US coastline are summarized in Table 4-4.

Table 4-4. Mean significant wave height and period at coastal localities of the United States (USACE 1984).

Location	Annual Mean Height m	(ft)	Period s	Location	Annual Mean Height m	(ft)	Period s
Atlantic Coast							
Cape Cod, MA	0.85	(2.8)	6.3	Seacrest, NC	0.85	(2.8)	7.5
Misquamicut, RI	0.49	(1.6)	8.3	Kill Devil Hills, NC	0.52	(1.7)	6.3
Southampton, NY	0.7	(2.3)	7.9	Nags Head, NC	0.94	(3.1)	8.6
Fire Island, NY	0.67	(2.2)	7.6	Wrightsville Beach, NC	0.79	(2.6)	7.8
Brigantine, NJ	0.73	(2.4)	6.1	Holden Beach, NC	0.64	(2.1)	7.5
Atlantic City, NJ	0.85	(2.8)	8.3	Murrels Inlet, SC	0.91	(3.0)	7.2
Ludlam Island, NJ	0.55	(1.8)	6.6	Daytona Beach, FL	0.67	(2.2)	8.7
Assateague, VA	0.67	(2.2)	7.8	Palm Beach, FL	0.64	(2.1)	6.4
Virginia Beach, VA	0.64	(2.1)	8.2	Boca Raton, FL	0.58	(1.9)	4.9
Gulf of Mexico Coast							
Naples, FL	0.3	(1.0)	4.7	Galveston, TX	0.4	(1.3)	5.7
Destin, FL	0.49	(1.6)	5.7	Corpus Christi, TX	0.79	(2.6)	6.7
Pacific Coast							
Imperial Beach, CA	0.85	(2.8)	13.6	San Simeon, CA	0.94	(3.1)	12.2
Torrey Pines, CA	0.91	(3.0)	15.7	Natural Bridges, CA	1.01	(3.3)	14.6
San Clemente, CA	0.85	(2.8)	14.5	Rogue River, OR	1.62	(5.3)	8.1
Huntington Beach, CA	0.73	(2.4)	12.9	Port Orford, OR	1.25	(4.1)	13.5
Venice, CA	0.37	(1.2)	10.5	Coquille River, OR	1.28	(4.2)	11.9
Point Mugu, CA	1.01	(3.3)	10.7	Coos Bay, OR	1.28	(4.2)	10.9
Channel Is. Harb., CA	0.85	(2.8)	11.5	Yaquina Bay, OR	2.07	(6.8)	10.3

Currents or circulation in the littoral zone are mostly wind and wave-driven motions superposed on the oscillatory motion of water waves. The velocities are generally low but are the main cause of sediment movement. In the case of sediment transport, the waves are usually shallow water waves. Using linear wave theory, the horizontal length (2A) of the path covered by the water particles as the wave passes in shallow water is

$$2A = \frac{HT\sqrt{gd}}{2\pi d} \qquad \textbf{4-28}$$

and the maximum horizontal water velocity is

$$u_{max} = \frac{H\sqrt{gd}}{2d} \quad \textbf{4-29}$$

For conditions evaluated at the bottom, the average bottom mass transport velocity ($\dot{u}_d$) is

$$\dot{u}_d = \left(\frac{u_{max(-d)}}{2C} \right)^2 \quad \textbf{4-30}$$

The maximum particle velocity in breaking waves from solitary wave theory is given by

$$u_{b_{max}} = C = \sqrt{g(H+d)} \quad \textbf{4-31}$$

and H+d is the distance from the crest to the sea bottom.

Onshore-offshore Currents

The water in the nearshore zone is divided into two distinct water masses that are delineated by the breaker line and there is only limited exchange between the two. Exchange mechanisms include shoaling waves, wind, wave setup, topographic currents, rip currents, and density currents. Rip currents (Figure 4-20) are the prominent exchange mechanism, and they are jets of water carried seaward through the breaker zone.

Longshore Currents

These currents flow parallel to the shoreline (Figure 4-20) and are typically confined to the area between the shore and the zone of breaking waves. The longshore current is due to the oblique approach of waves to the shore. The volume rate of flow depends primarily on the breaker height. The width of the surf zone increases as the breaker height increases since waves break in water depths nearly proportional to the water depth. The longshore current varies across the surf zone and in the longshore direction. To illustrate, the ratio of the longshore current at the breaker zone to the longshore current speed averaged across the surf zone varied from 0.4 at the start of the flow to 0.8 - 1.0 where the flow is fully developed.

Longuett-Higgins (1970) used radiation stress theory to develop an expression for the longshore current and adapted it to fit field data. The basic equation is

$$v_b = M_1 m (gH_b)^{0.5} \sin 2\alpha_b \quad \textbf{4-32}$$

where v_b is longshore current at the breaker position, m is beach slope, g is gravitational acceleration, H_b is breaker height, α_b is the angle between the breaker crest and shoreline, and

$$M_1 = \frac{0.694\Gamma(2\beta)^{-0.5}}{f} \quad \textbf{4-33}$$

where Γ is a mixing coefficient commonly used as 0.2, β is the depth to height ratio of breaking waves in shallow water and usually taken as 1.2, and f is a friction coefficient assumed to be 0.01. Using these values, M_1 is 9.0. Field data of the longshore current (Putnam, Munk, and Traylor 1949, Galvin and Eagleson 1965) have been used to modify the Longuet-Higgins equation and the result is

$$V = 20.7\,m\left(gH_b\right)^{0.5}\sin 2\alpha_b \qquad \textbf{4-34}$$

Longshore currents are most sensitive to breaker angle and to a lesser amount to breaker height. However, the volume rate of flow is most sensitive to breaker height. The above equation is recommended for the computation of longshore current (USACE 1984).

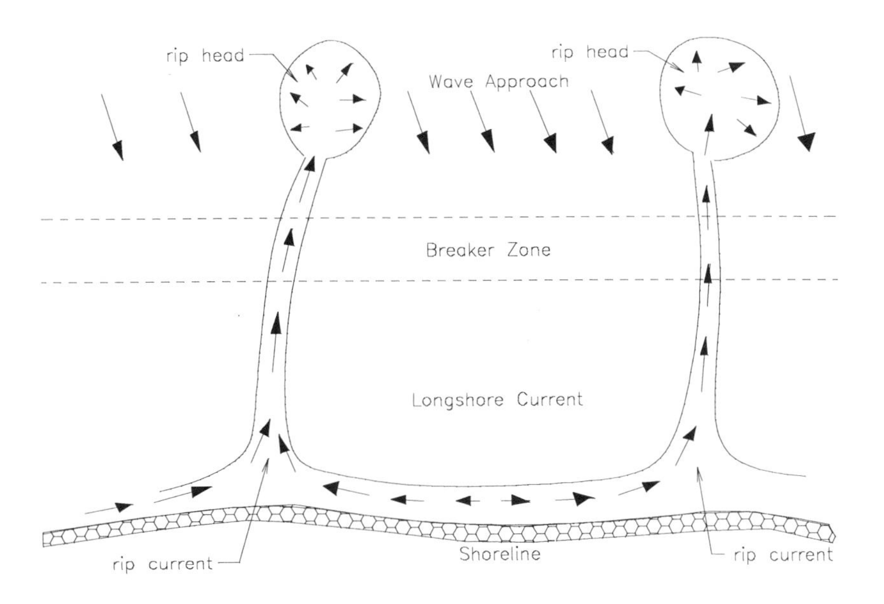

Figure 4-20. Nearshore circulation showing longshore and rip currents.

DESIGN OF RUBBLE MOUND STRUCTURES

Coastal protection structures such as breakwaters, jetties, revetments, and groins are constructed with layers of different size stone and a layer or two of large armor units that are either stone or manufactured concrete units. A schematic of the layering of a rubble mound breakwater is illustrated in Figure 4-21. The weight of the armor units (W) must be large enough to resist the effects of waves, but the entire structure can not be constructed with armor units because the high porosity would allow transmission of wave energy through the structure. Also the large quarry stone or armor units are expensive. Therefore, layers of different size stone are used and slight movement or partial failure can be repaired after extreme wave conditions have abated. Jetties and groins are similarly constructed, but are not as large and complex.

The significant wave height is commonly used as the design wave for a selected wave spectrum, but the H_{10} wave height is recommended by USACE (1984) as a more conservative approach. Additional considerations are to decide whether the wave breaks on the structure and if overtopping is permitted. Armor unit stability is increased by the interlocking that is attained with the concrete manufactured armor units known as tribar, dolos, and tetrapod. There are usually two layers of armor units, and the porosity is between 35 and 55 percent which is a function of the armor shape and placement method. The slope of the seaward breakwater typically varies between 1 on 1.5 to 1 on 3 while revetments may be as flat as 1 on 5. Flatter slopes give more stability, but the cost of the structure increases greatly. It is possible to consider allowing some damage in order to use smaller weight armor units. Thus, risk is considered and maintenance costs are delayed to a later time. These decisions lend themselves to trade-off studies based on costs and risk analysis.

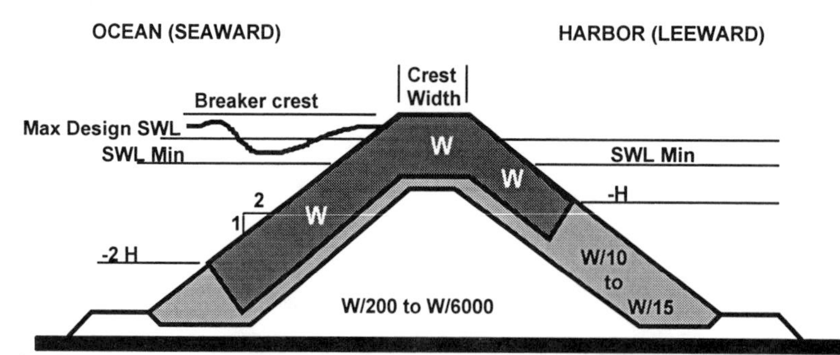

Figure 4-21. Sketch of common breakwater rubble mound structure where W is weight of armor unit.

The determination of the weight of the armor units to provide the desired stability under the selected environmental conditions is usually determined from the equation developed by Hudson (1959). This equation was developed from numerous physical modeling studies at the US Army Engineer Waterways Experiment Station in Vicksburg, MS. The resulting equation is

$$W = \frac{\gamma_a H^3}{K_D (S_a - 1)^3 \cot\alpha} \tag{4-35}$$

where W is the weight of the individual armor unit, γ_a and S_a are the armor unit specific weight and specific gravity respectively, α is the slope angle of the armor layer, and K_D is the experimentally determined stability coefficient for the particular armor unit. The stability coefficient is a function of armor shape, placement technique, wave breaking conditions, damage permitted, and amount of overtopping allowed. Most laboratory testing uses regular waves to evaluate the K_D, and it is common practice to use the significant wave height for an irregular wave train. Conservative approaches use the one-tenth wave height (H_{10}) for H in Equation 4-34. USACE (1984) suggests values for K_D as tabulated in Table 4-5.

DREDGING

Definitions and History

Dredging is the removal of bottom sediments from streams, rivers, lakes, coastal waters, and oceans, and the resulting dredged material is then transported by ship, barge, or pipeline to a designated disposal site on land or in the water where it is discharged. According to Herbich (1992), dredging is defined as raising material from the bottom of a water-covered area to the surface and pumping it over some distance. Webster defines dredging as "(1) to dig, gather, or pull out with a dredge; (2) to deepen (as a waterway) with a dredging machine" and defines a dredge as "a machine for removing earth usually by buckets on an endless chain or suction tube". Dredged material is sediment that has been excavated by a dredge and has been or is being transported to a disposal site. The term maintenance dredging involves the removal of sediments that have accumulated since a previous dredging operation and new work dredging involves the removal of materials that have not been previously dredged.

Table 4-5. Suggested armor unit stability coefficients (K_D) for rubble mound structures (USACE 1984).

Armor Unit	Number of Armor Unit Layers	Placement Technique	Trunk Structure		Head Structure		Slope
			K_D		K_D		
			Breaking Wave	Non Breaking Wave	Breaking Wave	Non Breaking Wave	Cot α
Quarrystone (smooth rounded)	2	random	1.2	2.4	1.1	1.9	1.5 to 3.0
Quarrystone (rough angular	1	random	Non recommended	2.9	Not recommended	2.3	1.5 to 3.0
Quarrystone (rough angular	2	random	2.0	4.0	1.9 1.6 1.3	3.2 2.8 2.3	1.5 2.0 3.0
Quarrystone (graded angular)		random	2.2	2.5			
Tetrapod	2	random	7.0	8.0	5.0 4.5 3.5	6.0 5.5 4.0	1.5 2.0 3.0
Tribar	2	random	9.0	10.0	8.3 7.8 6.0	9.0 8.5 6.5	1.5 2.0 3.0
Tribar	1	uniform	12.0	15.0	7.5	9.5	1.5 to 3.0
Dolos	2	random	15.8	31.8	8.0	16.0	2.0
Note: K_D values are for no damage and minor overtopping.							

Canal dredging was conducted as early as 4000 BC in the canals of Egypt using the labor of slaves, prisoners, and soldiers and with primitive tools such as spades and baskets. In 1435, a scraper dredge "Krabbelaar" (Figure 4-22) loosened the bottom sediments and the water current carried the sediment out to sea. This type of dredging is called agitation dredging and is still used occasionally. The grab dredge or clamshell dredge was developed in Italy and Holland in the 1500's. Lebby conceived the first hydraulic hopper dredge in 1855, and it was named the General Moultrie. The 365 ton dredge had a wooden hull that was 46 m (150 ft) long, 3.1 m (10.3 ft) deep and 8.1 m (26.7 ft) wide. It was equipped with a steam engine and a centrifugal pump with a 1.8 m (6 ft) impeller and 0.5 m (19 in) diameter suction pipe. The production of this dredge was about 251 cubic meters (328 cubic yards) of dredged material per working day. Bazin introduced the idea of suction dredging in 1867 using a rotating harrow under the bow of a ship and suction pipes under the stern and applied it to dredging in the Suez Canal. Modern dredges were developed in the 1900's, and these modern dredges operate either mechanically or hydraulically with most dredges using the hydraulic principle. These modern dredges have efficient pumps, heave compensating devices, electronic equipment for automatic controls, water jets, sophisticated navigation equipment, and advanced instrumentation. Environmental constraints continue to force the development of new advances in dredging and dredging equipment.

Figure 4-22. Scraper dredge *Krabbelaar*. Reprinted with permission from Herbich, 1992, *Handbook of Dredging Engineering*. (Full citing in references).

Dredging involves project planning, design, operation, and maintenance. Dredging, dredged material disposal, and other aspects of the overall navigation project should be considered as a total project. For example, the dredging and disposal equipment and procedures must be compatible. Navigation channel design dictates the dredging requirements, and refinement or modification to the navigation channels further influences the dredging requirements. Lake and cleanup dredging requires dredging over a dispersed area of "hot spots", precision removal of thin layers, and minimizing sediment resuspension during dredging.

Basic dredging requirements are determined by channel design and shoaling rates. The quantities of material to be dredged are determined from past records, and planning for dredging projects should be based on longterm requirements and hydrographic surveys. Horizontal positioning and depth measurements are conducted with electronic navigation and positioning equipment, and the data are usually reduced using computers. Accuracy and capabilities of positioning and surveying equipment are rapidly improving.

Dredges and Dredging Equipment

Excellent sources of information related to dredging and dredging equipment include Dredging and Dredged Material Disposal Engineering Manual, EM 1110-2-5025 (USACE 1983), Bray et al. (1997), de Heer and Rochmanhadi (1989), and Herbich (1992) are excellent sources of information related to dredging and dredging equipment. Dredges are classified as either mechanical or hydraulic as is shown in the Figure 4-23.

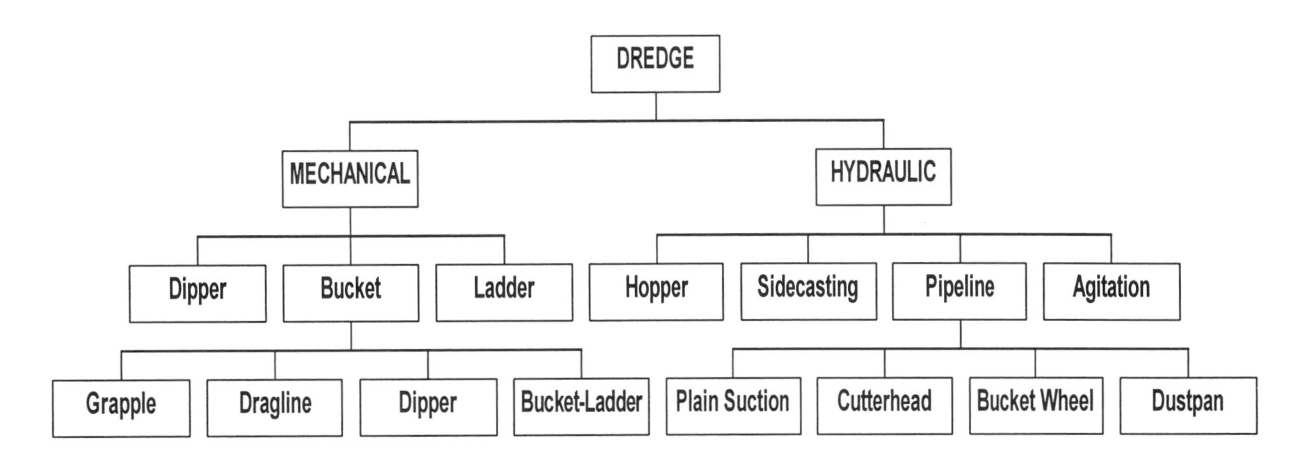

Figure 4-23. Classification of dredges.

Mechanical Dredges

A grapple dredge has a derrick mounted on a barge and is equipped with a clamshell bucket. It is best for working in soft sediments and works well in difficult to access areas. A dragline dredge has a steel bucket that is suspended from a movable crane as it is lowered to the bottom and then dragged toward the crane by a cable. A dipper dredge is a floating barge with a mechanically operated excavating shovel that works well in hard compact material. Photographs of a grapple and dipper dredge are shown in Figure 4-24.

Grabble Bucket Capacity	*23-38 m^3 (30-50 yd^3)*	*Dipper Bucket Capacity*	*14-23 m^3 (18-30 yd^3)*
Total Installed Power	*7,040 kw (9,430 hp)*	*Total Installed Power*	*7,040 kw (9,430 hp)*

Figure 4-24. Grapple and dipper dredges. (Reprinted with permission from Great Lakes Dredge and Dock).

A bucket-ladder dredge has an endless chain of buckets, and each bucket is thrust into the material to be dredged and brings its load to the surface. When used in sand and gravel operations, the buckets discharge their contents onto vibrating screens to separate the different sizes, and the separated material is placed on a barge for transport. The work cycle is continuous and more efficient than the grapple or dipper dredges. The buckets are attached to the chain that is guided and supported by a ladder. The lower end of the ladder is suspended from a hoisting gantry and lowered to the bottom for dredging and mining operations. The bottom sediment is cut by the rim of the bucket and then fills the bucket that travels up the ladder and discharges to a barge alongside the dredge. The dredge is swung from side to side with the aid of anchors and mooring lines. Tugs are used to push or pull the loaded barges to a disposal or offloading area. Bucket ladder dredges have been used in placer mining of gold and tin. Digging depth has increased from 15.2 m (50 ft) in 1905 to 53.4 m (175 ft) in 1973. Bucket sizes have increased from 0.2 to 1.5 m^3 (7 to 54 ft^3). A photograph of this type of dredge is shown in Figure 4-25.

Figure 4-25. Bucket-ladder dredge. Reprinted with permission from Herbich, 1992, *Handbook of Dredging Engineering*. (Full citing in references).

These mechanical dredges have limited ability to transport dredged material, no self-propulsion, and relatively low production. The advantages of a mechanical dredge are its ability to operate in restricted locations (e.g. docks, jetties, and piers), ability to treat and dewater dredged material in placer mining operations and that with special adaptations such as watertight clamshell buckets it can be used in working with contaminated sediments.

Hydraulic Dredges

Hydraulic dredges conduct both phases of the dredging operations (digging and disposing). Disposal is accomplished by pumping the dredged material through a floating pipeline to the disposal area or by storing the dredged material in hoppers that are emptied over a disposal area. Hydraulic dredges are more efficient, versatile, and economical. The dredged material is first loosened and mixed with ambient water by cutter heads or water jets and pumped as a fluid (slurry) through a long pipeline or to a hopper. The basic components of a hydraulic dredge are: dredge pumps, digging and agitation machinery, and hoisting and hauling equipment.

Hydraulic dredges are categorized as hopper (trailing suction), pipeline (plain suction, cutterhead, dustpan), bucket wheel, and sidecasting.

Hopper dredge

The development of self-propelled trailing suction hopper dredge revolutionized the dredging industry by reducing costs. These dredges are used extensively in Europe and the US, and they can work in all but hard materials. Hopper capacities of several hundred to 10,000 m^3 have been built. The hoppers are usually unloaded through the bottom doors and some have pump-out facilities. The maximum dredging depth is 18 to 21 m (59 to 69 ft), but it can be increased to 40 m (131 ft) with the addition of a submerged pump at the draghead. The typical hopper dredge components are illustrated in Figure 4-26. The dragarms and dragheads extend from both sides of the hull, and each is lowered to the sea bottom. The dredge moves slowly over the area to be dredged while the dredge pumps move the sediment and water mixture (slurry) through the dragarm into the hopper bins. When these hopper bins are full, the dredge moves under its own propulsion system to a designated disposal area and empties the dredged material through the hopper doors in the bottom of the hull, or it uses inboard pumps to transport the dredged material through pump-out lines to shore for beneficial uses such as beach nourishment. Some of the modern dredges have a single hopper in the midsection and the dredge is designed to split open and allow the dredged material to exit as the hull splits open (split-hull hopper dredge). The split-hull hopper dredge, Padre Island, is shown in Figure 4-27.

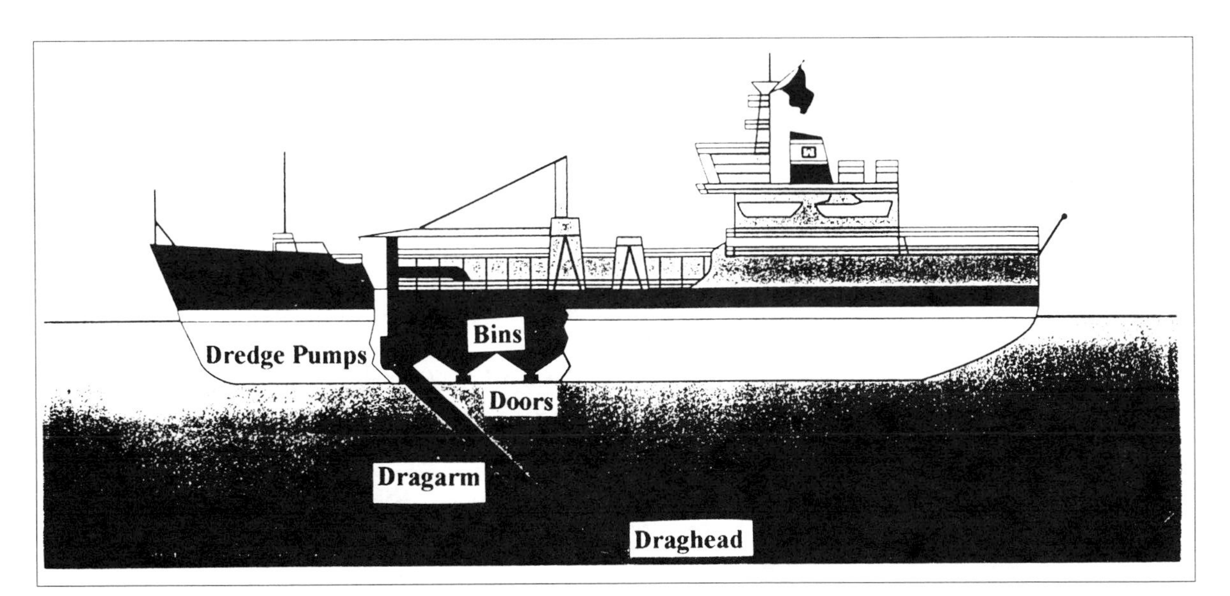

Figure 4-26. Typical hopper dredge components. (Reprinted with permission from US Army Corps of Engineers).

A hopper distribution system tries to minimize turbulence and allow for quick settling of the solid material. The slurry enters the hopper from some height entraining air that tends to keep material in suspension. Some distribution lines have been installed below the water line to reduce entrainment and turbulence. Overflow weirs are located at opposite end of the hopper from where the slurry enters in order to allow the sediment to settle to the bottom. The water is then allowed to overflow and return to the ambient water. Gratings are also installed in the hopper near the slurry inflow location to reduce entrainment and turbulence. The installation of submerged pump on the draghead or in the dragarm permits dredging at deeper depths. The addition of a submerged pump increases the dredging depth by reducing the chances of cavitation, and it permits the pumping of a higher specific gravity material. Most modern hopper dredges have dragarm-mounted pumps. The draghead is an important part of the hopper dredge. Several types of dragheads are used such as the California, Ambrose, Venturi, and IHC types. A grating is used to prevent large objects entering the suction pipe. A new type of draghead uses a rotating cylinder with knives. A Venturi draghead creates a negative pressure just above the seabed that results in a 30-40 % increase in the production of fine sand. An automatic draghead winch control system controls the movement of the suction pipe and draghead. It is coupled with a swell compensation system to maintain the correct pressure of the draghead on the bottom and the lateral position of the pipe hoist. The hopper dredge has the increased ability to operate in bad weather and minimizes risk of damage to equipment.

Hopper Capacity	*2,750 m^3 (3,600 yd^3)*
Total Installed Power	*5,620 kW (7,530 hp)*
Discharge Method	*Split-hull Placement*

Figure 4-27. Photograph of the split-hull hopper dredge *Padre Island*. (Reprinted with permission from Great Lakes Dredge and Dock).

Sidecasting dredge

Sidecasting dredges discharge the dredged material to the side of the channel and allow for continuous dredging. Some hopper dredges are equipped with a sidecasting boom as shown in Figure 4-28.

Figure 4-28. Sidecasting dredge *Schweitzer*. (Reprinted with permission from US Army Corps of Engineers).

Cutterhead dredge

The pipeline cutterhead dredge is a very versatile dredge, and its primary function is to excavate and move material hydraulically to a disposal location without rehandling. It is categorized by the size (diameter) of its floating discharge line. For example, a 61 cm (24 in) dredge has a floating discharge line that has a diameter of 61 cm (24 in). The components of a typical cutterhead dredge are shown in Figure 4-29, and an operating cutterhead dredge is illustrated in Figure 4-30. During a dredging operation, the floating discharge and shore pipeline are connected to the dredge. Additional equipment to support the operation is required such as a derrick, tugs, fuel and pipe barges, surveying boats, and other site specific special equipment. A cutter is connected at the forward end of the ladder and connected to the shaft of the cutter motor. There are generally two types of cutters that are classified as either straight arm or basket. Rotation of the shaft and cutter agitates soft or lose material and cuts hard material that is then picked up by the suction.

On the cutterhead dredge, the ladder supports the cutter, suction pipe, lubricating lines, and usually the cutter motor and reduction gear. The forward end of the ladder is supported by a A-frame with hoisting equipment to raise and lower the ladder. The length of the ladder determines the dredging depth and the dredging depth is typically considered to be 0.7 times the length of the ladder. Ladder lengths may be 7.6 m (25 ft) or less to 45.7 m (150 ft) or more and they can weigh as much as 400 tons. The suction pipe supported beneath the ladder transports the dredged material to the dredge pump. The discharge line is connected to the floating pipeline by a ball joint. The diameter of the discharge line depends on the pump size, and the suction pipe diameter is usually 1.25 to 1.5 times the pump discharge diameter. Practically, the suction pipe is one size larger than the pump discharge such as a 41 cm (16 in) pump discharge and a 46 cm (18 in) suction pipe.

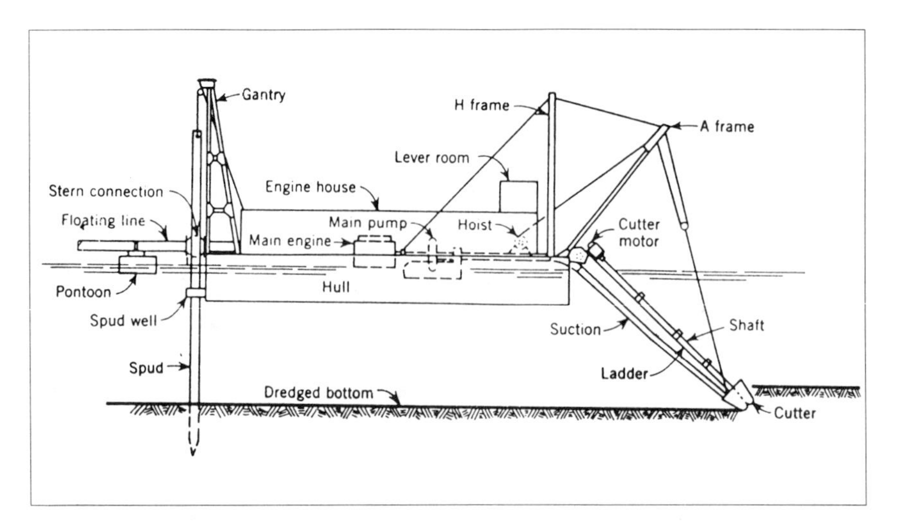

Figure 4-29. Cutterhead dredge components. Reprinted with permission from Houston, 1970, *Hydraulic Dredging*. (Full citing in references).

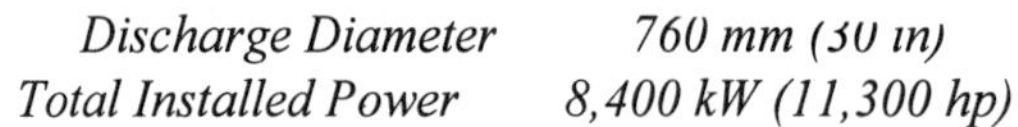
Discharge Diameter *760 mm (30 in)*
Total Installed Power *8,400 kW (11,300 hp)*

Figure 4-30. Photograph of cutterhead dredge *Illinois*. (Reprinted with permission from Great Lakes Dredge and Dock).

The dredge pump is located forward in the hull with its center near the loaded water line. A diesel engine, diesel electric motor, steam or gas turbine can be used to drive the pump. In some cases, shore electric power may be used to drive the pump. The horsepower required varies from 186 to more than 11,186 kw (250 to more than 15,000 horsepower). The pump rotative speed varies from about 300 to 900 RPM. The dredge is moved and held in position with spuds. These devices (usually two) are as big as 1.22 m (4 ft) in diameter, 30.5 m (100 ft) long and weigh as much as 30 tons. Spuds are typically spaced not less than one-tenth the distance between the stern of the dredge and the cutter so that moving ahead is not limited. A walking spud is a conventional spud placed in a groove such that it can move longitudinally along the centerline of the dredge.

The discharge line consists of three sections (pipe on dredge, floating line, and shore pipeline). The pipe on dredge runs along the dredge deck from the pump discharge to the stern, and it has a flap valve near the pump to prevent back flow. At the stern it is connected to a swivel elbow or ball joint. The floating line extends from the stern of the dredge to the shore. It is made of 9.1 to 15.2 m (30 to 50 ft) sections and each section is supported by floating pontoons. Strongbacks are used to connect the pontoons, and anchors are used where necessary to hold pontoons in place. A walkway is often constructed on top of the pipeline. Floating pipelines are used more often when waves are expected. The shore pipeline consists usually of shorter and lighter sections with a wye connection at the discharge end. A preliminary selection guide for cutterhead dredges is shown in Figure 4-31 below based upon production and length of discharge pipeline. Portable cutterhead dredges have been developed ranging from 20.3 cm (8 in) to a 50.8 cm (20 in) dredge. These are versatile and can undertake many small dredging projects.

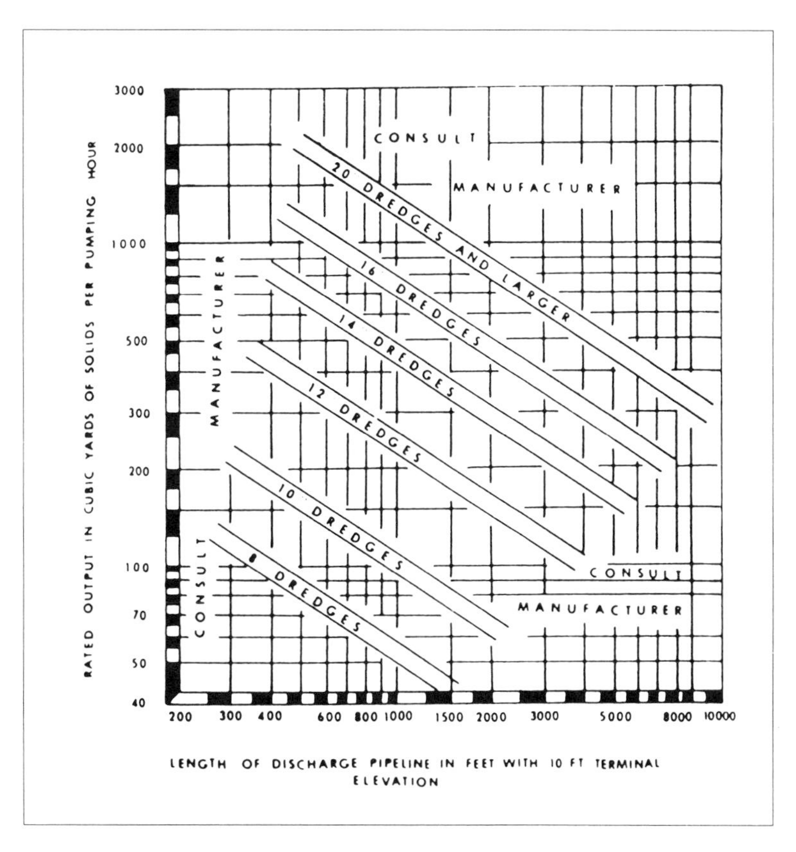

Figure 4-31. Preliminary selection guide for dredges. Reprinted with permission from Herbich,1992, *Handbook of Dredging Engineering.*, (Full citing in references).

Bucket wheel dredge

A rotating wheel equipped with bottomless buckets is used to cut or loosen soil that is then directed into the interior of the wheel and conveyed to the suction line. The bucket wheel is attached to the ladder as shown in Figure 4-32, but newer designs use a dual wheel concept.

Figure 4-32. Photograph of bucket wheel dredge. Reprinted with permission from Herbich, 1992, *Handbook of Dredging Engineering*. (Full citing in references).

Dustpan dredge

The dustpan dredge is a hydraulic, plain suction vessel. The wide (about same as hull width) vacuum cleaner-like head is lowered to the bottom by winches. It has a high velocity water jet to agitate and loosen material that is subsequently pumped through a floating pipeline to a disposal area. The dustpan dredge, *Burgess*, is shown in Figure 4-33.

Figure 4-33. Schematic of dustpan dredge. Reprinted with permission from Herbich, 1992, *Handbook of Dredging Engineering*. (Full citing in references).

REFERENCES

Battjes, J. A. "Surf Similarity." *Proceedings of the 14th Coastal Engineering Conference*, Copenhagen, Denmark, 1974.

Bray, R. N., A. D. Bates, and J. M. Land. *Dredging:A Handbook for Engineers*, Second Edition, London: Arnold, 1997.

Bretschneider, C. L. "The Generation and Decay of Wind Waves in Deep Water." *Transactions of the American Geophysical Union* 33(1952): 381-389.

de Heer, R. J., and Rochmanhadi. *Dredging and Dredging Equipment: Dredging Equipment.* International Institute for Hydraulic and Environmental Engineering, Delft, 1989.

Galvin, C. J., Jr., and Eagleson, P. S. "Experimental Study of Longshore Currents on a Plane Beach." TM-10, US Army Corps of Engineers, Coastal Engineering Research Center, Washington, 1965.

Gaythwaite, J. W. *Design of Marine Facilities*. New York: VanNostrand-Reinhold, 1990. Figure 4-11 reprinted with permission: "Source:Gaythwaite, *Design of Marine Facilities*. Copyright VanNostrand-Reinhold, 1990."

Hasselmann, K., T. P. Barnett, E. Bonws, H. Carlson, D. C. Cartwright, K. Enke, J. Ewing, H. Gienapp, D. E. Hasselmann, P. Kruseman, A. Meerburg, P. Muller, D. J. Olbers, K. Richter, W. Sell, and H. Walden. "Measurements of Wind-Wave Growth and Swell Decay During the Joint North Sea Wave Project (JONSWAP)." Deutches Hydrographisches Institute, Hamburg, 1973.

Herbich, J. B., *Handbook of Dredging Engineering*. New York: McGraw-Hill Co., 1992. Figures 4-19, 4-22, 4-25, 4-31, 4-32, and 4-33 reprinted with permission: "Source: Herbich, *Handbook of Dredging Engineering,* Copyright McGraw-Hill Co., 1992."

Horikawa, K. *Nearshore Dynamics and Coastal Processes*. University of Tokyo: Tokyo Press, 1988.

Huston, J. *Hydraulic Dredging*. Cambridge: Cornell Maritime Press. 1970. Figure 4-29 reprinted with permission: "Source: Huston *Hydraulic Dredging*. Copyright Cornell Maritime Press. 1970."

Hudson, R. Y. "Laboratory Investigations of Rubble-Mound Breakwaters and Jetties." *Proceedings of the American Society of Civil Engineers*, ASCE, Waterways and Harbors Division 1:85(1959) No. WW3, Paper No. 2171.

Hunt, I. A. "Winds, Wind Setup and Seiches on Lake Erie." Lake Survey, US Army Corps of Engineers, Detroit, 1959.

Ippen, A. T. *Estuary and Coastline Hydrodynamics*, New York: McGraw-Hill, 1966.

Leenknecht, D. A., A. Szuwalski and A. R. Sherlock. "Automated Coastal Engineering System: User's Guide." Coastal Engineering Research Center, U.S. Army Engineer Waterways Experiment Station, Vicksburg, 1992.

Longuet-Higgins, M. S. "Longshore Currents Generated by Obliquely Incident Sea Waves, 1 and 2." *Journal of Geophysical Research* 75.33(1970): 6788-6801.

Miche, R. "Le pouvoir reflechissant des ouvrages maritime exposes a l'action de la houle. *Annales Ponts et Chaussees*, 121(1951): 285-319.

Mitsuyasu, H. "On the Growth of the Spectrum of Wind-Generated Waves (I)." Reports of the Research Institute of Applied Mechanics, Kyushu University, Fukuoka, 16:55(1968): 459-482.

Pierson, W. J., Jr., and L. Moskowitz. "A Proposed Spectral Form for Fully Developed Wind Seas Based on the Similarity Theory of S. A. Kitaigorodskii." *Journal of Geophysical Research* 69:24(1964): 5181-5190.

Putnam, J. A., W. H. Munk and M. A. Traylor. "The Prediction of Longshore Currents." *Transactions of the American Geophysical Union* 30(1949): 337-345.

Richards, R. H. "Velocity of Galen and Quartz Falling in Water." *Transactions, AIME* 38(1908): 230-234.

Sverdrup, H. U., and W. H. Munk. "Wind, Sea, and Swell: Theory of Relations for Forecasting." Publication No. 601, U.S. Navy Hydrographic Office, Washington, 1947.

US Army Corps of Engineers (USACE). "Dredging and Dredged Material Disposal." Engineering Manual 1110-2-5025, Washington: Superintendent of Documents, 1983.

US Army Corps of Engineers (USACE). *Shore Protection Manual*. Washington: Superintendent of Documents, 1984. Figures 4-2 to 4-10, 4-12 to 4-14, and 4-16 to 4-17 reprinted with permission: "Source:*Shore Protection Manual*. US Army Corps of Engineers (USACE), Copyright US Government, 1984."

Wiegel, R. L. *Oceanographical Engineering*. Englewood Cliffs: Prentice-Hall, 1964.

PROBLEMS

4-1. A wave in deepwater has a height of 30 ft and period of 13 s. What is the height of the wave in 100 ft of water due to shoaling only?

4-2. An offshore platform is located in 1500 ft of water. An anemometer located 60 ft above the still water level measured the average wind speed as 50 mph for a duration of 8 hr, and the fetch length is estimated as 50 miles. Estimate the significant wave period and height. If the wind duration was 4 hr, what is the estimated wave height and period? What is the fetch length and duration for a fully developed sea?

4-3. Waves are approaching a beach which has straight and parallel depth contours and the beach slope is 0.02 (1:50). For a breaker wave height (H_b) of 3 and 6 ft, plot the longshore current for 10 degree increments of the breaker angle (α_b) between 10 and 50 degrees.

4-4. Beach material has a median grain size (d_{50}) of 1.5 mm. Determine the fall velocity in seawater at 15.6 °C. The typical specific gravity of sandy beach material is 2.65.

4-5. A 100 ft long wave of height 6 ft and period 6 s moves toward a coast from deep water. Determine the speed at which the wave approaches the coast. What is the total energy of the wave in ft-lb/ft of wave breadth? Determine the approximate depth at which the wave will begin to change form as a result of bottom friction. Evaluate the length, period, and height of this wave in a depth of 8 ft and whether or not it will break.

4-6. A 6.5 s wave with a height of 3.5 ft is approaching a rubble mound breakwater that has a slope of 1 on 3. Determine the reflection coefficient and the height of the reflected wave.

4-7. A 4 s wave is approaching a plane sloping impermeable beach that has a slope of 1 on 10. If the wave height is 3.0 ft, determine the runup on the beach.

4-8. The significant wave height of waves at the breaker zone approaching a 1 on 15 sloping beach is 3.8 ft and the angle of approach is 4.5 degree. Determine the estimated longshore current developed.

4-9. A rubble mound breakwater is planned for a depth of 35 ft. It has been determined that the significant wave height for the design is 3 ft with minimal breaking and no damage. Dolos armor units can be economically used at the proposed site. Assume the specific weight of a concrete Dolos is 140 lb/ft^3. Determine the weight of the armor unit.

4-10. A hurricane wave in deep water has a wave height of 58 ft and period of 12 s. Determine the height of this wave when it reaches a depth of 50 ft. Consider shoaling effects only. What is the energy per unit width of the when it reaches the 50 ft water depth?

4-11. Determine the terminal velocity of sand with a d_{50} of 0.9 mm in seawater at a temperature of 60 ^{o}F. Assume the specific weight of sand is 165.4 lb/ft^3.

4-12. A lake is 15 miles long and has an average depth of 22 ft. Determine the fundamental period of oscillation for the lake.

4-13. A rubble mound breakwater is being designed for 40 ft of water. The design is based upon the H_{10} wave height of 4.5 ft with no breaking and no damage. Compare the weight of smooth rounded quarrystone, tribar, dolos and tetrapod armor units for slopes of 1 on 1.5 and 1 on 3. Use a spreadsheet and plot the results.

4-14. A longshore current and breaker wave height are measured as 3.4 ft/s and 6 ft respectively. For a beach slope of 1:25 with straight and parallel depth contours, determine the angle of wave approach at the breaker zone.

4-15. The average longshore current and breaker angle are measured as 3.8 ft/s and 16° respectively. For a beach slope of 1:30 with straight and parallel depth contours, determine the height of the breaker at the breaker zone.

4-16. Determine the fall velocity for sand with a d_{50} of 0.5 mm in seawater with a specific gravity of 1.025. The specific gravity of sand is 2.65 and the specific weight of water is 9.8 kN/m^3.

4-17. Waves with a period of 11 s and a deep water wave height of 0.9 m propagate normal to the shore. An 150 m energy conversion device is in place parallel to the shore to convert the wave motion into electrical power. The device is located in a water depth of 8 m. For a 50% efficient device, evaluate the power produced in kilowatts.

CHAPTER 5: MATERIALS AND CORROSION

INTRODUCTION

Materials used in offshore and coastal waters must be specially selected to withstand the very harsh marine environment. Engineers are faced with selecting materials for offshore drilling and petroleum production platforms, coastal protection structures, subsea systems, ocean instrumentation, marina and harbor facilities, and moored systems. These structures or systems must endure wind, waves, currents, ice, temperature variations, biofouling, and a highly corrosive environment. Scour, hurricanes, earthquakes, typhoons, and mud slides are common extreme events that occur in the ocean environment, and ocean structures are designed with materials that best endure these environmental events. Since materials are in contact with ocean waters, they must not be toxic to the marine organisms living in the sea. Ocean structures may also experience excessive loads as a result of accidental collisions. Special materials are sometimes required to withstand the large hydrostatic pressures due to the extreme ocean depths. Selected sources of material information for ocean applications include ASM (1981, 1990), Battelle (1986), Dexter (1979), Meyers (1969), Reuben (1994), and Sheets and Boatwright (1970).

MATERIALS FOR OCEAN APPLICATIONS

Properties

Material strength is one of the most important mechanical properties used for material selection. The main mechanical properties are tensile strength, modulus of elasticity, yield strength, and elongation. Other important properties are toughness and fatigue strength. An important diagram describing the relationship between stress (load divided by the cross-sectional area) and strain (elongation divided by the initial length) is often used in the selection process, and typical stress-strain relationships are shown in Figure 5-1.

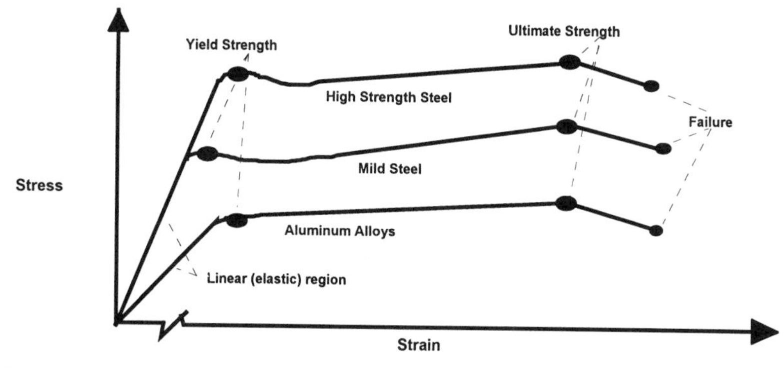

Figure 5-1. Typical stress-strain curve.

Elastic deformation occurs in the linear region and plastic deformation occurs beyond the yield point. Deformation is not permanent in the elastic region, but some permanent deformation occurs in the plastic region. The yield strength is normally the stress corresponding to the end of the linear region, or it is determined at the 0.2 percent offset strain. At some stress the material fails, and this stress level is called the ultimate strength. The modulus of elasticity is a measure of rigidity, or stiffness, of the material and is the proportionality constant between stress and strain in the elastic region. The modulus of elasticity is approximately 207 x 10^6 kPa (30 x 10^6 psi) for steel, 69 x 10^6 kPa (10 x 10^6 psi) for aluminum and 103 x 10^6 (15 x 10^6 psi) for titanium.

The ability of a material to deform without rupturing is determined by its ductility. A material that is not ductile is called brittle, and it fails quickly with little elongation or warning. Fatigue is the result of oscillating loads that can occur in waves, and consequently, it is important in ocean structures. Fatigue strength is normally determined experimentally, and it is generally defined on an S-N diagram that shows the fatigue strength as a function of the number of cycles that a specimen undergoes before failure. Another approach used in fracture mechanics gives the fatigue crack growth rate as a function of the change in toughness.

Metals

Steel

The ability to resist large forces, fatigue, impact, and be easy to fabricate and weld at a competitive price are desirable characteristics for materials used in the ocean. Steel has most of these characteristics, and consequently, it is one of the most frequently used materials for ocean applications. Low strength steels have yield strengths less than 414 MPa (60 ksi) as shown in Table 5-1, and these steels have good fatigue strength. Medium strength steels (Table 5-2) have yield strengths between 414 and 1034 MPa (60 and 150 ksi). Heat treatment, quenching, and tempering of these steels results in increased ductility and toughness. High strength steels have yield strengths of 1034 to 2068 MPa (150 to 300 ksi). Maraging steels are a sub-group of high strength steels that are more ductile and have improved characteristics in fatigue and energy absorption. Table 5-3 shows some common high strength steels used in ocean applications. A good source of characteristics of metals is the ASM International Handbook (ASM 1990). The classification and numbering of metals is accomplished by many organizations such as the American Iron and Steel Institute (AISI), American Society for Testing and Materials (ASTM), American Bureau of Ships (ABS), American Society of Mechanical Engineers (ASME) and others. ASTM and the Society of Automotive Engineers (SAE) publish a unified numbering system that includes a cross reference to all the numbering systems (ASTM/SAE 1989).

Table 5-1. Characteristics of low strengths steels.

Material	Yield Strength		Ultimate Strength		Elongation	Approx. Cost Rel. to
	MPa	ksi	MPa	ksi	%	Low Strength Steels*
Wrought Iron	186	27	331	48	14	1
ASTM A-242	317	46	462	67	19	1
ASTM A-441	317	46	490	71	19	1
ABS-class BH	324	47	490	71	19	1
AISI 1020	331	48	517	75	31	1
ASTM A-302	345	50	552	80	15	1

*Note: An approximate 1995 low strength structural steel cost is $0.35 per pound.

Table 5-2. Characteristics of medium strength steels for ocean applications.

Material	Yield Strength		Ultimate Strength		Elongation	Approx. Cost Rel. to Low Strength Steels
	MPa	ksi	MPa	ksi	%	
HY-80	552	80	689	100	20	3
ASTM A-543	586	85	724	105	16	2
HY-100	689	100	827	120	18	3
ASTM A-517-67	689	100	827	120	18	2
HY-140	965	140	1020	148	16	6
AISI 410	689	100	965	140	22	2
HY-180	1241	180	1310	190	18	6

Table 5-3. High strength steels for ocean applications.

Material	Yield Strength		Ultimate Strength		Elongation	Approx. Cost Rel. to Low Strength Steels
	MPa	ksi	MPa	ksi	%	
AISI H-11	1655	240	2034	295	12	11
Maraging 200	1379	200	1448	210	15	13
Maraging 250	1724	250	1793	260	12	13
Maraging 300	2068	300	2103	305	12	14
HY-TEN-B Alloy	1724	250	1999	290	28	15

Stainless Steels

Material known as stainless steel (Table 5-4) is iron based with greater than 12 % chromium added. It resists uniform rusting because of a thin oxide film that forms on the surface and isolates the material from the water. However, the complete submersion of stainless steel in seawater causes the oxide film to break down, and it is then susceptible to pitting, crevice, and intergranular corrosion. The type 300 series stainless steels (18 % chromium) are best suited for ocean applications and provide excellent corrosion protection for above waterline applications. Full submersion tends to lead to pitting and crevice corrosion, and type 304 and 316 are recommended for full submergence in seawater for only 4 to 6 months. One characteristic of the 300 series stainless steels is that they are nonmagnetic and the other series, such as the 400 series, are magnetic. Since stainless steels are susceptible to crevice corrosion, welded joints are preferred over mechanical fasteners. Special grades of stainless steels (e.g. type 304L and 316L) resist intergranular corrosion, but they are more expensive and have limited availability.

Table 5-4. Characteristics for selected stainless steels used in ocean applications.

Material	Yield Strength		Ultimate Strength		Elongation	Approx. Cost Rel. to Low Strength Steels
	MPa	ksi	MPa	ksi	%	
Type 302	276	40	621	90	60	6
Type 304	290	42	586	85	60	6
Type 304L	241	35	586	85	60	8
Type 316	290	42	621	90	60	8
Type 316L	290	42	558	81	55	10

Aluminum

Aluminum alloys have been used for ocean applications such as high-speed boats, hydrofoil craft, surface-effect ships, merchant and navy ships, marine propellers, SCUBA tanks, and submarines (Johnson-Sea-Link and Aluminaut). The specific weight of aluminum is 2.9 times lighter than similar steel, and the cost of an aluminum structure over its life is usually lower. The cost of steel is less, but when the cost of protective coatings are added, steel exceeds the cost of the aluminum that usually needs no coatings. Aluminum is known for its ease of fabrication. Certain alloys of aluminum resist corrosion in seawater to a great extent, and these alloys (Table 5-5) are frequently used for ocean applications. Improved mechanical properties can be attained through heat treatment and strain hardening processes. The thickness of aluminum structures are generally two or three times thicker than similar steel structures. Aluminum can be welded, but it is more difficult than welding steel. Since aluminum is lightweight and easy to fabricate, it is being used more and more in ocean applications.

Table 5-5. Characteristics of selected aluminum alloys for ocean applications.

Material	Yield Strength		Ultimate Strength		Elongation	Approx. Cost Rel. to Low Strength Steels
	MPa	ksi	MPa	ksi	%	
5083	228	33	317	46	16	7
5086	255	37	303	44	10	7
5454	248	36	303	44	10	8
5456	255	37	352	51	16	8
6061	276	40	310	45	12	8
7079	469	68	538	78	14	8

Titanium

Titanium is a high strength, lightweight, and expensive material (Table 5-6). It has been used in small submersibles (e.g. Alvin, Sea Cliff, and Turtle) and for the stress joint of a marine riser. It has excellent corrosion resistance and a moderate modulus of elasticity. A comparison of its use for a marine riser to that of steel and aluminum was studied by senior ocean engineering students (Buckeridge, et al. 1989), and they showed a significant reduction in cost was necessary to make it competitive. Titanium demonstrates excellent corrosion resistance and is nonmagnetic. Although titanium can be welded, it is more difficult than steel and weldable aluminum. The fatigue strength is relatively high.

Table 5-6. Characteristics of selected titanium alloys for ocean applications.

Material	Yield Strength		Ultimate Strength		Elongation	Approx. Cost Rel. to Low Strength Steels
	MPa	ksi	MPa	ksi	%	
5Al-2.5Sn	807	117	862	125	18	-
8Al-1Mo-1V	1034	150	1103	160	18	-
1AL-8V-5Fe	1524	221	1482	215	10	-
3Al-13V-11Cr	1207	175	1276	185	8	-
6Al-4V	1069	155	1172	170	8	20
7Al-4Mo	1207	175	1276	185	6	-

Other Non-ferrous Alloys

Copper-nickel alloys, monel, bronze, brass and aluminum bronze are also nonferrous alloys and their properties are illustrated in Table 5-7. Ships and other ocean systems frequently use seawater for various purposes such as ballast and cooling fluid for engines and condensers. These seawater systems are heat exchangers, valves, instruments, and piping that are subject to large variations in temperature, pressure and velocity and must be resistant to corrosion due to the seawater flowing through them. Copper and nickel alloys are excellent choices to best satisfy these requirements and have good fatigue properties, but they have been known to experience microbiologically influenced corrosion.

Table 5-7. Characteristics of selected nonferrous metals.

Material	Yield Strength		Ultimate Strength		Elongation	Approx. Cost Rel. to
	MPa	ksi	MPa	ksi	%	Low Strength Steels
Cupronickel 30	152	22	414	60	40	-
Cupronickel	255	37	469	68	28	-
K-500 Monel	448	65	724	105	25	-
Nickel-copper 400	310	45	586	85	35	-
Bronze (Cu90 Zn10)	69	10	276	40	45	14
Bronze (Cu85 Zn5 Sn5 Pb5)	117	17	255	37	30	-
Aluminum Bronze	186	27	517	75	25	-

Nonmetallic Materials

Thermoplastics

Thermoplastic material is a type of polymer that can be remelted and remolded to shape. This material can be reheated and remolded many times. The most common thermoplastics are polyethylenes (PE), polyvinyl chlorides (PVC), polypropylene (PP), and polystyrene (PS). The primary engineering thermoplastics include polyamides (nylons), polyethylene and polybutylene terephalates, polymethylmethacrylate (acrylic) and polyoxyemethylene (acetal). One of the well known polyamides is called Kevlar. Polymethylmethacrylate is better known as acrylic, perspex, or plexiglass and is frequently used as a replacement for glass because of its transparency. Table 5-8 summarizes some of the properties of thermoplastics.

Composites

Composites can be defined as material that is constructed from two or more components, separately manufactured, at the microscopic level. The composite consists of a matrix material that is polymeric, metallic, or ceramic and a reinforcing material that is fiber or particulate. The most used composites in the marine environment are polymeric or resin based. A primary marine application for composites has been its use in the construction of ship hulls. Future offshore applications include water piping systems, housings, buoys, and lifeboats.
Table 5-9 summarizes some of the characteristics of composites. High strength boron composites were used on America's Cup yachts. The disadvantage of these new composites is their expense. Graphite composites are relatively new and have been considered for use on new offshore platforms, but the cost is high.

Table 5-8. Summary of properties of selected thermoplastics.

Plastic	Elastic Modulus tension MPa (ksi)	Elastic Modulus flexure MPa(ksi)	Water Absorption (% in 24 hr)	Specific Gravity	Tensile Strength MPa(ksi)	Elongation %	Comments
PE polyethylene	140-6900 (20-1000)	69-970 (10-141)	<0.1 -<0.01	0.91-0.94	6-38 (0.9-5.5)	80-500	Resistant to marine exposure
PVC polyvinyl chloride plasticized	2.7-21 (0.4-3.0)		0.15-1	1.16-1.55	7-27 (1-3.9)	4.5-65	10% reduction in hardness due to water absorption over 5 years
PVC unplasticized	24-40 (3.5-5.8)	20-40 (2.9-5.8)	0.03-1.45	1.3-1.45	38-62 (5.5-9)	2-40	May be attacked by marine borers
PP polyproplene	689-1520 (100-220)	1100-1310 (160-190)	<0.03	0.89-0.91	29-34 (4.2-4.9)	115-350	Resistant to marine exposure, low water absorption
PS polystyrene	3100-3500 (450-508)	2750-3500 (400-508)	0.03-0.2	1.04-1.09	40-48 (5.8-7)	1-2.5	Slight strength loss due to sun, good resistance to sea water immersion, low water absorption
nylons polyamides	1400-3000 (203-435)		0.25-1.9	1.02-1.14	38-80 (5.5-11.6)	30-300	
acrylic polymethyl-methacrylate	2400-3000 (348-435)	2750 (400)	0.1-0.5	1.12-1.28	41-86 (5.9-12.5)	2-7	Some reduction of properties due to water absorption, sun resistant

Table 5-9. Summary of properties for unidirectional composites (ASM 1990).

Property	Boron-epoxy	Boron-polyamide	S-glass-epoxy	High-modulus graphite-epoxy	High-modulus graphite polyamide	High-strength graphite epoxy	Arimid epoxy
Reinforcement content (% Volume)	50	49	72	45	45	70	54
Specific Gravity	2.02	1.99	2.13	1.55	1.55	1.61	1.36
Tensile Strength (MPa)							
Longitudinal	1370	1040	1290	840	807	1500	1190
Transverse	56	11	46	42	15	40	11
Tensile Modulus (GPa)							
Longitudinal	201	221	61	190	216	154	84
Transverse	22	14	25	7	5	10	5
Compressive strength (MPa)							
Longitudinal	1600	1090	820	883	652	1700	290
Transverse	123	63	162	197	70	246	65
Shear modulus (GPa)	5.38	7.65	12.0	6.2	4.48	6.9	2.83
Intralaminar shear strength (MPa)	63	26	45	61	22	68	28
Poisson's ratio							
Major	0.17	0.16	0.23	0.10	0.25	0.28	0.32
Minor	0.02	0.02	0.09		0.02	0.01	0.02

Note: Divide MPa by 6.895 to convert to ksi (1 ksi = 6.895 kPa).

Fiberglass is a nonmetallic material that is frequently used in the construction of small boats, buoys, and submarines. Fiberglass is one example of an increasingly used type of composite material that consists of reinforcing fibers that are held together by a bonding material. The glass fibers are bonded together using epoxies, polyesters, phenolics, or silicones. The combination of glass fiber and polyester or epoxy binders is known as fiberglass and is a very common material in ocean or fresh water applications. These materials have a high strength to weight ratio and therefore are used in applications where lightweight is desirable. Fatigue strength of fiberglass is low, but it has excellent corrosion resistance. Good quality control is necessary to insure uniform qualities due to the processes involved in manufacturing. Table 5-10 presents a summary of properties for glass and fiberglass materials frequently used for ocean applications. Glass has high compressive strength, but very low tensile strength. It is also difficult to insure consistency in its physical characteristics. The cost of getting consistency through quality control is large.

Table 5-10. Summary of properties of glass and fiberglass materials.

Material	Yield Strength		Compressive Strength		Elastic Modulus $E \times 10^6$		Specific Weight lb/ft^3
	MPa	ksi	MPa	ksi	kPa	psi	
Fiberglass polyester mat	117	16.9	186	27	10.0	1.45	93.3
Fiberglass polyester cloth	365	53	290	42	18.6	2.7	112.3
Fiberglass epoxy cloth	586	85	448	65	31.7	4.6	113
Fiberglass epoxy filament	1034	150	827	120	55.2	8	129.6
Glass (tempered)	310	45	2758	400	82.7	12	207.4
Glass fiber	3103	450			86.2	12.5	

Cement and Concrete

Concrete has many marine applications that include foundations, dams, gravity offshore platforms, offshore pipeline coating, coastal protection structures (e.g. revetments, breakwater caps, piers, and docks). Concrete offshore fixed platforms are operational in the North Sea, and recently, concrete has been considered for floating structures such as semisubmersibles and offshore floating production systems. Concrete has excellent strength in compression but low strength in tension. Therefore, concrete structures need to be designed for minimum tension, bending, and shear stresses. Reinforcement using steel reinforcing members has greatly improved resistance to bending and tensile forces. In this case, the concrete must be sealed so that seawater does not permeate the concrete and contact the steel reinforcements that are subject to corrosion. Lightweight concrete has been developed and is finding uses in floating applications. Sharp corners are subject to cracking and abrasion and must be protected. The common ingredients in concrete mix are cement powder, water, sand, stones, various additives, and reinforcing steel bars. Typical properties of concrete are tabulated in Table 5-11.

Table 5-11. Properties of common concrete.

Material	Yield Strength		Compressive Strength		Elastic Modulus $E \times 10^6$		Specific Weight
	MPa	ksi	MPa	ksi	kPa	psi	lb/ft^3
Concrete lightweight	2.8	0.4	31.0	4.5	13.8	2.0	105.4
Concrete prestressed	4.8	0.7	48.3	7.0	31.0	4.5	155.5

Wood

The use of wood in marine applications has declined in recent years. Wood has higher strength along the grain than across the grain, and its strength is affected by moisture content, so treating the wood is often necessary to prevent water absorption and attacks by marine borers. There is a substantial influence of moisture content. Lamination is a common way of eliminating some of the anisotropy of properties. A summary of the mechanical properties for various woods is contained in Table 5-12. Marine grade plywood is made using weather and boil proof adhesive, and it is often used for the skins in small vessels. Applications of wood in the marine environment include quays, groins, small boats, and skids for launching offshore structures. A combination of wood and resins are used to produce wood laminates for marine applications. A wood laminate was used to construct the hull, propeller, and control surfaces for the human powered submarine *Aggie Ray* constructed by the ocean engineering students at Texas A&M University (Manlove et al. 1991).

Table 5-12. Summary of mechanical properties for wood with different moisture content.

Type	Moisture Content	Density	Static Bending			Compression parallel to grain		Com-pression normal to grain,	Shear parallel to grain,
	%	(t/m^3)	Propor-tional limit (MPa)	Ultimate strength (MPa)	Flexural modulus (MPa)	Propor-tional limit (MPa)	Ultimate Strength (MPa)	Propor-tional limit (MPa)	Ultimate strength (MPa)
Ash	43	0.54	36.5	65.5	9.7	23.2	28.0	5.9	9.3
	12	0.58	61.4	100.7	11.6	38.5	50.2	10.4	13.2
Beech	54	0.56	29.6	59.3	9.5	17.6	24.5	4.6	8.9
	12	0.64	60.0	102.7	11.9	33.6	50.3	8.6	13.9
Douglas fir	36	0.45	33.1	52.4	10.7	23.5	26.8	3.5	6.4
	12	0.48	55.8	80.7	13.2	44.5	51.2	6.3	7.9
Oak	70	0.59	32.4	55.8	8.3	20.3	24.3	5.9	8.8
	12	0.67	54.5	95.8	11.2	30.0	48.5	9.7	13.0
Pine	63	0.54	35.9	60.0	11.0	23.6	29.6	4.1	7.2
	12	0.58	64.1	101.4	13.7	42.4	58.2	8.2	10.3

Note: Divide MPa by 6.895 to convert to ksi (1 ksi = 6.895 kPa).

Buoyancy Material

Buoyancy material is used to provide buoyant force for various ocean applications, and it has a specific gravity less than water. These materials have been used on submarines, pipelines, marine risers and buoys to name a few. Gasoline with a specific gravity of 0.7 was used in the submarine Trieste during its record setting deep dive many years ago. Wood has a specific gravity of 0.5 and has been used as buoyancy material. However, wood absorbs water and eventually loses its buoyancy. Currently, buoyancy materials that do not absorb water and do not compress over the range of water depths are the most desirable. A common buoyancy material that is called syntactic foam has been developed that consists of hollow glass spheres mixed with a binding material such as a polyester or epoxy resin (Table 5-13). This material weighs between 1885 and 6598 N/m^3 (12 and 42 lb/ft^3) and has a buoyant force ranging between 98 and 231 N (22 and 52 lb).

Table 5-13. Properties of selected syntactic buoyancy material (Data from Emerson & Cuming).

Property	Shallow Depth	Moderate Depth	Moderate Depth	Deep Depth	Deep Depth	Deep Depth
Resin	epoxy	polyester or epoxy	polyester or epoxy	polyester	polyester	epoxy
Macrospheres	polystyrene	Dia = 7/16 in	none	none	none	none
Microspheres	none	Dia = 90 microns	Dia = 65 microns	Dia = 65 microns	Dia = 80 microns	Dia = 65 microns
Specific Weight N/m^3 (lb/ft^3)	1885 (12)	3770 (24)	5027 (32)	6598 (42)	6598 (42)	5498 (35)
Buoyancy in Sea Water N/m^3 (lb/ft^3)	8168 (52)	6283 (40)	5027 (32)	3456 (22)	3456 (22)	4555 (29)
Compressive Yield Strength MPa (ksi)	-	11.7 (1.7)	50.3 (7.3)	68.9 (10)	48.3 (7)	72.4 (10.5)
Hydrostatic Strength MPa (ksi)	1.03 (0.15)	12.4 (1.8)	48.3 (7.0)	103.4 (15. 0)	55.2 (8.0)	96.5 (14.0)
Failure Depth m (ft) of sea water	103 (338)	> 2058 >(6750)	6159 (20,200)	10976 (36,000)	7561 (24,800)	9604 (31,500)

MARINE CORROSION

Marine corrosion is the deterioration of metals in the marine environment due to electrochemical reaction. Ocean engineers must consider corrosion effects in the design process, and they must understand corrosion principles and learn to minimize or account for their effect. For design purposes, the engineer must realize that corrosion of the structure occurs, and as a consequence, the structural strength is reduced. It is through understanding corrosion principles that engineers can avoid poor designs and eventual failures. References for more detailed information on marine corrosion are AISI (1981), ASM (1987), NACE (1983), Fontana and Greene (1986), LaQue (1975), Reuben (1994), Schenck (1975), Schumacher (1979), Tuthill and Schillmoller (1971), and Uhlig and Revie (1985).

Electrochemical Reactions

For an electrochemical reaction to occur there must be a metal anode area that corrodes by oxidation and a cathode where the reduction reaction occurs. An electrolyte must be in contact with both the anode and cathode, and the ions are transferred through this path. Finally, there must be an electron flow between the anode and cathode, and the electrons flow from the anode to the cathode. The most common reactions at the cathode are illustrated in Table 5-14. M is the metal, Ze is the number of electrons lost, and Z is the metal valence. Reaction number three is the most common and important reaction in marine corrosion that indicates the reaction draws the needed oxygen from the surrounding environment.

Table 5-14. Common electrochemical reactions in corrosion.

Cathodic Type	Reaction Equation	Reduction Type
Acid	$2H^+ + 2e \rightarrow H_2$	hydrogen reduction
Oxygenated acid	$4H+ + O_2 + 4e \rightarrow 2H_2O$	oxygen reduction
Neutral or near-neutral solution	$O_2 + 2H_2O + 4e \rightarrow 4OH^-$	oxygen reduction
Metal reduction	$M^{Z+} + e \rightarrow M^{(Z-1)+}$	
Metal deposition	$M^{Z+} + Ze \rightarrow M^0$	
Anodic Type	**Reaction Equation**	**Reduction Type**
	$Fe \rightarrow Fe^{2+} + 2e$	
	$Al \rightarrow Al^{3+} + 3e$	

Galvanic Series of Metals

The galvanic series of metals in seawater are tabulated in Table 5-15. If there is a large potential difference between two metals, then there is a greater potential for corrosion. A small difference indicates a much lesser tendency to corrode. These differences do not indicate how fast the corrosion takes place. Electrode kinetics provide the ability to determine the corrosion current and thus, the speed of the reaction.

Typical Types of Corrosion Occurring in the Ocean Environment

Uniform and Galvanic Corrosion

General corrosion appears as a continuous layer of corrosion over an entire surface area. It occurs more often for objects exposed to air such as piping and plates on exposed structures such as offshore platforms and ships. This type of corrosion is not commonly found when objects are totally submerged. It is easier to design for this type of corrosion. Table 5-16 illustrates the corrosion rates for selected materials submerged in quiescent seawater. Titanium has no corrosion rate, and type 316 stainless steel has a corrosion rate of less than 0.1 mil/yr except for deep pitting. However, carbon steel corrodes at a rate of 4 to 7 mils/yr. For tubes and piping with seawater flowing inside (Table 5-17), titanium shows no corrosion rate for all flow velocities and carbon steel has a corrosion rate of 5 to 30 mils/yr for velocities from 0 to 4.57 m/s (0 to 15 ft/s) respectively. Type 316 stainless steel shows only deep pitting for velocities between 0 to 0.91 m/s (0 and 3 ft/s) and a corrosion rate of less than 0.1 mil/yr for velocities between 0.91 and 4.57 m/s (3 and 15 ft/s). The corrosion effects for seawater flowing in pumps and over hydrofoils are tabulated in Table 5-18. Pump parts and hydrofoils are expected to corrode at a rate of 30 to 300 mils/yr when velocities are between 6.1 (20) and 42.7 m/s (140 ft/s) and the material is carbon steel. Titanium again has no corrosion rate, and stainless steel type 316 has a corrosion rate of less than 1 mil/yr only at the highest flow velocity 36.6 to 42.7 m/s (120 to 140 ft/s).

Galvanic corrosion occurs when two dissimilar metals are connected directly or by a metallic path and are immersed in seawater or other liquid that acts as an electrolyte. The result is that one metal corrodes faster than normal, and the other more noble metal corrodes slower or ceases to corrode. The galvanic series in Table 5-15 can be used to estimate the potential for corrosion. Galvanic corrosion is also used to protect ship hulls by bolting zinc anodes to the steel hulls. It is best to avoid using dissimilar materials, but if dissimilar materials must be used, isolate the two materials by placing nonmetallic material between them. Also, making the anode larger than the cathode reduces galvanic corrosion.

Table 5-15. Galvanic series of metals and their alloys in seawater.

Material	Nobility	Electric Potential Range (V)
Graphite	Noble (cathode)	+0.3 to +0.20
Platinum		+0.35 to +0.20
Tantalum		+0.20
Stainless steel type 6X (passive)		+0.32 to -0.15
Hastelloy C		+0.10 to -0.04
Inconel 625		+0.10 to -0.04
Incoloy 825		+0.05 to -0.03
Titanium & titanium alloys		+0.06 to -0.05
Stainless steel (300 series)		-0.00 to -0.15
Monel 400 & K-500		-0.04 to -0.14
Silver		-0.09 to -0.14
Stainless steel 17-4 PH (passive)		-0.10 to -0.20
Copper		-0.14
Inconel 600 (passive)		-0.13 to -0.17
Molybdenum		-0.17
Copper-nickel (70-30)		-0.13 to -0.22
Lead		-0.19 to -0.25
Tungsten		-0.24
Stainless steel 430 & 431 (passive)		-0.20 to -0.28
Copper-nickel (80-20)		-0.21 to -0.27
Copper-nickel (90-10)		-0.21 to -0.28
Nickel silver		-0.23 to -0.28
Silicon bronze		-0.24 to -0.27
Manganese bronze		-0.25 to -0.33
Stainless steel 410 (passive)		-0.24 to -0.35
Lead-tin solder 50/50		-0.26 to -0.35
Red brass, CDA 230		-0.20 to -0.40
Stainless steel 17-4 PH & type 6X (active)		-0.20 to -0.40
Cast brasses and bronzes		-0.24 to -0.40
Naval brass		-0.30 to -0.40
Aluminum bronze		-0.30 to -0.40
Inconel 600 (active)		-0.30 to -0.42
Stainless steel 300 series (active)		-0.35 to -0.57
Stainless steel 400 series (active)		-0.45 to -0.57
Maraging steels		-0.57 to -0.58
Alloy steels 4130		-0.60
High strength steels (HY80 & above)		-0.60 to -0.63
Low strength alloy steels		-0.57 to -0.63
Plain carbon steels		-0.60 to -0.70
Cast irons		-0.60 to -0.72
Aluminum alloys		-0.70 to -0.90
Zinc		-0.98 to-1.03
Aluminum		-1.25 to -1.50
Magnesium	Active (anode)	-1.60 to -1.63

Table 5-16. Corrosion effects on materials submerged in quiescent seawater (Petersen and Soltz 1975).

Material	Usual Average Corrosion Rates (mils/yr)
Nickel-chromium-high molybdenum alloys	None
Titanium	None
Stainless steel type 316	< 0.1 except for deep pitting
Stainless steel type 304	< 0.1 except for deep pitting
Nickel-chromium alloys	< 0.1 except for deep pitting
Nickel-copper alloy	< 1.0 except for pitting
Nickel	< 1.0 except for deep pitting
Copper-nickel 70-30	0.1 to 0.5
Copper-nickel 90-10	0.1 to 0.5
Copper	0.5 to 3.0
Aluminum brass	0.5 to 1.5
Nickel-aluminum bronze	1.0 to 2.0
Nickel-aluminum-manganese bronze	1.0 to 2.0
Manganese bronze	1.0 to 3.0 dezincifies
Cast iron	2.0 to 3.0
Carbon steel	4.0 to 7.0

Table 5-17. Corrosion effects in pipes and tubes with seawater flowing (Petersen and Soltz 1975).

Material	Usual Average Corrosion Rates (mils/yr)				
Nickel-chromium-high molybdenum alloys	None				
Titanium	None				
Copper-nickel alloy 70-30	< 1.0				
Copper-nickel alloy 90-10	< 1.0				
Aluminum brass	< 1.0				> 5.0
Copper	< 3.0			> 5.0	
Carbon steel	5.0	>	>	>	30.0
Nickel-copper alloy 400	May pit		< 1.0		
Stainless steel type 316	Deep pitting		< 1.0		
Nickel-chromium alloys	Deep pitting		< 1.0		
Stainless steel type 304	Deep pitting		< 1.0		
Water Velocity ft/s	0–3	3–6	6–9	9–12	12–15
m/s	0–0.91	0.91–1.83	1.83–2.74	2.74–3.66	3.66–4.57

Intergranular

Intergranular corrosion is a microscopic form of corrosion that is caused by the potential difference between the grain boundaries of the metal and grain bodies. When the grain body is anodic to the grain boundaries, corrosion occurs along the boundaries, and this is frequently the case along weld zones. Dealloying corrosion occurs in some metal alloys that are susceptible to corrosion by attacking elements in the alloy. The more active element is removed, and the more noble element remains. The result is a porous and weakened structure. This type of corrosion is common for cast iron immersed in seawater, and it occurs in brass having more than 15 % zinc.

Table 5-18. Corrosion effects for seawater flowing in pumps and over hydrofoils (Petersen and Soltz 1975)

Material	Usual Average Corrosion Rates (mils/yr)											
Nickel-chromium-high molybdenum alloys	None											
Titanium	None											
Stainless steel type 316	None										< 1	
Stainless steel type 304	None										< 1	
Nickel-chromium alloys	None										< 1	
Nickel-copper alloy 400	None										<1	
Nickel	None										< 1	
Copper-nickel 70-30 (5 Fe)	None										7	
Carbon steel	30	>		>				>			300	
Nickel-aluminum bronze	< 10										> 30	
Nickel-aluminum-manganese bronze	< 10										> 30	
Austenitic nickel cast iron	< 10										>30	
Manganese bronze	dezincifies											
Copper-nickel alloy 70-30 (0.5 Fe)	< 10										>50	
		Pump impellers & ship propellers										
	Pump & valve body					Hydrofoils						
ft/s	20		40		60		80		100		120	140
m/s	6.1		12.2		18.3		24.4		30.5		36.6	42.7

Typical range of velocities for equipment item

Crevice and Pitting

Crevice corrosion occurs where there is a limited availability of oxygen such as that which occurs at slightly open joints (crevices). Other examples are under nuts, bolt heads, and washers. The localized nature of the corrosion can cause mechanical failures. Pitting corrosion is similar to crevice corrosion, but it does not require an existing pit or crevice to cause the pits to occur. These pits can penetrate a hull or weaken a structure.

Erosion

Erosion corrosion occurs when seawater is flowing, and it is often found in bends and elbows of pipes. Corrosion due to cavitation is also caused by the velocity of seawater, but the mechanism is different. As the fluid flows over the metal surface, vapor bubbles form at the interface (cavitation) as a result of the reduced local pressure. When these vapor cavities collapse in regions of higher pressure, the forces of the water on the metal surface can damage the metal surface or protective coating. This type of corrosion commonly occurs on propellers, hydrofoils, and pump impeller blades.

Stress Corrosion Cracking

Stress corrosion cracking is a result of the combined action of the ocean environment and mechanical stress that results in cracking of the material, and it can go undetected until the structure fails. Another type of corrosion is hydrogen embrittlement that is caused by atomic hydrogen penetrating the metal and combining either with the metal to form a brittle phase or by creating a large pressure. The hydrogen can be introduced by corrosion or by a cathodic protection system.

Preventing Corrosion

The optimum time to consider corrosion prevention is during the design stage, and the worst time is after the existence of corrosion has been discovered. Several methods of preventing corrosion are now briefly described that include material selection, good design, paint or coating, cathodic protection, and inhibitors. Selecting the right material for a particular application can avoid or minimize corrosion. However, the selection of the most corrosion resistant material doesn't always work because there are other requirements that the design must satisfy. Some typical requirements and a few procedures for making proper material choices are tabulated in Table 5-19.

Table 5-19. Typical requirements and procedures for material selection for corrosion prevention.

Requirements	Procedures
Properties (corrosion, mechanical, physical)	Define life of system
Fabrication (constructability, weldability, machinability)	Material life
Compatibility	Reliability (safety, failure consequences)
Maintainability	Availability/delivery time
Data availability	Costs (material, maintenance, inspection)
	Comparison with other corrosion protection possibilities

Design

Good design incorporates corrosion protection methods during the design stage when the engineer can view drawings and determine the possibility of eliminating geometric configurations that are known to cause or accelerate the corrosion process. Examples are the elimination of crevices, stagnant areas, and stress risers. The design stage is a good time to consider effects of other corrosion protection methods, but they can also be incorporated at a later time.

Coating

Painting metal surfaces is a very common type of corrosion control. The life of corrosion protection from paint is relatively short (several years), and the repair of worn, scratched, or chipped spots is routine maintenance. Some coatings (long-life paints) have been developed that have a useful life of over 10 years in the marine environment. Of course, coatings are very expensive on a per gallon basis, and they require extensive surface preparation that is also costly. Inorganic zinc is one of the better coatings for above water applications, and its use underwater requires a top-coating. Other types of coatings are tabulated in Table 5-20 along with a brief description of their characteristics.

Table 5-20. Characteristics of paint and coatings for marine protection.

Paints and Coatings	Characteristics
Water based paints	• Easy application • Nearly odorless • Easy clean-up
Oil based paints	• Easy application • Relatively inexpensive • Permeable • Recommended for mild atmospheric conditions
Alkyd paints	• Must be baked to dry • Better corrosion resistance than oil based paints • Not suitable for resistance to chemicals
Urethane paints	• Good resistance to abrasion • Corrosion resistance approaches vinyls and epoxies
Vinyl paints	• Better corrosion resistance than oils or alkyds • Adherence and wetting often poor • Good resistance to aqueous acids and alkalines • Maximum temperature of 150 °F
Chlorinated rubber	• Poor wetting • Quick drying • Good resistance to water and inorganics • Maximum temperature of 150 °F • Painting often improves protection
Epoxy paints	• Coal tar-epoxy offers good resistance to water soil and inorganic acids • Polyamide-hardened more resistant to moisture but less resistant to acids • Amine-hardened more resistant to chemicals • Epoxy-ester easier to apply but less corrosion resistant
Silicone paints	• Excellent water repellent • Maximum temperature of 1200 °F • Poor chemical resistance
Zinc paints (organic & inorganic)	• Used for galvanic protection • Organic requires less surface preparation • Inorganic easier to topcoat • Used effectively in neutral and slightly alkaline solutions • Inorganic more heat resistant
Coal tars	• Applied hot • Good for underground applications

Cathodic Protection

Cathodic protection is the most common form of corrosion protection for submerged material, and it is best used in conjunction with paints and coatings. Two types of cathodic protection are called impressed and galvanic cathodic protection. The impressed current system is a more permanent protection system and requires the use of external electrical power. It is much more complex than the galvanic system and has been used for offshore structures and mooring applications. Galvanic protection employs aluminum, magnesium, or zinc anodes that are attached to the steel material in seawater. The cathodic protection principle is that when a metal receives electrons, it becomes the cathode and can no longer corrode. Aluminum and aluminum-zinc anodes are commonly used for offshore applications such as for offshore oil platforms, marine pipelines, and instrument mooring chains. The aluminum anodes have a better current/weight ratio than zinc and are very durable. Aluminum anodes do not corrode reliably in fresh or brackish water, so zinc anodes are the more common material used as a sacrificial material to protect steel in fresh or brackish waters. When the anodes deteriorate, they are replaced, and continued protection of the steel is accomplished. The most common use of cathodic protection is on ship hulls, but it is also used on the submerged part of offshore structures and offshore pipelines.

The design of a cathodic protection system depends on whether the anodes can supply the required external current for the design life of the system. The general sequence used to design a cathodic protection system is

- Select proper maintenance current density based on geographical location.
- Calculate surface areas to be protected.
- Compute total anode material (number of anodes) required for selected life of structure.
- Determine anode geometry and initial current density assuming appropriate driving potential (i.e. 0.45 V between steel and aluminum alloy anodes).
- Determine life of anode for polarized material (i.e. 0.25 V potential for polarized steel).
- Distribute anodes evenly over material to be protected.

First, the total number of anodes is determined. The amount of maintenance current density (i_m) needed for protecting selected material in seawater or saline mud is known. The known surface area (A) to be protected is used to determine the total current (I_{total})

$$I_{total} = i_m A \qquad \textbf{5-1}$$

The total weight (W_{total}) of anode material is evaluated by

$$W_{total} = I_{total}(\text{Design Life})SC \qquad \textbf{5-2}$$

where S is a safety factor and C is the anode current capacity. The total number (N) of anodes is determined by

$$N = \frac{I_{total}}{C} \qquad \textbf{5-3}$$

Equal spacing of these anodes over the protected surface is required. Next, the anode resistance is found using

$$R = \frac{P}{2\pi L}\left[\ln\left(\frac{4L}{r}\right) - 1\right] \qquad \textbf{5-4}$$

where R is the anode resistance, P is the resistivity of the electrolyte, r is the radius of a cylinder and L is the cylinder length. Other anode geometry uses an effective radius as defined by

$$r = \sqrt{\frac{\text{cross sectional area}}{\pi}} \quad \textbf{5-5}$$

The driving potential (E) between the anode (E_{anode}) and protected material (E_{steel}) is determined by

$$E = E_{steel} - E_{anode} \quad \textbf{5-6}$$

Typically, the driving potential between unpolarized steel and aluminum alloy anodes is 0.45 V (i.e. E_{steel} = -0.60 and E_{anode} = -1.05). The average initial current at the anode (I) is found from

$$I = \frac{E}{R} \quad \textbf{5-7}$$

The initial structure current density (i) is

$$i = \frac{I\,N}{A} \quad \textbf{5-8}$$

This initial current fully polarizes the material to be protected (e.g. steel), and it causes the driving potential to be reduced (e.g. the driving potential is reduced to 0.25 V between steel and aluminum alloy anode). To calculate the life of the anode (T), the current output is recalculated using Equation 5-7 with the reduced driving potential and a reduced radius or effective anode radius. The lifetime (T) of an anode is determined by

$$T = \frac{W\,C}{I\,S} \quad \textbf{5-9}$$

where W is the weight of the anode, C is the current capacity of the anode as shown in Table 5-21 (ASM 1987), and S is a safety factor. If the life of the anode just exceeds the desired life of the structure, then the design is satisfactory. Otherwise, the type and size of the anode must be adjusted and design steps repeated until a satisfactory design is found.

Table 5-21. Current capacity of sacrificial anodes (ASM 1987).

Alloy	Current Capacity (A-h/lb)
Aluminum-zinc-mercury	1250-1290
Aluminum-zinc-indium	760-1090
Aluminum-zinc-tin	410-1180
Zinc	370
Magnesium	500

Example problem 5-1

An example of determining a cathodic protection system that has a twenty year design life for an offshore platform in the Gulf of Mexico is outlined in Table 5-22. The cylindrical anodes are to be aluminum alloy weighing 1050 lb (8 ft long and 0.4 ft radius). The water resistivity is 20 ohm-cm and a safety factor of 1.25 is required. The calculated platform surface area is 35,000 and 50,000 square feet in the water zone and mud zone, respectively. The anode capacity is 1250 A-hr/lb and the maintenance current is 6 mA/ft^2. The life and number of anodes is to be determined.

Table 5-22. Example calculation for cathodic protection system.

Given:	Design life	= 20 yr
	Calculated surface area (A)	= 35,000 ft^2 (water zone), 50,000 ft^2 (mud zone)
	Anode capacity (C)	= 1250 A-hr/lb
	Safety factor (S)	= 1.25
	Water resistivity (P)	= 20 ohm-cm
	Maintenance current density (i_m)	= 6 mA/ft^2
	Anode material	= Aluminum alloy
	Assumed anode parameters	= 1050 lb net weight, 8 ft long, 0.4 ft radius
Find:	Number of anodes (N) and Life of anodes (T)	
Soln:	1. Total current (I_{total})	
	$Current_{water}$ = (6 mA/ft^2)(35,000 ft^2)/1000 = 210 A	
	$Current_{mud}$ = (6 mA/ft^2)(50,000 ft^2)/1000 = 300 A	
	2. Total weight of required anode material	
	$Total\ weight_{water}$ = (210 A)(8760 hr/yr)(20 yr)(1.25)/(1250A-hr/lb) = 36,792 lb	
	$Total\ weight_{mud}$ = (300 A)(8760 hr/yr)(20)(1.25)/(1250 A-hr/lb) = 52,654 lb	
	3. The number of anodes, N, needed	
	N_{water} = (36,792 lb)/(1050 lb/anode) = 35.0	
	N_{mud} = (52,654 lb)/(1050 lb/anode) = 50.1	
	N_{total} = 35 + 50 = 85	
	4. Resistance (R) of a single anode	
	$R = \frac{P}{2\pi L}\left[\ln\left(\frac{4L}{r}\right) - 1\right]$	
	where L is anode length (in), P is water resistivity (ohm-cm) and r is an effective radius (in).	
	$R = \frac{20}{2\pi(96)2.54}\left[\ln\left(\frac{4(96)}{4.8}\right) - 1\right] = 0.0441$ ohm	
	5. Assuming a driving potential of 0.45 V, the initial current input is computed as	
	I = E/R = (0.45)/0.0441 = 10.2 A	
	6. The initial current density for the entire structure is	
	Current density = (10.2 A/anode)(85 anodes)/(35,000) = 0.025A/ft^2	
	This is above the maintenance current density of .006 A/ft^2.	
	7. Next, the anode resistance is computed for a 10 % reduced radius of 4.3 in	
	$R = \frac{20}{2\pi(96)2.54}\left[\ln\left(\frac{4(96)}{4.3}\right) - 1\right] = 0.0456$ ohm	
	8. For a potential drop of 0.25 V, the total reduced current I is	
	I = (0.25)/(0.0456) = 5.48 A	
	9. The design life in years of the cathodic protection system is	
	$T = \frac{WC}{IS}$	
	Life = (1050 lb/anode)(1250 A-hr/lb)/[(8760 hr/yr)(5.48 A/anode)(1.25)]	
	Life = 21.8 yr (Slightly exceeds desired 20 yr, so design is acceptable)	

Inhibitors

The protection of closed systems such as engines, boilers, or tanks is usually accomplished with the use of chemical inhibitors. Five classes of inhibitors are: absorption (affects anodic and cathodic reactions), hydrogen evolution poisons (affects hydrogen evolution), scavengers (removes oxygen needed for the cathodic reaction), oxidizers (works with iron only), and vapor phase inhibitor. The first four methods use a solution to protect the metal. For example, rust inhibitors are used in automobile coolants to protect the engine and radiator.

REFERENCES

Handbook of Corrosion Protection for Steel Structures in Marine Environments. American Iron and Steel Institute (AISI), 1981.

ASM Metals Reference Book, American Society of Metals (ASM). Metals Park: American Society of Metals, 1981.

Metals Handbook, Ninth Edition, Volume 13, Corrosion. American Society of Metals (ASM). Metals Park: ASM International, 1987.

Metals Handbook, Tenth Edition, Volume 1 & 2. American Society of Metals (ASM). Metals Park: ASM International, 1990.

Metal & Alloys in the Unified Numbering System. American Standards and Testing Manual/Society of Automotive Engineers (ASTM/SAE). Warrendale: Society of Automotive Engineers, Inc., 1989.

Seawater Corrosion of Metals and Alloys. Battelle. Columbus: Battelle, 1986.

Buckeridge, S., S. Campbell, D. Hagen, H. Hagen, H. Hvide, E. Kaiser, G. Luckman and M. Phillips. "Preliminary Design of a Deepwater Tension Leg Platform Production Riser." Ocean Engineering Senior Design Class, Ocean Engineering Program, Texas A&M University, College Station, Texas, May 1989.

Dexter, S. C. *Handbook of Oceanographic Engineering Materials.* New York: Wiley-Interscience, 1979.

Fontana, M., and N. Greene. *Corrosion Engineering*, 2nd Ed. New York: McGraw-Hill, 1986.

LaQue, F. L. *Marine Corrosion.* New York: Wiley-Interscience, 1975.

Manlove, T. L., J. K. Longridge and R. E. Randall. "Performance of a Human Powered Submarine: The Aggie Ray." *MTS '91 Conference Proceedings*, New Orleans, Louisiana, November 11-13, 1991.

Meyers, J. J., Editor. *Handbook of Ocean and Underwater Engineering.* New York: McGraw, 1969.

Petersen, C., and G. Soltz. Ocean Corrosion, Chapter 5, *Introduction to Ocean Engineering.* Edited by H. Schenck, New York: McGraw, 1975.

"Recommended Practice: Corrosion Control of Steel, Fixed Offshore Platforms Associated with Petroleum Production." National Association of Corrosion Engineers (NACE), NACE RP-01-76, 1983 Revision, National Association of Corrosion Engineers, 1983.

Reuben, R. L. *Materials in Marine Technology.* London: Springer-Verlag, 1994.

Schenck, H., Editor. *Introduction to Ocean Engineering.* New York: McGraw, 1975.

Schumacher, M., Editor. *Seawater Corrosion Handbook.* Noyes Data Corporation, 1979.

Sheets, H. E., and V. T. Boatwright. *Hydronautics*, Chapter VI. New York: Academic, 1970.

Tuthill A. H., and C. M. Schillmoller. *Guidelines for Selection of Marine Materials*, 2nd Edition. New York: The International Nickel Co. Inc., 1971.

Uhlig, H. H., and R. W. Revie. *Corrosion and Corrosion Control*, 3rd Edition. New York: Wiley-Interscience, 1985.

PROBLEMS

5-1. A swimmer assist vehicle requires 100 lb of buoyancy at depths of up to 300 ft. The vehicle is to be compact and will remain submerged for weeks at depth outside an underwater habitat. Discuss selection of buoyancy material.

5-2. A barge of 100 ft length and 18 ft beam is to be constructed. Discuss the use of fiberglass, steel, and ferroconcrete as the construction material.

5-3. A designer plans to rivet two pieces of metal together. Discuss whether the rivets should be anodic or cathodic.

5-4. If you paint a metal structure, discuss the need to paint anode or cathode.

5-5. Determine the number of zinc anodes required to protect the 1000 ft chain mooring line for a taut mooring system for 5 years. The surface area of the low strength steel chain mooring line is 1500 ft^2. The seawater resistivity is assumed to be 30 ohm-cm and the safety factor is 1.5. The anode weight and current capacity are assumed as 1135 lb and 370 A-hr/lb. The anode is rectangular in shape with a length 0.5 ft, width of 2 in, and a thickness of 0.5 in.

5-6. Determine the number of aluminum-zinc-tin anodes necessary to protect 50,000 ft^2 of carbon steel plate (0.5 in thick) submerged in seawater for 15 years. The seawater resistivity is assumed to be 30 ohm-cm. The anode weight is 1135 lb and the current capacity is 1180 A-hr/lb. The cylindrical anode is 8 in long with a 3 in radius.

CHAPTER 6: FLOATING AND SUBMERGED BODY HYDRODYNAMICS

TERMINOLOGY

The analysis and design of ocean structures includes ocean systems that are fixed, floating, and submerged. Ship structure systems require a means of propelling them through the ocean waters, and the design and analysis of ships has been conducted by naval architects for decades. The navies of the world operate ships and other propelled vessels over the oceans of the world. The terminology associated with ship design is quite different from that encountered in traditional design of land structures, and some of this terminology is also used in the field of ocean engineering. Some of the common nautical terminology used aboard naval vessels and merchant ships is tabulated in Table 6-1 and that related to naval architecture, marine and ocean engineering for the design and analysis of floating bodies is defined in Table 6-2.

Table 6-1. Nautical terminology.

TERM	DEFINITION
Aft	Rear
Athwartship	Across ship from side to side
Beam	Extreme width of ship at widest point
Below deck	Below the main deck
Bow	Front of vessel
Bulkhead	Wall
Capstan	Electric device for winding mooring lines
Deck	Floor
Fathom	Six feet of water depth
Focsle, Forecastle	Forward part of ship above main deck
Forward	Front
Head	Bathroom, restroom, lavatory
Keel	Bottom of center portion of vessel
Knot	Nautical mile per hour
Lee	Side of ship sheltered from wind
Leeward	Being in or facing the direction toward which wind is blowing
Passageway	Hallway
Port	Left side of vessel when facing bow
Starboard	Right side of vessel when facing bow
Stern	Rear of vessel
Topside	Above the main deck
Windlass	Electric device used to raise anchor
Windward	Being in or facing the direction from which wind is blowing

Table 6-2. Terminology used in naval architecture, marine and ocean engineering.

Term	Definition
∇	Displacement
Δ	Mass displacement
AP	After perpendicular
B, CB	Center of buoyancy
BM	Distance between center of buoyancy and metacenter
CG, G	Center of gravity
DWL	Designed load water line
FP	Forward perpendicular
GM	Distance between CG and metacenter
GZ	Righting arm
I	Moment of inertia
K	Keel
KB	Distance between keel and center of buoyancy
KG	Distance between keel and center of gravity
KM	Distance between keel and metacenter
LCB, LCF	Longitudinal center of buoyancy, flotation
LCG	Longitudinal center of gravity
M	Transverse metacenter
SM	Simpson multiplier
TCG	Transverse center of gravity
V	Volume of displacement
VCG, VCB	Vertical center of gravity, buoyancy
Waterplane	Horizontal planes parallel to designed load waterplane
WL	Intersection of waterplane with vessel's form
WL	Waterline

Ship Geometry

The geometry of a ship is illustrated in what is known as the lines drawing and an example is shown in Figure 6-1. A set of lines drawings consist of the sheer, half-breadth, and body plans. The sheer plan consists of vertical planes with the centerplane through the ship centerline and buttock planes at specified distances from the centerline. The half-breadth, or waterlines, plan shows planes that are parallel to the baseplane and intersecting the hull, and all these planes are called waterplanes. The body plan illustrates planes that are perpendicular to the buttock and waterline planes and intersect the hull.

A vertical line is drawn in the sheer plan at the intersection of the designed load waterline (DWL), and it is called the forward perpendicular (FP). A similar vertical line is drawn at the stern and is known as the after perpendicular (AP). The distance between these two perpendiculars is the length of the ship and is often referred to as the length between perpendiculars (LBP). This length is used in the American Bureau of Ships' Rules for Building and Classing Steel Vessels.

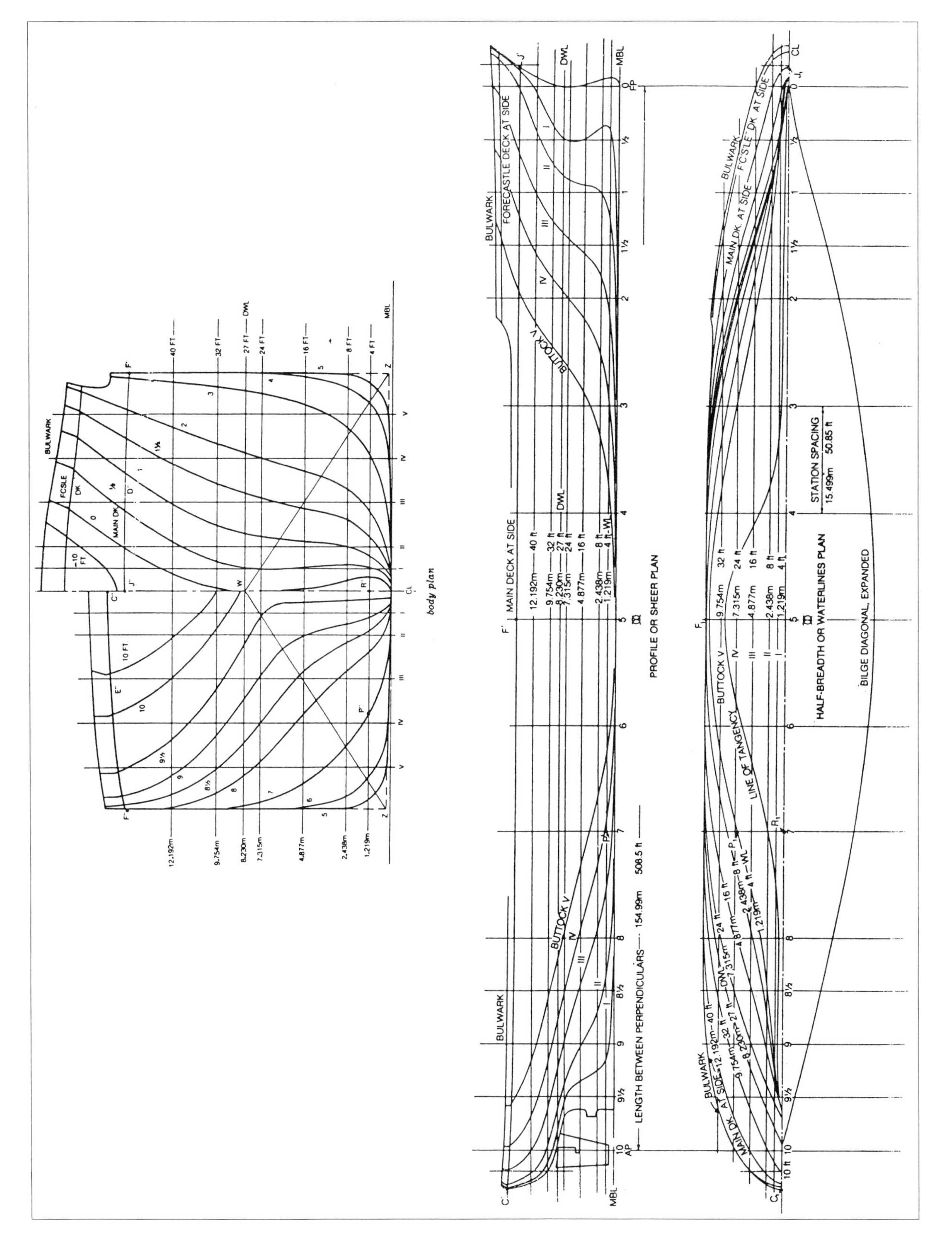

Figure 6-1. Example of ship lines drawing. Reprinted with permission from Lewis, 1988, *Principles of Naval Architecture*. (Full citing in references).

The length between perpendiculars is divided into 10 to 40 intervals by body plan planes that are called body plan stations, or simply stations. These stations are usually numbered from the bow with station "0" being the location of the FP. The molded base line is a straight horizontal line that is used as a reference datum for design and construction purposes. It is also the bottom of the vessel's molded surface and coincides with the top surface of the keel.

The draft of a vessel is the vertical distance between the waterline at which the vessel is floating and the bottom of the vessel. The molded draft is measured vertically between the waterline and the molded baseline, and the keel draft is measured between the waterline and the bottom of the keel. The average of the forward and aft drafts is the mean draft. Most ships have draft marks located amidships and at both ends as close to the forward and aft perpendiculars as possible. The difference between the forward and aft draft readings is known as the trim. Vessel offsets are measured from the vessel centerline to the waterline at each station, and a complete set of offsets for the various waterlines is known as a table of offsets. Computer programs, such as HULDEF and SHIPHUL, have been developed to generate ship lines and offsets.

HYDROSTATICS AND STABILITY

Displacement

The static behavior of floating and submerged bodies is governed by Archimedes' principle that says a body immersed in a fluid is buoyed up by a force equal to the weight of the fluid it displaces. This means the weight of the vessel is a downward force that is proportional to the mass of the body, and the equal upward buoyant force is proportional to mass of the displaced fluid. Assume that a floating body is partially immersed in a fluid such as water at rest as shown in Figure 6-2. At any point or depth (D), the mass of fluid above the depth is ρAz where ρ is the fluid density, A is the cross-sectional area parallel to the free surface of the column of fluid, and z is the distance from the free surface. Since a fluid does not support shear forces, static equilibrium requires that equal forces occur in all directions at that point. The pressure force experienced at this point (D) is ρgAz which is the weight of the column of fluid above D. If the floating body is rigid, then the integration of the vertical components of the pressure force over the surface of the body is the buoyant force. Thus, the buoyant force (F_B) is

$$F_B = \int_S \rho gz \cos \alpha \, dS \qquad \textbf{6-1}$$

where α is the angle of inclination of any portion of the body surface (S) from the horizontal. For equilibrium, the buoyant force must equal the weight of the body. Thus, the weight of the floating body and its contents are equal to the weight of the displaced fluid or the displacement. Similarly, the mass of a floating body and its contents are equal to the mass of the displaced fluid. For a totally submerged body, such as a submarine, the upward buoyant force must equal the weight of the fluid it displaces.

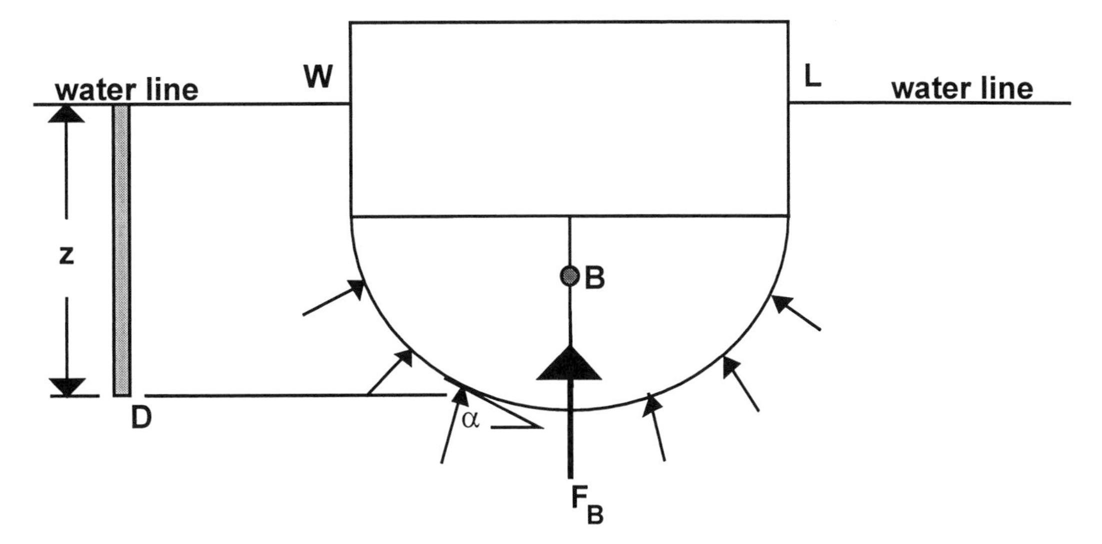

Figure 6-2. Buoyant forces acting on a floating body.

The volume of the underwater, or submerged, portion of a floating body can be calculated using numerical techniques such as Simpson's or trapezoidal rules. If the submerged portion is a simple shape it can be calculated directly. This underwater volume is called the volume of displacement (∇). Therefore, the weight of the displaced fluid, or the displacement weight (W) is

$$W = \rho g \nabla = \gamma \nabla \qquad \textbf{6-2}$$

where ρ is the fluid density, g is the gravitational acceleration, and γ is the specific weight of the fluid. Using Archimedes' principle, this displacement weight is equal to the weight of the floating body and its contents. In English units, W is usually expressed in long tons (2240 lb per ton), and in the SI system, it is usually expressed in metric tons (1000 kg per metric ton).

The centroid of the submerged portion of a floating body is called the center of buoyancy (B), and it represents the point through which the buoyant force acts. The density of the fluid medium affects the displaced volume of fluid and hence the draft of the vessel. Since seawater is more dense than freshwater, the draft of a vessel increases as it travels from ocean to inland waters that are less saline or to lakes (e.g. Great Lakes) that are freshwater.

Coefficients of Form

A number of coefficients are used to express characteristics of a vessel's shape, body plan sections, and waterlines. The Block Coefficient (C_B) is defined as the ratio of the volume of displacement (∇) of the molded form up to any waterline to the volume of a rectangular prism of length, breadth, and depth equal to the length, breadth, and mean draft of the vessel at the waterline. In equation form, it is

$$C_B = \frac{\nabla}{L\,B\,T} \qquad \textbf{6-3}$$

where L is length, B is breadth, and T is mean molded draft. The values used for L, B, and T vary so there can be slight differences. For example, the length between perpendiculars may be used for L, or the length at the waterline may be used. Typical values of C_B vary between 0.36 for high speed vessels to 0.92 for slow speed cargo vessels (Lewis 1988). The Midship Coefficient (C_M) is the ratio of the immersed area of the midship station to that of a rectangle of equal area to the product of the molded breadth and molded draft at the midship section. The common range of C_M values is between 0.75 to 0.995 (Lewis 1988). It is expressed as

$$C_M = \frac{\text{Immersed area at midship section}}{B\,T} \tag{6-4}$$

The Prismatic Coefficient (C_P) is the ratio of the volume of displacement (∇) to a prism with a length equal to the length of the ship and a cross section equal the midship section area. The usual range of values is 0.5 to 0.9 (Lewis 1988). It is expressed in equation form as

$$C_P = \frac{\nabla}{L \text{ x Immersed area at midship section}} \tag{6-5}$$

or

$$C_P = \frac{\nabla}{L\,B\,T\,C_M} = \frac{C_B}{C_M} \tag{6-6}$$

The Waterplane Coefficient (C_{WP}) is the ratio of the waterplane area (A_{WP}) to the area of a circumscribing rectangle and is expressed as

$$C_{WP} = \frac{A_{WP}}{L\,B} \tag{6-7}$$

This coefficient may be calculated at any draft and typical values at the DWL range between 0.65 to 0.95 (Lewis 1988). The Vertical Prismatic Coefficient (C_{VP}) is the ratio of the displacement volume (∇) to the volume of a cylindrical solid with a depth equal to the molded mean draft and with a uniform horizontal cross section equal to the waterplane area at that draft, and it is defined as

$$C_{VP} = \frac{\nabla}{C_{WP}\,L\,BT} = \frac{C_B}{C_{WP}} \tag{6-8}$$

Finally, there is the Volumetric Coefficient (C_V) that is defined as the displacement volume divided by the cube of one tenth of the length of the vessel and is expressed as

$$C_V = \frac{\nabla}{\left(\frac{L}{10}\right)^3} \tag{6-9}$$

Common values of C_V range from 1.0 for light and long ships (e.g. destroyer) to 15 for short heavy vessels (e.g. trawler). Certain ratios of the principal vessel dimensions are also frequently used in ratio form such as the length to breadth ratio (L/B, 3.5 to 10), length to draft ratio (L/T, 10 to 30) and breadth to draft ratio (B/T, 1.8 to 5). Since different definitions of length can be used in these ratios and coefficients, it is important to define which definition is employed.

It is necessary to calculate areas, centroids, volumes, and other geometrical characteristics for a vessel floating at a prescribed waterline (draft). Because the waterplane areas are not simple shapes, these characteristics usually require numerical techniques to accomplish the calculations. In most cases, Simpson's rules for integration are used, and these rules are defined in Table 6-3 and further explained by Lewis (1988). Mechanical integrators have also been used to evaluate plane areas, and the planimeter is an example of such an instrument.

Table 6-3. Integration Rules using the combined Simpson's first and second rules.

Station	Ordinate	Simpson Multiplier	Product
	Simpson's First Rule Primary		
0	y_o	1	y_o
1	y_1	4	$4y_1$
2	y_2	2	$2y_2$
3	y_3	4	$4y_3$
4	y_4	17/8	$17y_4/8$
5	y_5	27/8	$27\ y_5/8$
6	y_6	27/8	$27\ y_6/8$
7	y_7	9/8	$9\ y_7/8$
			$\sum$ product
	$\text{Area} = \frac{\Delta y}{3}(\sum \text{product})$, where Δy is station spacing		
	Simpson's Second Rule Primary		
0	y_o	1	y_o
1	y_1	3	$3y_1$
2	y_2	3	$3y_2$
3	y_3	17/9	$17\ y_3/9$
4	y_4	32/9	$32\ y_4/9$
5	y_5	16/9	$16\ y_5/9$
6	y_6	32/9	$32\ y_6/9$
7	y_7	8/9	$8\ y_7/9$
	$\text{Area} = \frac{3\Delta y}{8}(\sum \text{product})$, where Δy is station spacing		

Curves of Form

In the design of vessels, the curves of form are normally produced for the vessel at a series of drafts. The curves of form consist of a number of hydrostatic properties of the vessel as a function of the different drafts. These curves are used by the vessel operation personnel and are also known as the curves of form, or hydrostatic curves (Figure 6-3). The range of drafts should extend from below the lightest possible operational draft to the deepest possible draft.

Hydrostatic Calculations

Hydrostatic calculations include computation of the waterplane area (A_{WP}), tons per unit immersion (TPI or TP cm), longitudinal center of flotation (LCF), transverse metacentric radius (BM), height of the transverse metacenter (KM), longitudinal metacentric radius (BML), height of longitudinal metacenter (KML), molded displacement, total displacement (∇), displacement (Δ), longitudinal center of buoyancy (LCB), and vertical center of buoyancy (KB). Table 6-4 shows example calculations for the waterplane characteristics and uses Simpson's first rule with half multipliers in these computations. Similar computations for displacement, longitudinal center of buoyancy, volume of displacement, and height of center of buoyancy are illustrated in Table 6-5 and Table 6-6.

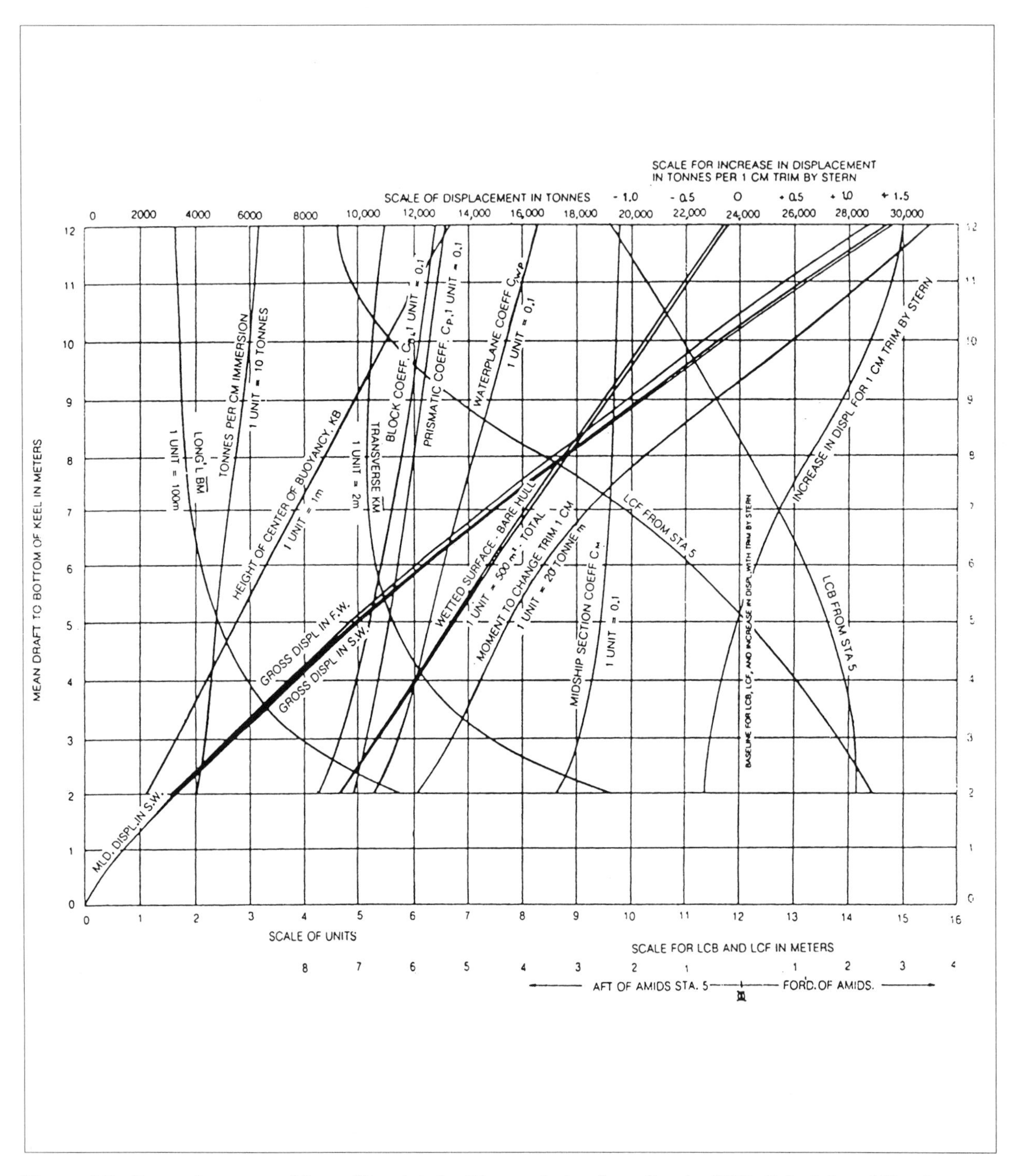

Figure 6-3. Example curves of form. Reprinted with permission from Lewis, 1988, *Principles of Naval Architecture*. (Full citing in references).

Table 6-4. Waterplane characteristics for a particular waterline (Lewis 1988).

Station	Half-breadth (m)	1/2 SM	Product	Moment Arm	Product	Moment Arm Squared	Product	(Half-breadth) cubed	Product
0	0	0.25	0	5.0	0	25.0	0	0	0
0.5	1.245	1.0	1.245	4.5	5.603	20.25	25.211	1.93	1.93
1	3.140	0.5	1.570	4.0	60280	16.0	25.120	30.96	15.48
1.5	5.359	1.0	5.359	3.5	18.757	12.25	65.648	153.90	153.90
2	7.597	0.75	5.698	3.0	17.094	9.0	51.282	438.46	328.84
3	10.956	2.0	21.912	2.0	43.824	4.0	87.648	1315.09	2630.18
4	12.007	1.0	12.007	1.0	12.007	1.0	12.007	1731.03	1731.03
5	12.039	2.0	24.078	0	0	0	0	1744.90	3489.8
6	12.039	1.0	12.039	-1.0	-12.039	1.0	12.039	1744.90	1744.90
7	11.899	2.0	23.798	-2.0	-47.596	4.0	95.192	1684.73	3369.46
8	10.271	0.75	7.703	-3.0	-23.109	9.0	69.327	1083.52	812.64
8.5	8.417	1.0	8.417	-3.5	-29.460	12.25	103.108	596.31	596.31
9	5.962	0.5	2.981	-4.0	-11.924	16.0	47.696	211.92	105.96
9.5	3.057	1.0	3.057	-4.5	-13.756	20.25	61.904	28.57	28.57
10	0	0.25	0	-5.0	0	0	0	0	0
			S_1=129.8		S_2= -34.3		S_3=656.2		S_4=15009

Station spacing (s), $s = \frac{L}{10} = \frac{154.99}{10} = 15.499$ m

Waterplane area, $A_{WP} = S_1 \frac{4}{3} s = [(129.9)(20.67)] = 2{,}683.8\ m^2$

Waterplane coefficient, $C_{WP} = \frac{A_{WP}}{LB} = \frac{2683.8}{[(154.99)(24.078)]} = 0.719$

Tonnes per centimeter immersion $= \frac{(2683.8)(1.025)}{100} = 27.51$

Longitudinal Center of Flotation , $LCF = \left(\frac{S_2}{S_1}\right) s = \left(\frac{-34.3}{129.9}\right) 15.499 = 4.10\,m$ abaft Station 5

Transverse moment of inertia, $I_T = S_4 \left(\frac{9}{4}\right) s = (15{,}009)(6.8884) = 103{,}390\ m^4$

Volume of displacement, $\nabla = 17{,}845$, from displacement curve Figure 6-3

Transverse metacentric radius, $BM = \frac{I_T}{\nabla} = \frac{103{,}390}{17{,}845}\ 5.79\,m$

Two approximate expressions for determining the height of the center of buoyancy for conventional ship forms are due to Morrish (1892) (Equation 6-10) and Posdunine (1925) (Equation 6-11),

$$KB = \frac{1}{3}\left[\frac{5T}{2} - \frac{\nabla}{A_{WP}}\right] \qquad \textbf{6-10}$$

$$KB = T\left[\frac{A_{WP}}{A_{WP} + \frac{\nabla}{T}}\right] \qquad \textbf{6-11}$$

The hydrostatic calculations for the curves of form are repetitious and consequently computers (mainframe and personal computers) are now used extensively. The US Navy's Ship Hull

Characteristics Program (SHCP) is used widely in shipyards and design offices throughout the US (NAVSEA, 1976), but other commercial software are available.

Table 6-5. Example calculation of displacement and LCB for a particular waterline (Data from Lewis 1988).

Station	Area (m^2)	1/2 Simpson Multiplier (SM)	Product	Non-dimensional Moment Arm	Product
-0.07	0	0.0175	0	5.07	0
-0.035	3.0	0.07	0.21	5.035	1.1
0	4.2	0.2675	1.12	5.0	5.6
0.5	12.7	1.00	12.70	4.5	57.2
1	22.6	0.50	11.30	4.0	45.2
1.5	35.1	1.00	35.10	3.5	122.9
2	50.6	0.75	37.95	3.0	113.9
3	83.3	2.0	166.60	2.0	333.2
4	106.1	1.0	106.1	1.0	106.1
5	113.7	2.0	227.40	0	0
6	107.6	1.0	107.6	-1.0	-107.6
7	81.4	2.0	162.80	-2.0	-325.6
8	44.0	0.7813	34.38	-3.0	-103.1
8.5	29.1	0.8438	24.3	-3.5	-85.1
9	17.4	0.8438	14.68	-4.0	-58.7
9.5	5.3	0.3138	1.66	-4.5	-7.5
9.565	3.3	0.13	0.43	-4.57	-2.0
9.63	0	0.0325	0	-4.63	0
			$S_1 = 944.33$		$S_2 = 95.6$

Sectional area curve extended beyond Stations 0 and 9.5 to extremities, as shown by Figure 6-1 and read at midpoint between last station and extremity. Simpson's Multipliers proportioned accordingly. At station -0.035, 1/2 SM= 0.5(4)(0.035) = 0.07; at station 0, 1/2 SM = 0.25+0.0175 = 0.2675; at station 8, 1/2 SM = 0.5[1.0+(9/8)(0.5)] = 0.7813 (First and Second Rules); at station 8.5 and 9, 1/2 SM = [(0.5)(9/8)(3/2)] = 0.8438; at station 9.5, 1/2 SM = 0.5[(9/8)(1/2)+(0.65/1.0)] = 0.3138; at station 9.565, 1/2 SM = [(1/2)(4)(0.065/1.0)] = 0.13.

Volume of displacement, $\nabla = S_1\,(2/3)\,s = 944.33\,(2/3)\,15/499 = 9{,}757\ \text{m}^3$

Displacement, $\Delta = 1.025(9757) = 10{,}000\ \text{t}\ (\text{seawater})$

Longitudinal center of buoyancy, $LCB = \frac{S_2}{S_1}(s) = \left(\frac{95.6}{944.33}\right)(15.499) = 1.57\ \text{m}$ forward of station 5

Static Stability

The stability of submerged and floating vessels is usually classified as being either static or dynamic. The concepts of static stability are discussed in this section, and the subject of dynamic stability is described in other texts such as Lewis (1988), Tupper (1996), and Zubaly (1996). Some dynamic effects such as forces due to wind can be considered as static when steady conditions occur. Another stability problem occurs when vessels go aground or when they are damaged and flooding occurs.

Floating and submerged vessels are usually rigid bodies, and for a state of equilibrium to exist, the resultant of all forces and moments acting on the vessel must equal zero. Static equilibrium of a floating vessel is concerned with the vessel in an upright position in still water. This means that the resultant of all gravity (weight) forces acting downward and buoyant forces acting upward on the vessel are of equal magnitude and act through the same vertical line.

Table 6-6. Example calculation of the volume of displacement and height of the center of buoyancy.

Height above baseline (m)	Waterplane area (m^2)	Multiplier for volume	Product	Multiplier for moment	Product
0	194	5	970	3	582
1	1714	8	13712	10	17140
2	1976	-1	-1976	-1	-1976
			$S_1 = 12706$		$S_2 = 15746$
1	1714	5	8570	3	5142
2	1976	8	15808	10	19760
3	2137	-1	-2137	-1	-2137
			$S_3 = 22241$		$S_4 = 22765$

Values for 1 m draft

Volume of displacement, $\nabla = \left[\frac{s}{12}\right] S_1 = \frac{1}{12}(12{,}706)$

Moment of volume about baseline, $M_\nabla = \left(\frac{s^2}{24}\right) S_2 = \left(\frac{1^2}{24}\right) 15{,}746 = 656.1\ \text{m}^4$

Height of center of buoyancy, $KB = \frac{M_\nabla}{\nabla} = \frac{656.1}{1059} = 0.62\ \text{m}$

Values for 2 m draft

Added volume of displacement, $\delta\nabla(1\ \text{to}\ 2\ \text{m}) = \left[\frac{s}{12}\right](S_3) = \frac{1}{12}(22{,}241) = 1853\ \text{m}^3$

Total volume, $\Sigma\nabla = \nabla + d\nabla = 1059 + 1853 = 2912\ \text{m}^3$

Moment of added volume, $\delta M_\nabla (1\ \text{to}\ 2\ \text{m})$ about 1m waterline $= \frac{s^2}{24}(S_4) = \frac{1^2}{24}(22765)\ \text{m}^4 = 948.5\ \text{m}^4$

Moment of added volume about baseline $= 948.5 + 1853(1 - 0) = 2801.5\ \text{m}^4$

Moment of total volume about baseline, $\Sigma M_\nabla = 656.1 + 2801.5 = 3457.6\ \text{m}^4$

Height of center of buoyancy, $KB = \frac{\Sigma M_\nabla}{\Sigma\nabla} = \frac{3457.6}{2912} = 1.19\ \text{m}$

The three conditions of stability for floating or submerged vessels are stable, neutral, and unstable. Stable means that the vessel returns to its original equilibrium state after an external force or moment has been applied and removed. The external moment or force causes the vessel to change its angular attitude and when the force or moment is removed, the vessel returns to its original equilibrium position. When a force or moment is applied to a vessel causing it to change its angular orientation and this new orientation is maintained after the force or moment is removed, it is called neutral stability. An unstable condition exists when a force or moment is applied to a vessel and the change in orientation continues even after the force or moment is removed (i. e. the vessel goes from upright to upside down).

Floating and submerged vessels may be inclined either athwartship (heel or list) or longitudinally (trim). Athwartship or transverse stability and longitudinal stability are usually discussed separately, and the transverse stability is of the most concern because it usually determines whether a vessel will or will not capsize, or turn upside down. In the English system, the displacement, weight, and buoyant forces are expressed in long tons (or lb), but in the SI system, the displacement (Δ) is expressed as mass in metric tons (or kg). The righting and heeling moments are expressed as metric ton-meters (t-m) or long ton-feet (ft-t). As previously discussed, the weight or displacement is determined from the curves of form and draft marks.

The center of gravity (G) location is determined by experiment or calculation. Calculating the location of G requires knowledge of the weight of all items on a vessel and their location with respect to a selected coordinate system. The total weight of the vessel is assumed to act through the center of gravity. An inclining experiment is conducted to experimentally determine the center of gravity location (G) that is established relative to three reference planes. The vertical location of the center of gravity (VCG) is referenced to a horizontal plane passing through the baseline (keel). A vertical transverse plane passing through the midship location or through the forward perpendicular is the reference for the longitudinal center of gravity (LCG). The transverse center of gravity (TCG) is referenced to a vertical plane passing through the vessel centerline. Usually, G is located very near to the centerline plane.

The buoyant force acting on the vessel is equal to the weight of the displaced fluid, and it acts vertically upward through the center of buoyancy (B). The submerged volume is normally converted to weight or mass of the displaced fluid and is termed displacement (W or Δ). The orientation or attitude of a floating vessel is determined by the interaction of the forces of buoyancy and weight. When no other forces are acting, the vessel immerses until the forces of buoyancy and weight are equal, and it rotates until the centers of gravity (G) and buoyancy (B) are in the same vertical line. A submerged vessel rotates until the center of gravity is directly below the center of buoyancy. An important difference between floating and submerged objects is that for a floating body the center of buoyancy location changes when the object is rotated, and it has a fixed location for the submerged object.

There are two types of hydrostatic moments acting on floating and submerged vessels that determine the vessel's position. The righting moment occurs when the vessel inclines to a position where the forces of weight and buoyancy cause the vessel to return to an upright and vertical position. Heeling moments occur at any angle of inclination, and the forces of weight and buoyancy cause the vessel to move away from the upright position.

The effect of the location of the center of gravity in floating and submerged vessels is illustrated in Figure 6-4. Lowering the center of gravity increases the stability by increasing the separation of the forces of weight and buoyancy. Lowering the center of gravity can also change a heeling moment to a righting moment. The longitudinal separation of B and G effects the draft and trim of the vessel. For the submerged body, the center of buoyancy does not move, and positive stability requires that G remain below B. An unstable condition occurs when G moves above B.

Upsetting forces may act on a vessel causing it to heel, and the forces of weight and buoyancy must create righting moments to offset the heeling moments in order to prevent capsizing or excessive heel. Examples of upsetting forces are wind, lifting of a heavy weight over the side, high-speed turns, grounding, mooring lines, towlines, shifting of onboard weights, and entrapped water on deck.

In design, it is important to catalog all weights and their location so that the overall weight and location of the center of gravity can be estimated. The height of the center of gravity and weight estimates are critical in determining the adequacy of the vessel stability. The weight and longitudinal center of gravity determine the drafts at which the vessel floats. The distance G is offset from the vessel centerplane determines the list of the vessel. An example of the calculation of the weight and height of G above the keel is shown in Table 6-7.

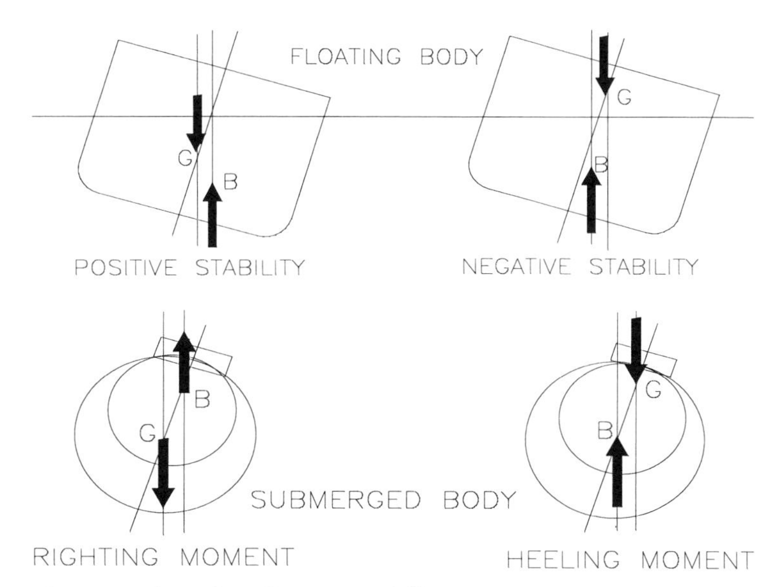

Figure 6-4. Location of center of gravity effects on stability.

Metacentric Height

When a symmetric vessel is heeled to a very small angle ϕ, the center of buoyancy moves off the vessel centerline as a consequence of the inclination (Figure 6-5). The resultant forces due to weight (F_w) and buoyancy (F_B) are separated by a distance GZ called the righting arm. A vertical line is drawn through the new center of buoyancy (B) and intersects the original vertical line drawn through the initial center of buoyancy. The intersection of these two lines is called the transverse metacenter (M). The location of the metacenter does vary with displacement and trim, but it is constant for a given draft. The metacenter is essentially stationary for small vessel angles of inclination (less than 7 to 10 degrees).

If the locations of G and M are known, then the calculation of the righting arm is accomplished by

$$GZ = GM \sin\phi \qquad \textbf{6-12}$$

The distance GM is an important index of the transverse stability at small angles of heel and is termed the transverse metacentric height. GM is positive when M is above G and negative when M is below G. If GM is positive, then the vessel is stable, and it is unstable if GM is negative. The GM is used to measure the initial stability of the vessel. Other useful relationships are

$$BM = \frac{I_T}{\nabla} \qquad \textbf{6-13}$$

$$GM = KM - KG \qquad \textbf{6-14}$$

$$GM = KB + BM - KG \qquad \textbf{6-15}$$

Table 6-7. Example calculation of vessel weight and vertical center of gravity.

Item	Weight (lb)	Vertical Location (ft)	Moment (ft-lb)
1	400	2	800
2	900	1.5	1350
3	600	3.0	1800
4	750	0.5	375
5	500	2.5	1250
6	1200	0.75	900
7	300	4.0	1200
8	150	5.0	750
9	400	1.75	750

$$VCG = \frac{\sum \text{Moments}}{\sum \text{Weights}} = \frac{\sum \text{Weight x Vertical Location}}{\sum \text{Weights}}$$

$$VCG = \frac{9125}{5200} = 1.75 \text{ ft}$$

$KG = 1.75$ ft

Similar procedures are used to evaluate transverse and longitudinal CGs.

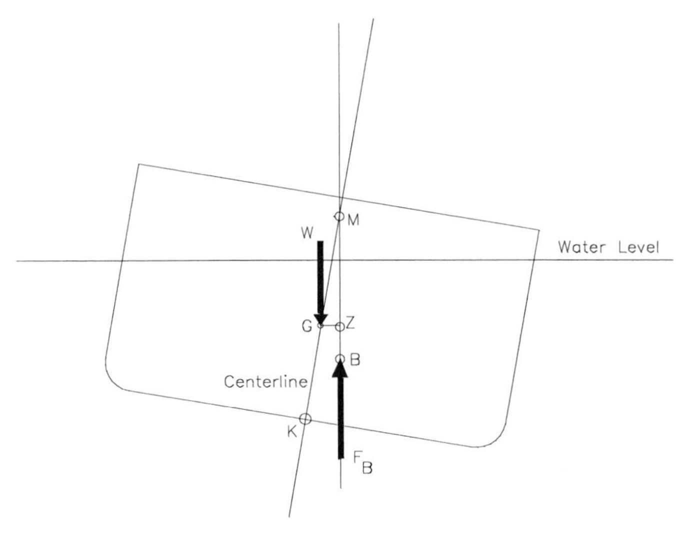

Figure 6-5. Transverse metacenter and righting arm.

For submerged submarine, the center of buoyancy is stationary with respect to any angle of inclination. When the submarine is submerged, B and M coincide. Therefore, the initial stability of a submerged body is GB instead of GM. For floating bodies, the metacentric height is also used to determine the moment to heel one degree, moment to trim one centimeter, and moment to trim one inch. Expressions for these parameters are

$$\text{Moment to heel one degree } = \Delta \text{ GM} \sin(1) \tag{6-16}$$

$$\text{Moment to trim one centimeter} = \frac{\Delta \ GM_L}{100 \ L} \ (t - m) \qquad \textbf{6-17}$$

$$\text{Moment to trim one inch} = \frac{W \ GM_L}{12 \ L} \ (ton - ft) \qquad \textbf{6-18}$$

where L is ship length and GM_L is longitudinal metacentric height. The period of roll in still water is estimated by

$$T = \frac{C \text{ x Beam}}{\sqrt{GM}} \qquad \textbf{6-19}$$

where C is a constant obtained from measured data. The value of C doesn't vary much and can be reasonably assumed as 0.8 for surface ships and 0.67 for submarines (Lewis 1988). This expression can be used to estimate GM when the period of roll is measured.

Cross Curves of Stability

The moment of weight and buoyancy that is necessary to return a vessel to its upright position is determined by knowing the distance between the center of gravity and the vertical line through the center of buoyancy for a particular inclination angle and displacement. This distance is called the righting arm (GZ) and is illustrated in Figure 6-6.

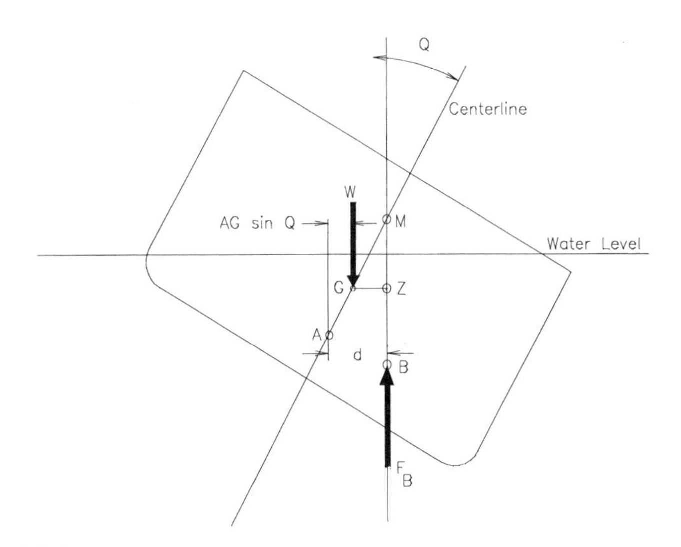

Figure 6-6. Transverse righting arms.

The cross curves of stability offer a way to illustrate the righting arm for any probable value of displacement and for several angles of heel. It is convenient to assume a location for the center of gravity such as point A on Figure 6-7 or the baseline. The distance (d) is calculated for several waterlines (displacements) and various angles of heel. If G is above A, the actual values of GZ are obtained from

$$GZ = d - AG \sin Q \qquad \textbf{6-20}$$

If G is below A, then

$$GZ = d + AG \sin Q \qquad \textbf{6-21}$$

Often A is taken at the baseline (K) and the value KG sin ϕ is always subtracted from d. A sample set of cross curves of stability are illustrated in Figure 6-7 that shows the distance (d) for several displacements and angles of heel.

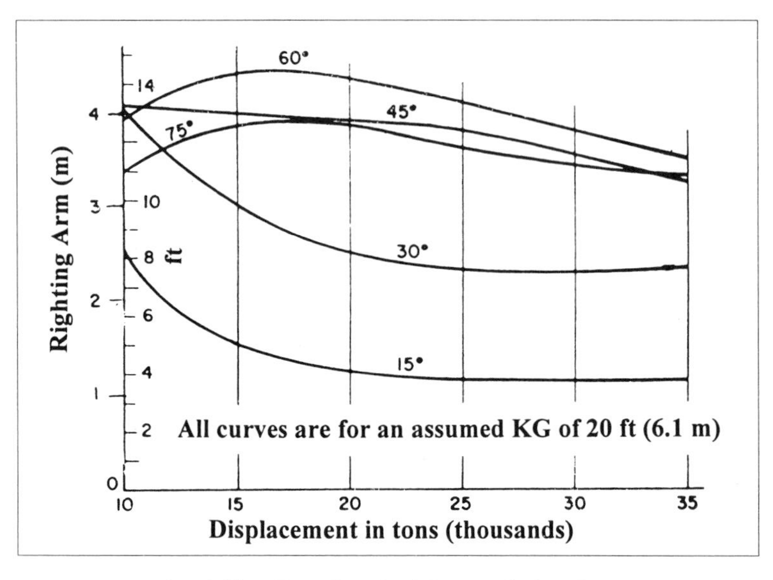

Figure 6-7. Sample cross curves of stability. Reprinted with permission from Lewis, 1988, *Principles of Naval Architecture*.. (Full citing in references).

Methods of obtaining the cross curves of stability include the manual method that uses transverse sections of the vessel. The most common method is to use digital computers along with inputs of the vessel offsets such as the Navy's program SHCP mentioned earlier. Another method is to use mechanical integrators. All these methods are described by Lewis (1988).

Curves of Static Stability

The static stability curve is a graph of the righting arm versus the angle of inclination (heel) for a given vessel load condition. An example curve is illustrated in Figure 6-8 that was obtained from the cross curves of stability. The calculations of the righting arms using values from Figure 6-7 for a displacement of 30,000 metric tons and KG of 8.8 m are tabulated in Table 6-8.

Table 6-8. Example calculation of righting arm from cross curves of stability.

Angle of Inclination, ϕ, degrees	0	15	30	45	60	75
Righting Arms, m, from Figure 6-8	0	1.17	2.44	3.57	3.79	3.41
Adjustment of actual KG, (6.1-8.8) sin ϕ	0	-0.70	-1.35	-1.91	-2.34	-2.61
Free surface effect	0	-0.05	-0.06	-0.05	-0.05	-0.04
Righting Arms, m	0	0.13	0.77	1.40	1.25	0.68

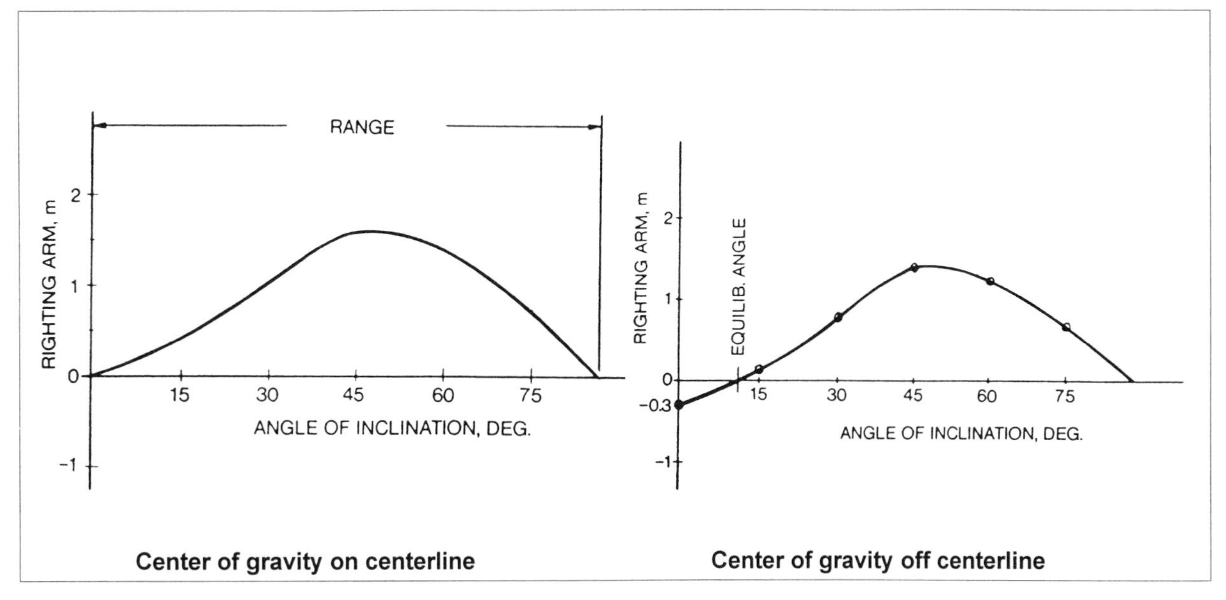

Figure 6-8. Typical static stability curve with center of gravity on and off the centerline.

The stability criteria for vessels is governed by several international and national organizations such as the International Maritime Organization (IMO), US Coast Guard (USCG), American Bureau of Shipping (ABS), and US Navy (USN). The metacentric height is an approximate indication of the stability, but the more accurate indication is the comparison of the righting arm curve (static stability curve) and the heeling arm curve as illustrated in Figure 6-9. The residual dynamic stability is represented by the cross-hatched area in between the righting and heeling arm curves. Satisfactory stability is attained if (1) heeling arm at the intersection of the righting arm and heeling arm curves (point C) is not greater than 0.6 of the maximum righting arm and (2) area A_1 is not less than 1.4 A_2 where area A_2 extends 25 degrees to windward from point C (Lewis 1988).

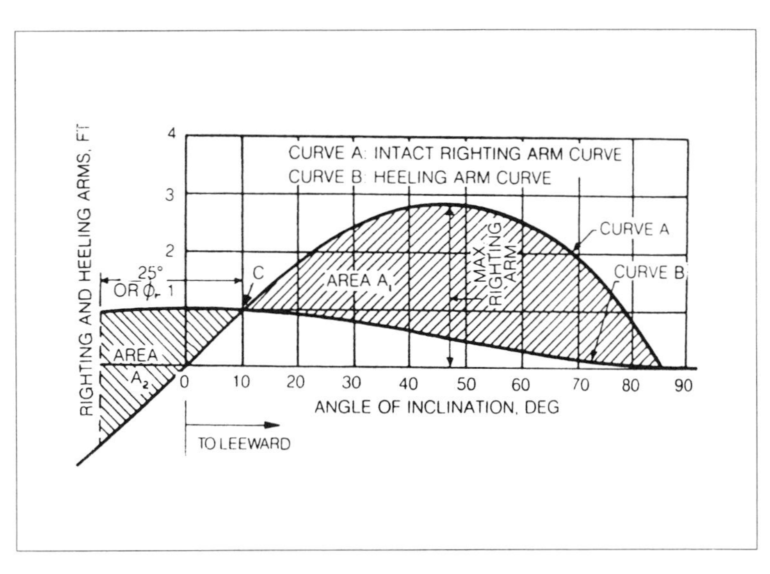

Figure 6-9. Example of US Navy criterion for stability in wind and waves. Reprinted with permission from Lewis, 1988, *Principles of Naval Architecture*. (Full citing in references).

RESISTANCE AND PROPULSION

Resistance

Vessels that move through ocean waters, or any other fluid, must have sufficient thrust from a propulsion device to overcome the resistance to movement of the vessel through the water, or fluid. The resistance is different for a completely submerged vessel and a vessel that moves through the water at the surface. Resistance forces are usually distributed over the entire wetted surface of the vehicle. Propulsor forces are applied at the stern of the vehicle where the propeller is located.

In real fluids, the relative motion between the marine vehicle and fluid results in the development of a boundary layer. It is assumed that no relative motion exists at the surface of the vehicle or in other words the velocity of the fluid and the vehicle are the same, otherwise known as the no-slip condition. Assuming the vehicle is moving through a fluid at rest, then the velocity of the fluid near the vehicle varies from the velocity of the vehicle to zero far away from the vehicle. If the vehicle is stationary and the fluid is moving by it, then the velocity varies from zero at the vehicle surface to the fluid velocity at some distance away from the vehicle. The layer of fluid between the vehicle surface and the location where it approaches the freestream velocity is called a boundary layer. Since the velocity in the boundary layer approaches the freestream velocity asymptotically, the thickness is not easily defined but is usually assumed to be the distance at which the velocity in the boundary reaches 95 to 99 % of the freestream velocity. In many cases the fluid separates from the rear of the vehicle and forms a wake, and as a result, there is a pressure difference between the bow and stern of the vehicle. For surface vehicles, there is an additional phenomena due the generation of waves developed at the air-water interface.

If steady motion is assumed and the effects of lift and side forces are neglected, then the only horizontal force is that due to drag or resistance. The weight and buoyancy forces are in the vertical direction. For the submerged vehicle, the total resistance (drag) is the summation of the forces resulting from the summation of the pressure and shear stress distribution over the wetted hull surface. Therefore, the total resistance (R_T) is the sum of the pressure resistance (R_P) and skin-friction resistance (R_F), and it is expressed as

$$R_T = R_P + R_F \qquad \textbf{6-22}$$

A surface vehicle moving at the air-water interface results in the added complication of including the resistance due to wave making. In this case the total resistance must include the wave making contribution to the resistance (R_W) and is written as

$$R_T = R_P + R_F + R_W \qquad \textbf{6-23}$$

For the vehicle, a dimensionless resistance coefficient is defined as

$$C_T = \frac{R_T}{\frac{1}{2}\rho U^2 A} \qquad \textbf{6-24}$$

where $(1/2)\rho\, U^2 A$ is the dynamic pressure and A is a characteristic area of the body. The total resistance equation is now expressed in dimensionless form as

$$C_T = C_P + C_F + C_W \qquad \textbf{6-25}$$

and C_P is the pressure coefficient and C_F is the skin-friction coefficient that are defined as

$$C_P = \frac{R_P}{\frac{1}{2}\rho U^2 A} \quad ; \quad C_F = \frac{R_F}{\frac{1}{2}\rho U^2 A} \quad ; \quad C_W = \frac{R_W}{\frac{1}{2}\rho U^2 A} \qquad \textbf{6-26}$$

For bluff bodies such as a sphere, cylinder, or disc normal to flow, the boundary layer separation produces a large wake behind the body that results in a large pressure drag, and consequently, C_P is much greater than C_F. When a body is streamlined, then the boundary layer separation occurs near the rear or not at all and the skin friction resistance dominates.

Determining the total resistance of a vessel (R_T) is very difficult because there is no general relationship between the coefficient of resistance (C_T) and Reynolds (Re) and Froude (Fr) numbers. The Reynolds number is the ratio of the inertia to viscous forces, and the Froude number is the ratio of the inertial to gravitational forces. Expressions for the Reynolds and Froude numbers are

$$Re = UL/\nu, \qquad Fr = U/\sqrt{g\,L} \qquad \textbf{6-27}$$

Equivalency of the prototype and model Re and Fr insures dynamic similarity between the model and the prototype. Thus, laboratory modeling has been used to determine the resistance of the model vessel, and scaling laws are used to evaluate the prototype resistance. The C_T is split into its components and each component is evaluated using the appropriate scaling law. The Froude procedure is:

1. Determine R_T for the model when $(Fr)_M = (Fr)_P$ where subscripts M and P refer to model and prototype.
2. Evaluate the skin friction resistance R_F for the model using data from submerged planks of equal surface area, length and finish and model Reynolds number.
3. Determine the residual resistance $(R_R)_M$ from $(R_R)_M = (R_T)_M - (R_F)_M$ and determine $(C_R)_M$. C_R is assumed to be a function of Froude number only so that $(C_R)_M = (C_R)_P$.
4. Compute frictional resistance R_F for the plank corresponding to Reynolds number of the prototype and deduce $(C_F)_P$.
5. Evaluate prototype resistance from $(C_T)_P = (C_F)_P + (C_R)_M$.

Once C_T is determined then the resistance can be evaluated from

$$R_T = \frac{1}{2} C_T \rho A U^2 \qquad \textbf{6-28}$$

The assumptions are good for streamline hull forms and Froude numbers greater than 0.35. Also, the use of very fast digital computers such as work stations and Cray computers allow numerical computations of resistance that give very good results. More detailed descriptions of the resistance determination can be found in Lewis (1988) and Clayton and Bishop (1991).

Propulsion

Marine vessels are propelled through the fluid medium by some type of propulsion device that generates a thrust to overcome the vessel resistance. In the early days, thrust was developed by the wind (sailing vessels) and by oars (human powered). Later steam was used as the energy source to drive propulsors. Currently, the more common energy source used to drive propulsors is diesel fuel, gasoline, or natural gas. Some of the different types of propulsors used on marine vessels are tabulated in Table 6-9.

Table 6-9. Summary of typical marine propulsors (Clayton and Bishop 1991).

Propulsor	Advantage	Disadvantage	Comments
Oars, paddles	human powered	small, unsteady thrust	small boat use only
Paddle wheel	efficient	requires low speed engine	used with steam power
Vertical axis propellers	efficient, high degree of speed control and thrust	weight and cost	used on ferries
Sails	free power	ineffective without wind	used for sport and recreation
Hydraulic Jets	no parts external to vessel	large duct losses	efficiency increases at high speeds
Pump jet	reduced duct length	costly and difficult to install	used in shallow draft vessels
Kort nozzle	high thrust at zero speed	cavitation occurs earlier than pump jet	used on tugs & large tankers
Screw propeller	reliable, simple and efficient	cavitation limits performance	most common propulsor

For steady flow of a fluid through a screw propeller, the propeller is replaced by a thin actuator disc, and the fluid passing through the disc is assumed to receive a sudden increase in pressure while the fluid axial velocity remains continuous. The magnitude of the thrust developed by the disc is determined by evaluating the change in the axial momentum of the fluid (Froude 1889). The flow relative to an actuator disc in open water is illustrated in Figure 6-10. The disc advances forward at a steady velocity (V_a), called the velocity of advance, that is parallel to the axis of rotation. The fluid is assumed to be stationary, constant density, inviscid, and infinite in extent. The propeller disc is assumed to uniformly accelerate all the fluid passing through it, and therefore the thrust generated is uniform over the disc. The axial velocity relative to the disc increases from V_a far upstream of disc to $V_a(1+a)$ at the disc to $V_a(1+b)$ far downstream of the disc. The term "a" is called the axial inflow factor.

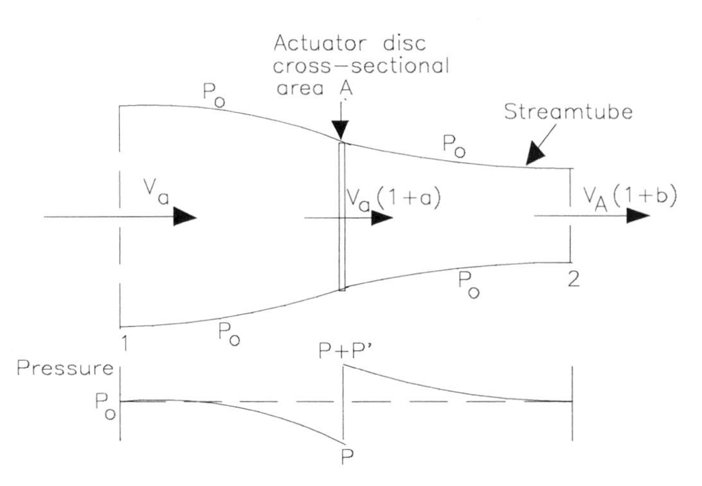

Figure 6-10. Flow relative to an actuator disk in open water.

As shown in Figure 6-10, the uniform pressure on the upstream face of the disc is p and on the downstream face it is p + p'. Using the Bernoulli theorem between the upstream location and upstream side of the actuator disc yields

$$p_o + \frac{1}{2}\rho V_a^2 = p + \frac{1}{2}\rho\left[V_a(1+a)\right]^2 \qquad \textbf{6-29}$$

and on the downstream side is

$$p + p' + \frac{1}{2}\rho\left[V_a(1+a)\right]^2 = p_o + \frac{1}{2}\rho\left[V_a(1+b)\right]^2 \qquad \textbf{6-30}$$

Combining Equations 6-58 and 6-59 and solving for p' yields

$$p' = \frac{1}{2}\rho\, b(2+b)\, V_a^2 \qquad \textbf{6-31}$$

The thrust of the fluid on the disc of area (A) is

$$T = Ap' = \frac{1}{2}\rho A b(2+b)\, V_a^2 \qquad \textbf{6-32}$$

The force of the disc on the fluid is written as

$$T = (\text{mass flow rate})\left[V_a(1+b) - V_a\right] \qquad \textbf{6-33}$$

Using Equations 6-61 and 6-62, it is shown that b equals 2a, and thus the thrust is expressed as

$$T = 2\rho A a(1+a) V_a^2 \qquad \textbf{6-34}$$

The velocity at the disc is the arithmetic mean of the velocities well upstream and downstream of the disc. Using energy concepts, the ideal efficiency η_i for the actuator disc model is

$$\eta_i = \frac{1}{1+a} \qquad \textbf{6-35}$$

The equation for thrust can be used to define a thrust coefficient C_T as

$$C_T = \frac{T}{\frac{1}{2}\rho A V_a^2} = 4a(1+a) \qquad \textbf{6-36}$$

and the inflow factor is determined from

$$a = \frac{1}{2}\left[(1+C_T)^{1/2} - 1\right] \qquad \textbf{6-37}$$

The ideal efficiency is now expressed as

$$\eta_i = \frac{2}{1 + (1+C_T)^{\frac{1}{2}}} \qquad \textbf{6-38}$$

The above equation is illustrated in Figure 6-11 and represents the highest possible propeller efficiency. An ideal efficiency of 0.7 is common.

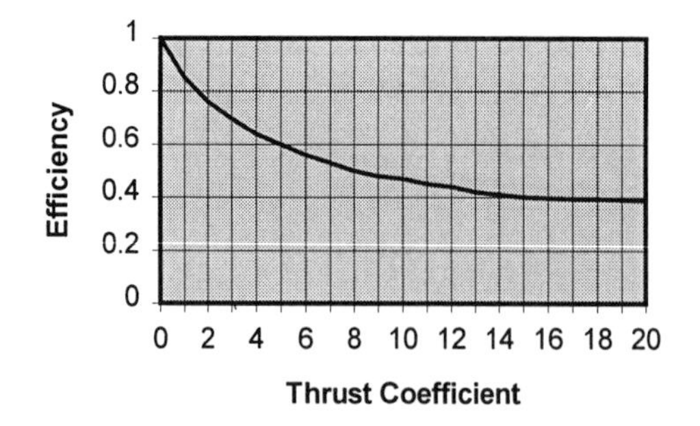

Figure 6-11. Relationship between ideal efficiency (η_i) and thrust coefficient (C_T). Reprinted with permission from Clayton and Bishop, 1991, *Mechanics of Marine Vehicles.* (Full citing in references).

When a propeller is placed at the stern of the vessel, there is an effect due to the hull-propeller interaction. The difference between the bare hull resistance and thrust as a result of adding the propeller is expressed as a thrust deduction fraction defined as

$$t = \frac{T - R_T}{T} \tag{6-39}$$

and

$$R_T = T(1 - t) \tag{6-40}$$

where the thrust deduction factor is always positive, and thus R_T is less than T. There is also an effect due to the wake behind the vessel. The overall effect is that the velocity of advance is slightly less than the forward velocity (U) of the vessel. A wake fraction is defined to account for the wake effect and is defined as

$$w = \frac{U - V_a}{U} \tag{6-41}$$

or

$$V_a = U(1 - w) \tag{6-42}$$

The total propeller efficiency (η_T) is usually expressed as the product of the open water efficiency (η_o), hull efficiency (η_H), and relative rotative efficiency (η_R). The open water efficiency is defined as

$$\eta_o = \frac{T_o V_a}{2 \pi n Q_o} \tag{6-43}$$

where T_0 is the thrust delivered at a torque Q_0 for a relative rotative speed of n (revolutions per unit time) when the velocity of advance is V_a. The hull efficiency is expressed as

$$\eta_H = \frac{1 - t}{1 - w} \tag{6-44}$$

The overall efficiency is then expressed as

$$\eta_T = \eta_o \, \eta_H \, \eta_R \qquad \textbf{6-45}$$

Some representative values of the wake fraction, thrust deduction fraction, relative rotative efficiency, hull efficiency, and overall efficiency are tabulated in Table 6-10.

Table 6-10. Representative values of efficiency, wake fraction, and thrust deduction factor (Clayton and Bishop 1991).

Vehicle Type	1-ω	1-t	η_H	η_R	η_T
Single propeller submarine	0.50-0.80	0.80-0.90	1.10-1.45	1.00-1.10	0.72-0.93
Single propeller merchant ship	0.55-0.85	0.75-0.85	1.10-1.30	0.99-1.09	0.72-0.86
Dual propeller merchant ship	0.83-0.95	0.82-0.89	0.96-0.97	-	0.62-0.70
Dual propeller destroyer	0.93-1.01	0.9-0.99	-	-	0.61-0.70
Four propeller ship	0.87-0.99	0.82-0.95	-	-	0.60-0.67
Ducted propeller cargo ship	0.66	0.8	1.21	0.98	0.80

A propeller converts the power supplied by the propulsion motor into thrust as efficiently as possible, but this process always produces losses. The power delivered by the propulsion device (electric or hydraulic motor) to the propeller shaft is the shaft horsepower (SHP). The effective horsepower (EHP) is the product of the vessel speed and the hull resistance expressed in the units of horsepower as

$$EHP = \frac{R_T V}{550} \qquad \textbf{6-46}$$

where the constant 550 is the conversion of ft-lb/s to horsepower. The thrust horsepower (THP) is the product of vessel speed and the thrust delivered by the propeller. The difference between EHP and THP is a result of the thrust deduction (1-t) and the wake fraction (1-w), and it is expressed as

$$EHP = \left(\frac{1-t}{1-w}\right) THP = \eta_H (THP) \qquad \textbf{6-47}$$

The relationship between the SHP and EHP is illustrated in Figure 6-12 and expressed in equation form by

$$\eta_T = \frac{EHP}{SHP} \qquad \textbf{6-48}$$

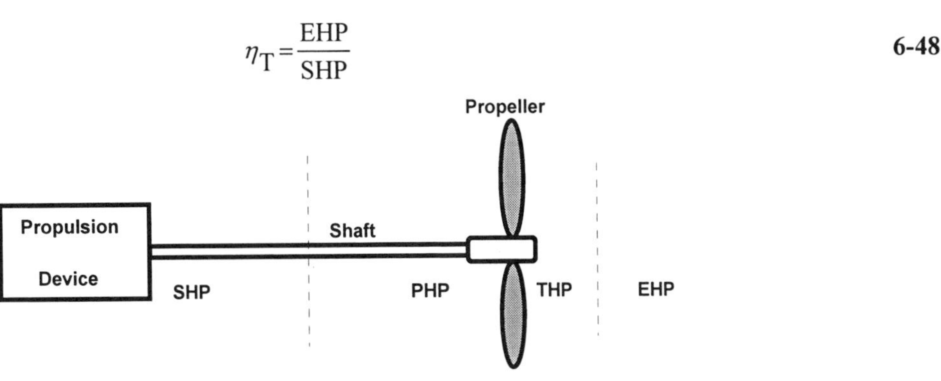

Figure 6-12. Propeller power relationships.

The total efficiency η_T does not include power reductions due to reduction gears or bearing losses which must be included if these devices are being used. In many cases these are not present in submersibles (Allmendinger 1990). Typical values of the thrust deduction factor (1-t) and the wake fraction (1-w) that are used in evaluating the hull efficiency are illustrated in Figure 6-13. The determination of the hull resistance requires model testing or calculation using Equation 6-28 if the total resistance coefficient (C_T) is known. Values for streamlined hull shapes typical of a submarine hull are shown in Figure 6-14.

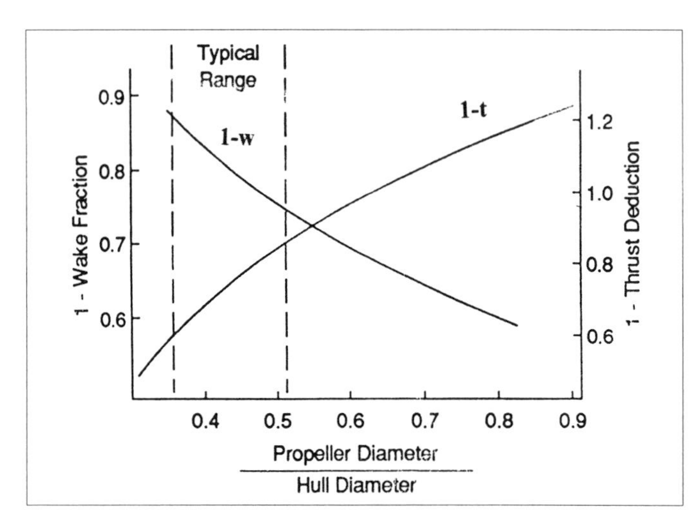

Figure 6-13. Typical wake fraction and thrust deduction for range of propeller diameter to hull diameter ratio. Reprinted with permission from Allmendinger, 1990, *Submersible Vehicle Systems Design*, (Full citing in references).

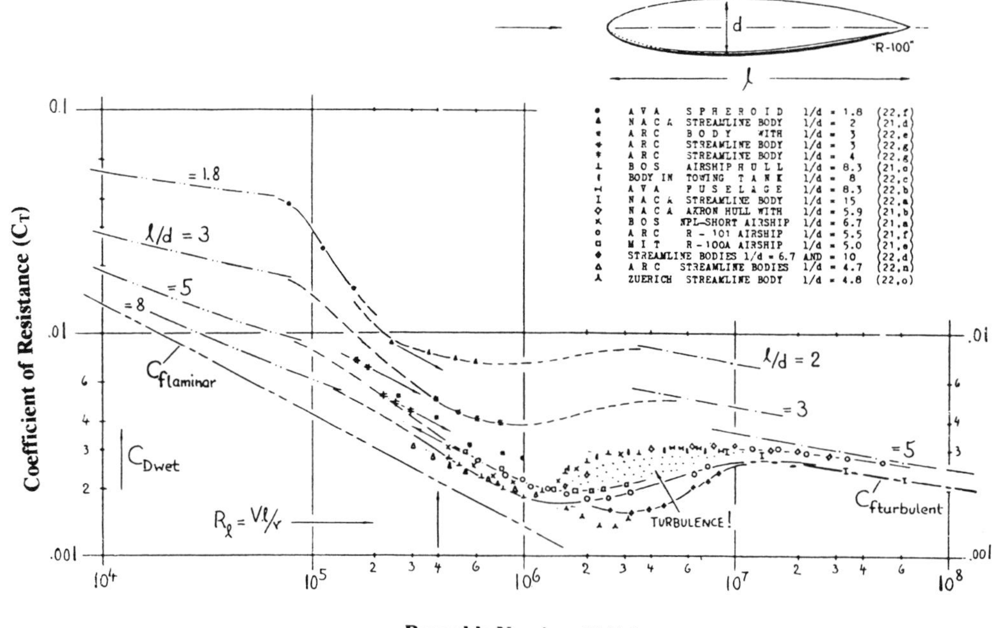

Figure 6-14. Resistance coefficients for streamlined hull shapes. Reprinted with permission from Allmendinger, 1990, *Submersible Vehicle Systems Design*, (Full citing in references).

Propeller model tests are usually conducted to assess the performance of marine propellers. The first test is used to determine the open water characteristics of the propeller. The model propeller is placed on a long shaft in either a water tunnel or beneath a towing carriage in a towing tank, and measurements of torque and thrust are made for various velocities of advance and rotative speeds. For the second test, the model propeller is installed in its proper position on a model hull and similar measurements are conducted as in the first test. Dimensional analysis is applied using the parameters of torque, thrust, velocity of advance, diameter, rotative speed, fluid density and viscosity, pressure, and vapor pressure of fluid. This analysis shows that

$$K_T \text{ and } K_Q = \text{function}(J, Re_D, \sigma_N) \tag{6-49}$$

where

$$\text{thrust coefficient} = K_T = \frac{T}{\rho n^2 D^4} \tag{6-50}$$

$$\text{torque coefficient} = K_Q = \frac{Q}{\rho n^2 D^5} \tag{6-51}$$

$$\text{advance coefficient} = J = \frac{V_A}{nD} \tag{6-52}$$

$$\text{propeller Reynolds number} = Re_D = \frac{\rho V_a D}{\mu} \tag{6-53}$$

$$\text{nominal cavitation index} = \sigma_N = \frac{\overline{p} - p_v}{\frac{1}{2}\rho V_a^2} \tag{6-54}$$

and $\overline{p}$ is the upstream static pressure, D is propeller diameter, n is the propeller rotative speed, p_v is fluid vapor pressure, and μ is fluid dynamic viscosity. Typical open water performance characteristics curves for a non-cavitating propeller are shown in Figure 6-15.

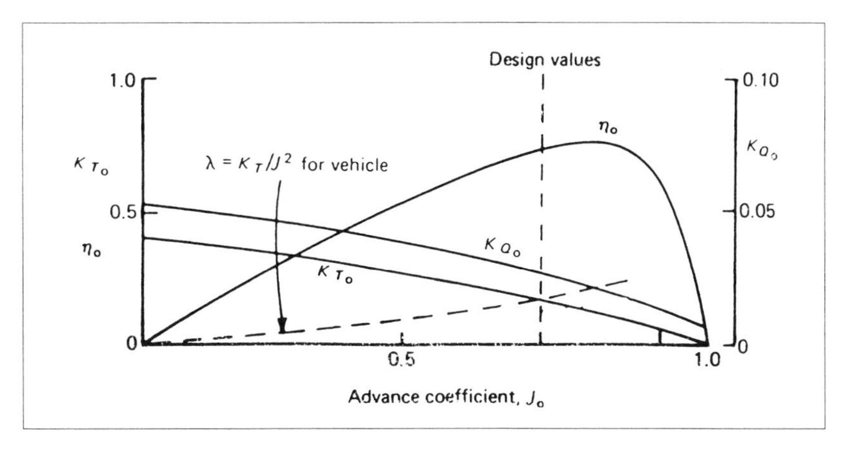

Figure 6-15. Representative propeller performance curves. Reprinted with permission from Clayton and Bishop, 1991, *Mechanics of Marine Vehicles*. (Full citing in references).

BUOY SYSTEMS

There are numerous types of buoy systems placed in inland, coastal, and ocean waters, and there are also a variety of needs for these different types of buoys. The objective of this section is to discuss the various types and uses of buoy systems and the static analysis of these systems. Buoys, like other vessels, are placed in a very dynamic ocean environment and consequently have dynamic responses induced by ocean waves and currents. However, the dynamic analysis of buoy systems is beyond the intended scope of this text, and the reader is referred to a more advanced publication such as Berteaux (1991).

Buoy Types and Uses

Buoys are floating or submerged objects which are typically cylindrical, spherical, disc, or toroidal in shape and moored to the seafloor with some type of anchor and mooring line (i. e. wire rope, chain, synthetic line, or combination). A major use of buoys is for aids to navigation that mark ship channels, obstacles such as wrecks or other underwater hazards, and port entrances. In the US alone, it is estimated that some 24,000 buoys are deployed and maintained as aids to navigation by the US Coast Guard (Berteaux 1991). These buoys are often equipped with bells, whistles, and lights and sometimes depend on wave induced motion to produce the sounds. Heavy chain and clump anchors are frequently used to slack moor the buoy system in depths usually not exceeding 45.7 m (150 ft). The offshore industry employs precise navigation aids for exploration and seismic surveys. Buoys may be moored with elastic tethers in water depths up to 305 m (1000 ft), and the elastic tethers are kept in high tension that limits the buoy lateral movement.

Buoys are also used for markers and ship moorings. Marker buoys are used to locate and retrieve objects on the seafloor such as anchors, lobster traps, sunken ships, and offshore pipelines. Large cylindrical buoys are used to mark rig anchors for an offshore drilling platform mooring. Buoys also identify the location of ship moorings that are sometimes used in bays and harbors to moor ships and boats when dock space is limited or when water depth is too shallow near the shore. Very large single point mooring systems are used to moor oil tankers while transferring oil to the tanker for eventual transport to land.

Weather stations and oceanic platforms are another use for buoys. The US National Data Buoy Center has a network of some 50 moored weather stations that measure and report wind speed and direction, air temperature and pressure, wave conditions, and visibility. These systems benefit the maritime, fishing, offshore industries, and the general public. Oceanic buoy platforms are either moored or free drifting, and they are also used to support sensors to measure a variety of oceanic parameters. The scientific and engineering community use buoy systems to attain measurements of wind, atmospheric pressure, air and surface temperature, wave evolution and noise, ocean currents, water temperature, salinity and dissolved oxygen, distribution of trace metals, sedimentation rates, ambient acoustic noise, light transmission and turbidity, and other parameters. The data are transmitted using satellites and radio transmittersthrough the atmosphere and acoustical systems for transmission through the water column. Data are stored internally on tapes or in computer memory. Examples of these buoy systems are shown in Figure 6-16.

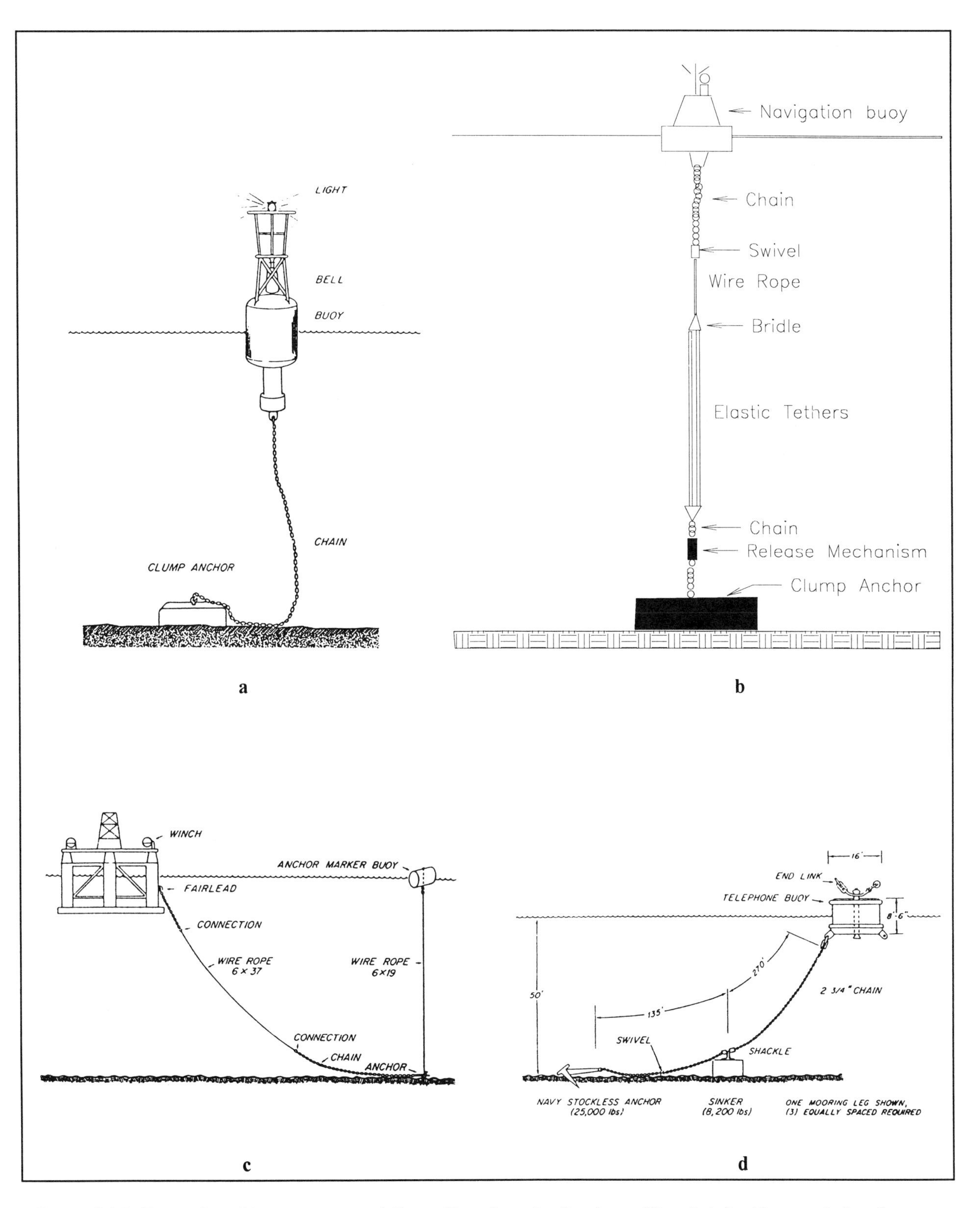

Figure 6-16. Examples of buoy systems. a)Coast Guard navigation buoy (Reprinted with permission from USCG, 1979, *Aids to Navigation Manual.* (Full citing in references), b) small watch circle marker buoy, c) offshore semisubmersible mooring with anchor marker buoy, d) US Navy free swinging buoy (Reprinted with permission from NAVFAC, 1990, *Design Manual, Harbor and Coastal Facilities.* (Full citing in references].

The basic hull forms for surface buoy systems are a disc (toroid, flat disc), sphere, cone, boat hull, and spar as shown in Figure 6-17. Disc buoys are easier to design and build and are more cost efficient, but they follow the waves in heave and slope. Sphere and cone buoys respond to the wave amplitude but not the slope. Boat hull buoys align themselves with the direction of the waves, current, and wind, but still heave and roll considerably. Spar buoys have the advantage of having good stability. The common techniques for mooring surface buoy systems in deep and shallow water are illustrated in Figure 6-18, and these include slack, taut, and three point mooring techniques.

Typical subsurface buoy and mooring systems are shown in Figure 6-19. These systems usually have single point subsurface mooring that is recovered by disconnecting the mooring line near the anchor using a remote release mechanism (acoustic release is common). This configuration is economical and easy to deploy. When the subsurface buoy is well below the free surface the effect of wave motion is greatly reduced, and the stability and endurance of the mooring are improved. The multileg subsurface systems provide advantages of reduced motion, ample space below the buoy for instruments, and reliability as well as disadvantages of increased cost and more difficult deployment.

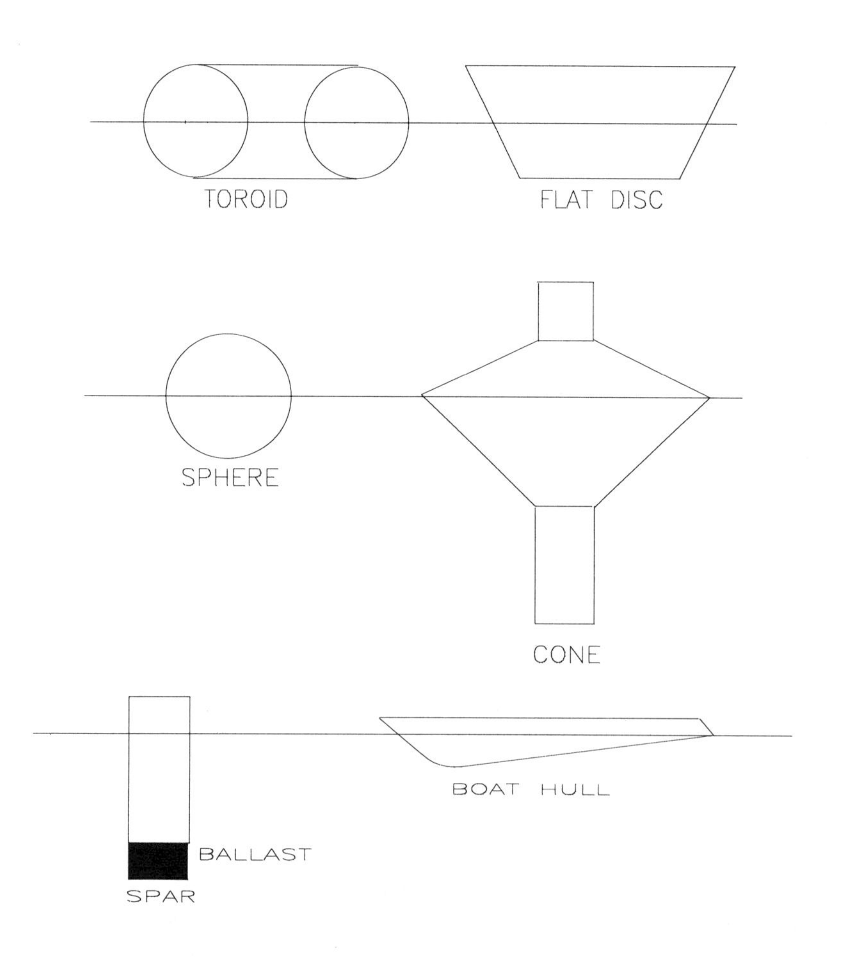

Figure 6-17. Examples of buoy hull forms.

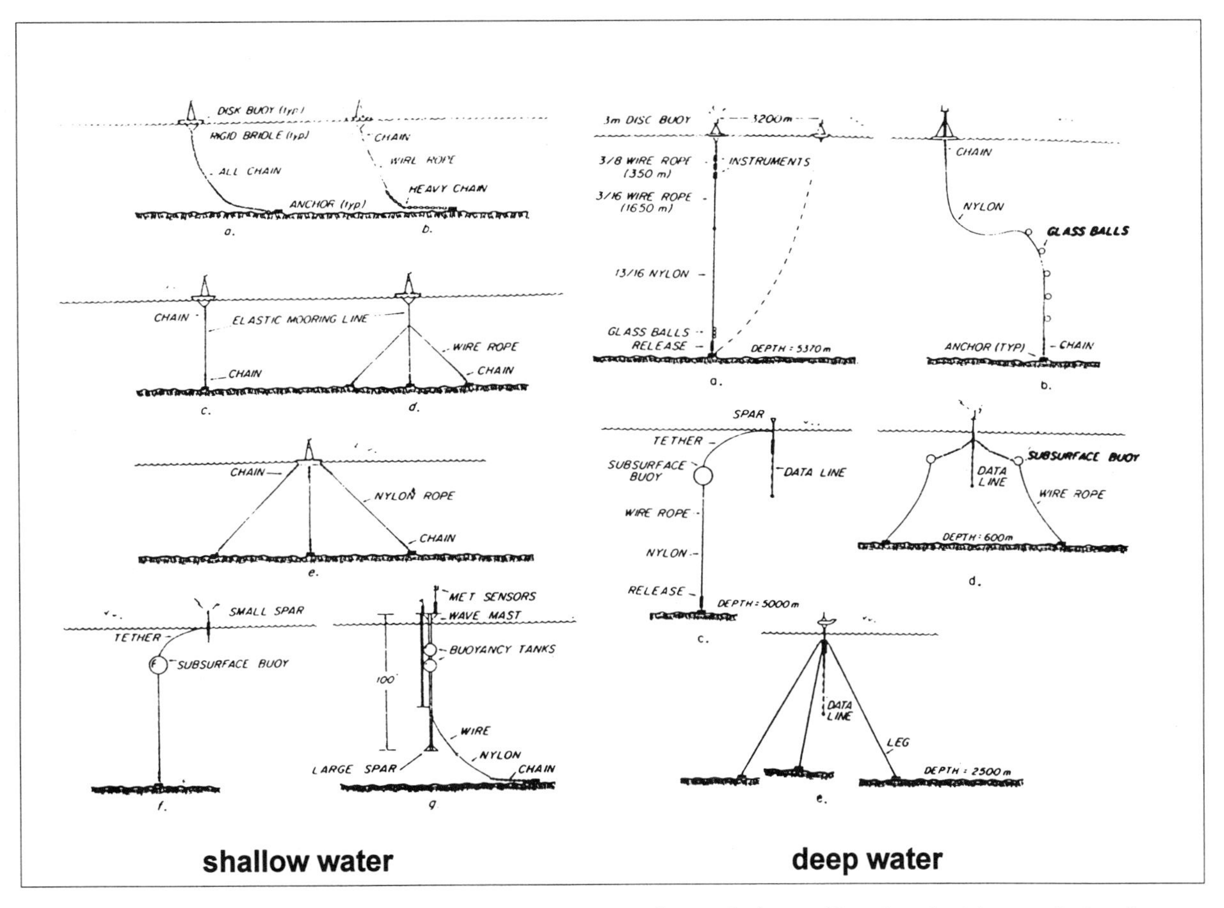

Figure 6-18. Example shallow water and deep water mooring techniques. Reprinted with permission from Berteaux, 1991, *Coastal and Oceanic Buoy Engineering*. (Full citing in references).

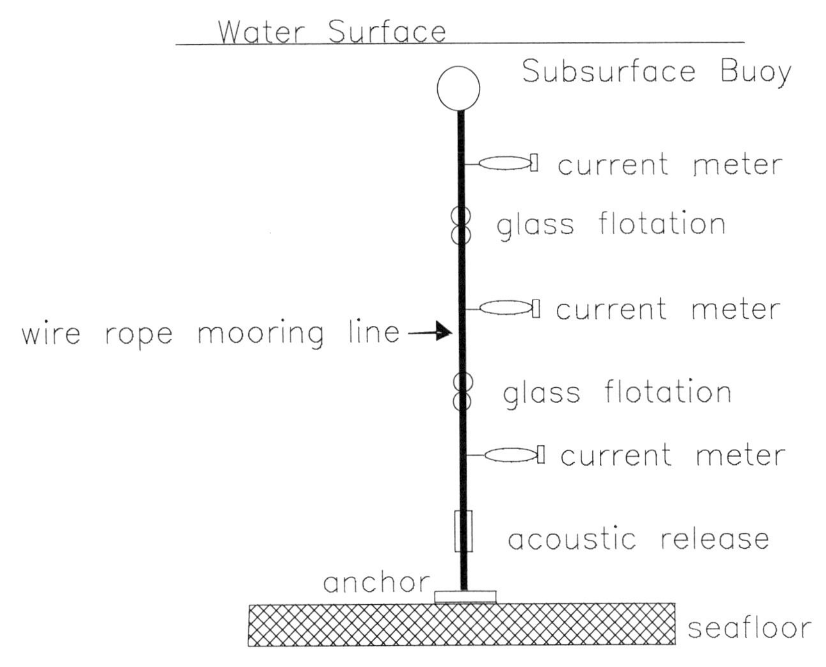

Figure 6-19. Typical subsurface mooring configurations.

Static Analysis of Buoy Systems

The buoy weight (W) acts vertically downward through the buoy's center of gravity (G). The buoyancy of the buoy is equal to the weight of the displaced fluid and this buoyant force (F_B) is directed vertically upward through the center of buoyancy (B). The center of buoyancy is also the centroid of the buoy's submerged volume. Since the buoy is either floating or submerged, it is also exposed to the flow of water (ocean currents) past the buoy surface that results in drag and lift forces acting on the buoy. The drag force (F_D) consists of friction and pressure forces as a result of tangential and normal stresses acting on the buoy surface. If the flow past the buoy is laminar, then the shear (tangential), or friction, stresses predominate. When the flow is turbulent and the buoy shape is blunt the normal stresses (pressure) dominate. The resultant force is obtained by integrating the shear and pressure stresses over the area of the buoy and is called the resistance force. This force has two components called drag in the direction of flow and lift in the direction normal to the flow. The common "Drag Law" is used to determine the drag force (F_D) and the similar "Lift Law" is used to compute the lift force (F_L). The equations for drag and lift are

$$F_D = \frac{1}{2}\rho C_D A_N U^2 \quad \text{"Drag Law"} \qquad \textbf{6-55}$$

$$F_L = \frac{1}{2}\rho C_L A_L U^2 \quad \text{"Lift Law"} \qquad \textbf{6-56}$$

where ρ is the fluid density, C_D and C_L are the drag and lift coefficients, A_N is the area projected on a plane normal to the fluid flow, A_L is the characteristic area used in determination of C_L, and U is the average fluid velocity. As discussed previously, drag coefficients are a function of Reynolds number (UD/ν) where D is the characteristic buoy dimension and ν is the fluid kinematic viscosity. The drag force is assumed to act through the centroid of the projected area (A_N). Drag coefficients for 2- and 3-dimensional body shapes are tabulated in Table 6-9. Waves acting on buoys also add a resistance force, but this force is usually small and is neglected.

Moored Subsurface Buoy

The free body diagram for a subsurface buoy moored in a current is illustrated in Figure 6-20. For static equilibrium to exist, the sum of the forces and moments must be zero.

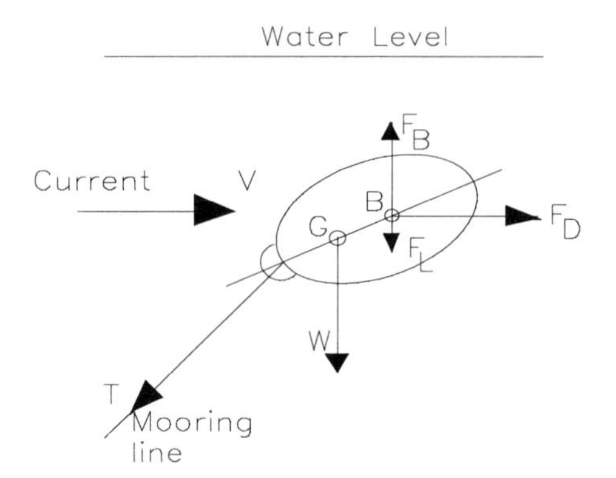

Figure 6-20. Free body diagram of a moored subsurface buoy in a current.

By summing the forces at the attachment point, the result is

$$T_H = F_D \tag{6-57}$$

and

$$T_V = F_B - W - F_D \tag{6-58}$$

where T_H and T_V are the horizontal and vertical tension in mooring cable. The forces F_D, F_B, and F_L are the drag, buoyant, and lift forces respectively, and W is the weight of the buoy. Equations 6-25 and 6-26 are used to estimate the tension in the mooring cable. The resultant cable tension is determined by

$$T = \sqrt{T_H^2 + T_V^2} \tag{6-59}$$

and the angle between the cable and the horizontal is found by

$$\theta = \tan^{-1}\frac{T_V}{T_H} \tag{6-60}$$

Static Cable Analysis

Buoy systems are moored with flexible cables that are subjected to ocean currents. As a result, cable tension increases and strong anchoring systems are needed. In this description, only two dimensional cases with coplanar currents are considered. The cable immersed weight (W_I) is defined as

$$W_I = F_B - W \tag{6-61}$$

If W_I is positive, the cable is positively buoyant (polypropylene rope), and if it is negative, the cable is negatively buoyant (wire rope). The hydrodynamic force due to the current is expressed as

$$Rds = \frac{1}{2}\rho C_{DN} D U^2 ds \tag{6-62}$$

where C_{DN} is the normal drag coefficient. A schematic of the hydrodynamic forces are illustrated in Figure 6-21.

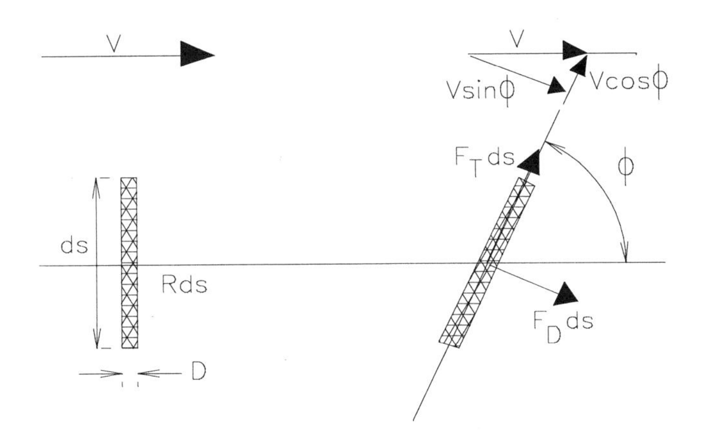

Figure 6-21. Schematic of forces on a cable element.

Table 6-11. Drag coefficients for two and three dimensional body shapes.

Shape: Description	Shape: Schematic	Reynolds Number (R)	Characteristic Length Ratio (L/D, L/H)	Drag Coefficient (C_D)
Rectangular plate normal to flow		$>10^3$	L/H = 1	1.16
			5	1.20
			10	1.50
			∞	1.90
Circular plate normal to flow		$>10^3$	-	1.12
Circular cylinder with axis parallel to flow		$>10^3$	L/D = 0	1.12
			1	0.91
			2	0.85
			4	0.87
			7	1.00
Circular cylinder with axis perpendicular to flow		$\sim10^5$	L/D = 1	0.63
			2	0.68
			5	0.74
			10	0.82
			20	0.90
			40	0.98
			∞	1.20
		$>5 \times 10^5$	L/D = 5	0.35
			∞	0.34
Sphere		$\sim10^5$	-	0.5
		$\sim3\times10^5$		0.2
Hemisphere concave to flow		$>10^3$	-	1.33
Hemisphere convex to flow		$>10^3$	-	0.34
Ellipsoid with major axis perpendicular to flow		$<5 \times 10^5$	L/D = 0.75	0.6
		$>5 \times 10^5$		0.21
Ellipsoid with major axis parallel to flow		$>2 \times 10^5$	L/D = 1.8	0.07
Airship hull		$>2 \times 10^5$	-	0.05
Solid cone (apex pointing in opposite direction of velocity vector)		-	-	0.34
Solid cone (apex pointing in same direction as velocity vector)		-	-	0.51

When the cable is at an angle ϕ with the horizontal direction of the current, then the resistance is expressed as two components, normal and tangential to the cable. The normal component $F_D ds$ and tangential component $F_T ds$ are expressed as

$$F_D\, ds = \frac{1}{2}\rho\, C_{DN}\, D U^2 \sin^2\phi ds = R \sin^2\phi ds \tag{6-63}$$

$$F_T\, ds = \frac{1}{2}\rho\, \gamma\, C_{DN}\, (\pi D) U^2 \cos^2\phi ds = \pi\gamma\, R \cos^2\phi ds \tag{6-64}$$

where γ is the ratio between tangential and normal drag coefficients ($C_{DT} = \gamma\ C_{DN}$), C_{DN} is the normal drag coefficient, C_{DT} is the tangential drag coefficient, and $(\pi D)ds$ is the surface area of the cable element (ds). Normal (ϕ=90°) and tangential (ϕ=0°) drag coefficients for several cables are shown in Figure 6-22. The vector summation of the forces in the normal and tangential directions yields the classical cable equilibrium equations

$$T d\phi = (F_D + W_I \cos\phi) ds \tag{6-65}$$

$$dT = (W_I \sin\phi - F_T) ds \tag{6-66}$$

These equations must be integrated numerically, and the result permits evaluation of tension and cable geometry with known boundary conditions.

If heavy cables such as chain or wire rope are assumed, then the resistance terms can be neglected. The above equations reduce to

$$T\, d\phi = W_I \cos\phi\ ds \tag{6-67}$$

$$dT = W_I \sin\phi\ ds \tag{6-68}$$

Integrating these equations from the origin where $\phi = 0$ to any point on the cable where the cable angle is ϕ yields

$$T\cos\phi = T_o \tag{6-69}$$

$$T\sin\phi = W_I\, s \tag{6-70}$$

where T_o is the tension at the origin and s is the cable length to the origin. The coordinates x and z of a point along the cable at an angle ϕ and the cable length (s) from the origin are expressed as

$$x = \frac{T_o}{W_I} \sinh^{-1}\left(\frac{W_I\, s}{T_o}\right) \tag{6-71}$$

$$z = \frac{T_o}{W_I} \cosh\left[\frac{W_I\, x}{T_o} - 1\right] \tag{6-72}$$

$$s = \frac{T_o}{W_I} \sinh\left(\frac{W_I\, x}{T_o}\right) \tag{6-73}$$

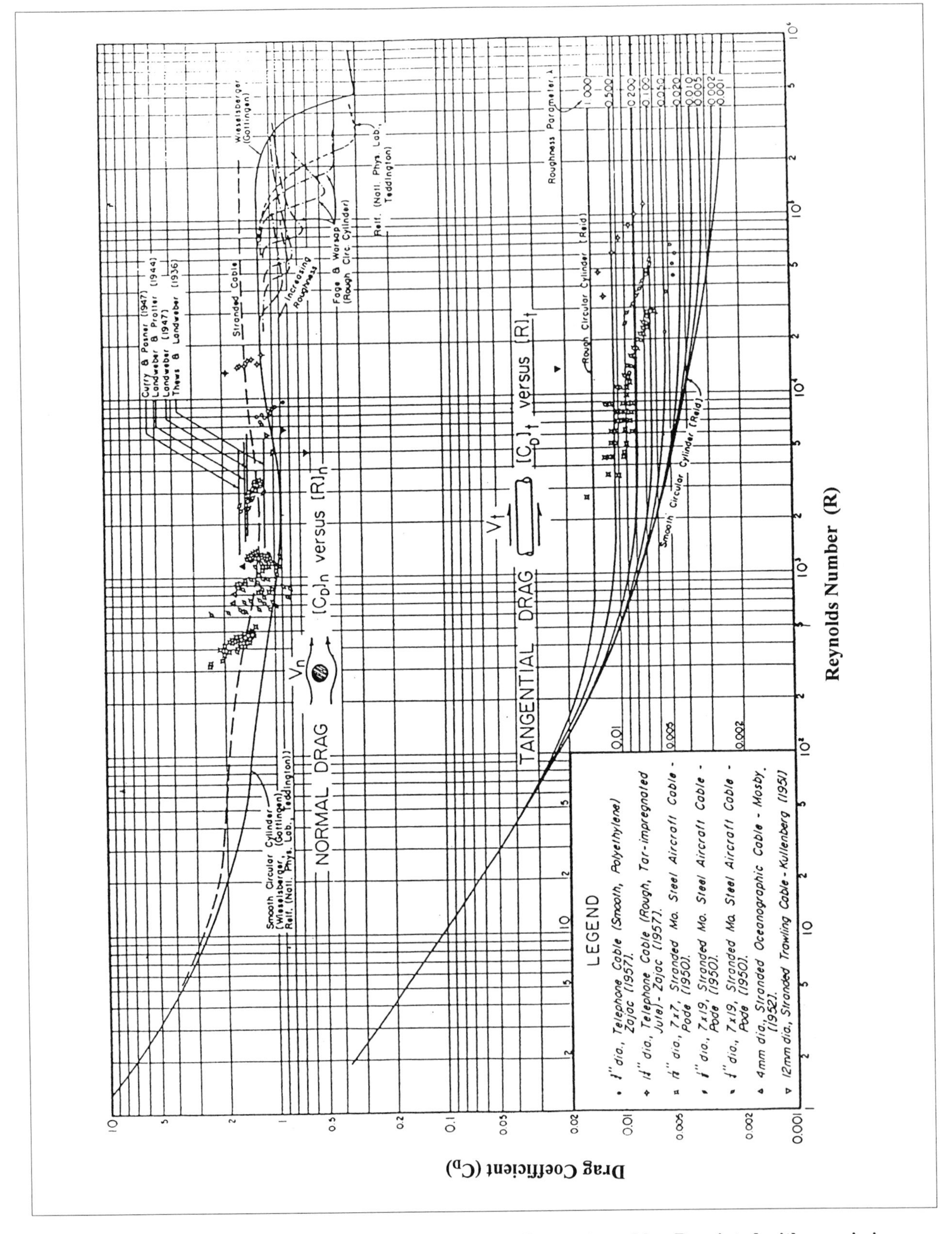

Figure 6-22. Normal and tangential drag coefficients for different size cables. Reprinted with permission from Berteaux, 1991, *Coastal and Oceanic Buoy Engineering*. (Full citing in references).

Example Problem 6-1

A surface buoy (Figure 6-23) is moored in 185 ft of water with a chain weighing 2.8 lb/ft (immersed weight), and the drag force on the buoy is 275 lb. Determine the length of chain necessary to maintain an angle of 25° or less at the anchor. Estimate the distance that the buoy moves downstream of the anchor, and determine the tension at the buoy and anchor. The solution to example problem 8-1 is tabulated in Table 6-12.

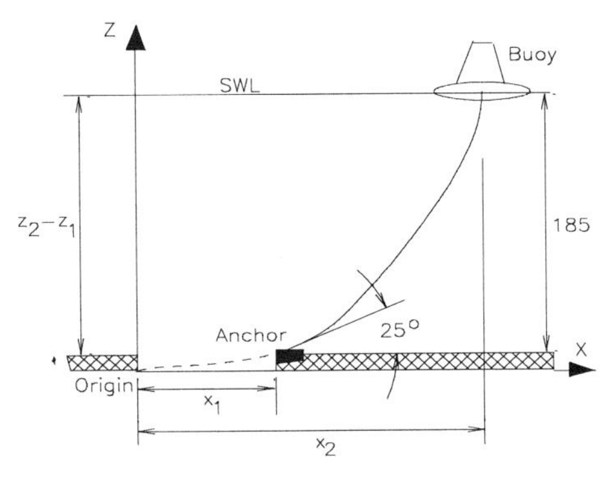

Figure 6-23. Heavy chain mooring of buoy.

Table 6-12. Results of example problem 6-1 solution.

Given	$W_I = 2.8$ lb/ft, Water depth = 185 ft, Buoy drag force =275 lb, cable angle at buoy = 25°.
Find	T_A, T_B, x_2
Solution	Horizontal component of tension at anchor (T_{HA}) equals the buoy drag: $T_{HA} = 275$ lb Vertical component of tension at anchor (T_{VA}) equals weight of fictitious chain length (s_1) between the anchor and origin. Tension at anchor (T_A): $\mathbf{T_A = 275/\cos 25 = 303.4\ lb}$ $T_{VA} = T_{HA} \tan 25 = 128.2$ lb $s_1 = 128.3/2.8 = 45.8$ ft $x = \frac{T_o}{W_I} \sinh^{-1}\left(\frac{W_I s}{T_o}\right)$ $x_1 = \frac{275}{2.8} \sinh^{-1}\left(\frac{2.8(45.8)}{275}\right) = 44.3$ ft $z = \frac{T_o}{W_I} \cosh\left[\frac{W_I x}{T_o} - 1\right]$ $z_2 - z_1 = 185 = \frac{275}{2.8}\left[\cosh\left(\frac{2.8}{275} x_2\right) - \cosh\left(\frac{2.8(44.3)}{275}\right)\right]$ Solving for x_2, yields: $\mathbf{x_2 = 172.4\ ft}$ $s = \frac{T_o}{W_I} \sinh\left(\frac{W_I x}{T_o}\right)$ $s_2 = \frac{275}{2.8} \sinh\left(\frac{(2.8)172.4}{275}\right) = 275.6$ ft The length of chain required is: $s_2 - s_1 = 275.6 - 45.8 = 229.8$ ft The distance downstream is: $x_2 - x_1 = 172.4 - 44.3 = 128.1$ ft The vertical tension at the buoy T_{VB} equals the weight of the chain plus the vertical component of tension at the anchor, $T_{VB} = 2.8(229.8) + 128.2 = 773.4$ lb Tension at the buoy is: $\mathbf{T_B = \sqrt{275^2 + 773.4^2} = 820.9\ lb}$

Neutrally Buoyant Cables

A neutrally buoyant towing line for a towed body is illustrated in Figure 6-24.

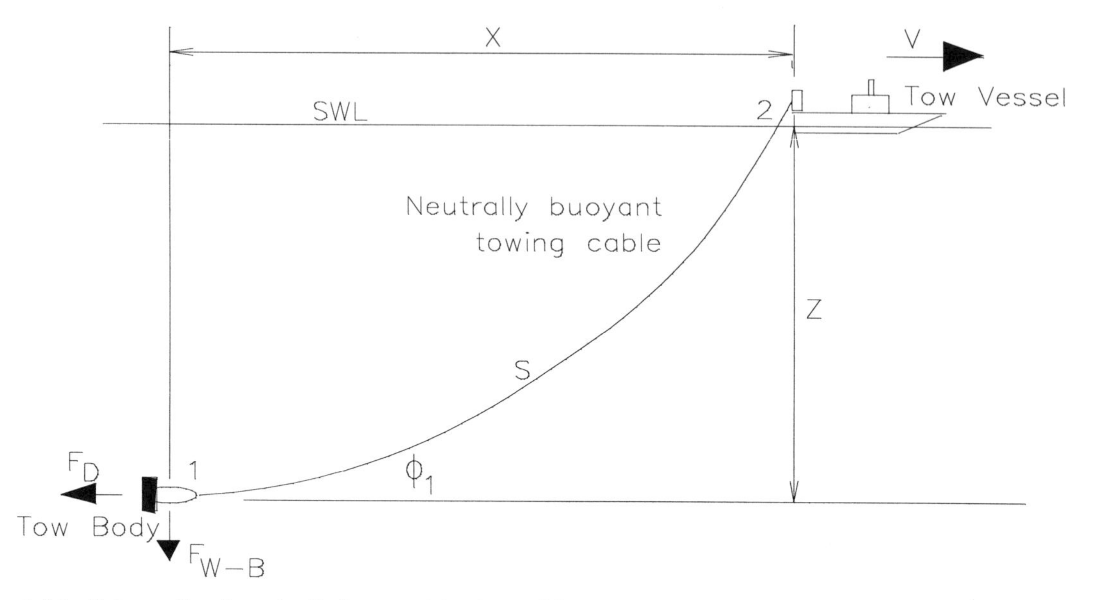

Figure 6-24. Schematic of neutrally buoyant towing cable.

In this case, the immersed weight (W_I) and the tangential drag force can be neglected for relatively short lengths (hundreds but not thousands of ft) of neutrally or nearly neutral buoyant cables such as synthetic fiber ropes. Previous expressions are simplified to

$$T\, d\phi = R \sin^2\phi \, ds \qquad \textbf{6-74}$$

$$\frac{dT}{ds} = 0 \qquad \textbf{6-75}$$

Integrating Equation 6-75 yields that $T = T_o$ = constant. Thus, the tension in the cable is the same at both ends of the towing line. Substituting T_o for T in Equation 6-74 and solving for ds yields

$$ds = \frac{T_o d\phi}{R\sin^2\phi} \qquad \textbf{6-76}$$

Also,

$$dx = ds \cos\phi = \frac{T_o}{R}\frac{\cos\phi}{\sin^2\phi} d\phi \qquad \textbf{6-77}$$

$$dz = ds \sin\phi = \frac{T_o}{R}\frac{d\phi}{\sin\phi} \qquad \textbf{6-78}$$

For a constant current, or boat speed, R is constant and the differential equations can be integrated between any two points along the cable [i.e. $P(\phi_1)$ and $P(\phi_2)$]. The result gives the cable length and the horizontal and vertical distances between the two points as

$$s = \frac{T_o}{R}\left[\cot\phi_1 - \cot\phi_2\right] \tag{6-79}$$

$$x = \frac{T_o}{R}\left[\csc\phi_2 - \csc\phi_1\right] \tag{6-80}$$

$$z = \frac{T_o}{R}\left[\ln\tan\frac{\phi_2}{2} - \ln\tan\frac{\phi_1}{2}\right] \tag{6-81}$$

REFERENCES

Allmendinger, E. E., Editor. *Submersible Vehicle Systems Design*. Jersey City: Society of Naval Architects and Marine Engineers, 1990. Figures 6-13 and 6-14 reprinted with permission: "Source: Allmendinger, *Submersible Vehicle Systems Design*, Copyright Society of Naval Architects and Marine Engineers, 1990."

Berteaux, H. O. *Coastal and Oceanic Buoy Engineering*. Woods Hole: Berteaux, 1991. Figure 6-18 and 6-22 reprinted with permission: "Source: Berteaux, *Coastal and Oceanic Buoy Engineering,* Copyright Berteaux, 1991."

Clayton, B. R. and R. E. D. Bishop. *Mechanics of Marine Vehicles*. Houston: Gulf Publishing Co., 1982. Figures 6-11 and 6-15 reprinted with permission: "Source: Clayton and Bishop, *Mechanics of Marine Vehicles*, Copyright Gulf Publishing Co., 1982."

Lewis, E. V., Editor. *Principles of Naval Architecture*, Second Revision. Jersey City: The Society of Naval Architects and Marine Engineers, 1988. Figures 6-1, 6-3, 6-7, and 6-9 reprinted with permission: "Source: Lewis, *Principles of Naval Architecture*, Second Revision. Copyright The Society of Naval Architects and Marine Engineers, 1988."

Morrish, S. W. F. "Approximate Rule for the Vertical Position of the Center of Buoyancy." *Transactions*, INA, now RINA, 1892.

Naval Sea Systems Command (NAVSEA). "Ship Hull Characteristics Program." NAVSEA SHCP User's Manual, CASDAC No. 231072, 1976.

Navy Facilities Engineering Command (NAVFAC). Design Manual, Harbor and Coastal Facilities, DM26, 1968. Figure 6-16 reprinted with permission: "Source: Design Manual, Harbor and Coastal Facilities, Copyright NAVFAC, 1968."

Posdunine, V. "Some Approximate Formulae Useful in Ship Design." *The Shipbuilder*. April 1925.

Tupper, E. C. *Introduction to Naval Architecture*, Third Edition. Jersey City: Society of Naval Architects and Marine Engineers, 1996.

U. S. Coast Guard (USCG). Aids to Navigation Manual-Technical. Department of Transportation. COMDINST MI 65003, 1990. Figure 6-16 reprinted with permission: "Source: Aids to Navigation Manual-Technical. U.S. Coast Guard (USCG), 1990."

Zubaly, R. B. *Applied Naval Architecture.* Jersey City: Society of Naval Architects and Marine Engineers, 1996.

PROBLEMS

6-1. A weight distribution for a vessel is given in the table below. Determine the location of the vertical center of gravity from the keel.

Item	Weight (lb)	Vertical Location Above the Keel (ft)
1	300	3
2	1000	2.5
3	500	4.0
4	950	0.75
5	400	3.5
6	1800	0.5
7	300	4.0
8	250	4.0
9	650	3.8
10	425	3.7
11	75	5.8
12	25	6.5
13	500	1.25

6-2. A diving bell has an outside diameter of 6 ft and is constructed of 0.75 inch thick steel (specific weight = 490 lb/ft^3). The diving bell is a vertical cylinder with a hatch in the bottom for the divers to enter and exit. How much weight must be added for the bell to float with a draft of 3 ft? Determine how much weight will have to be added with two 180 lb persons inside in order for the bell to just submerge.

6-3. A rectangular barge is 100 ft long, 35 ft wide and 6 ft deep. The draft of the barge is 3.5 ft. Determine the displaced volume and the location of the center of buoyancy. If the center of gravity is 4 ft above the bottom of the vessel, what is the condition of stability? Determine the location of the metacenter assuming the second moment of area of the waterplane (I_T) is given by $LB^3/12$. What is the magnitude of the righting arm when the barge is heeled over 2 degrees?

6-4. A spherical buoy is fully immersed in a tidal flow. Calculate the drag force on the buoy knowing that the buoy diameter is 6 ft, the current is 3 kts, and the water temperature is 60°F. Assume the γ_{sw} = 64 lb/ft^3 and $\nu = 1.26 \times 10^{-5}$ ft^2/s.

6-5. The spherical buoy in problem 4 weighs 2000 lb. Find the tension in the mooring line which would hold the buoy on station.

6-6. An aid to navigation buoy is moored in 80 ft of water. The anchor chain immersed weight is 3 lb/ft and the drag force on the buoy is 300 lb. Compute the length of chain required to maintain an angle of 25 degrees or less at the anchor. Compute the tension at the buoy and anchor. What is the distance of the buoy downstream of the anchor?

6-7. A barge has a rectangular cross-section that is 15 ft high and 30 ft wide. Its length is 300 ft. Ten items in the barge are vertically located as shown in the table below. The weight of the barge without the items is 6000 lb. Determine the draft of the barge, the center of buoyancy of the displaced volume and the distance between the center of gravity and center of buoyancy. If the metacenter is 30 ft above the keel, what is the righting moment developed for roll angle of 5 degrees?

Item	Weight (lb)	Vertical Location Above the Keel (ft)
1	200	1.5
2	350	2.5
3	500	1.5
4	150	3.0
5	325	2.7
6	180	4.5
7	400	5.2
8	275	6.1
9	400	3.1
10	175	7.1

6-8. A submarine weighs 20,000 lb in air. Determine the displaced volume of the submarine in sea water.

6-9. Find the angle of tilt of a cylindrical discus buoy shown below,. The mooring line tension is 5000 lb and is applied at the apex of a rigid bridle 6 ft below the buoy keel. The drag on the buoy is 868 lb and is applied at the buoy center of buoyancy when the buoy is on an even keel. The buoy weight is 3000 lb which acts through the center of gravity located at the geometric center of the buoy.

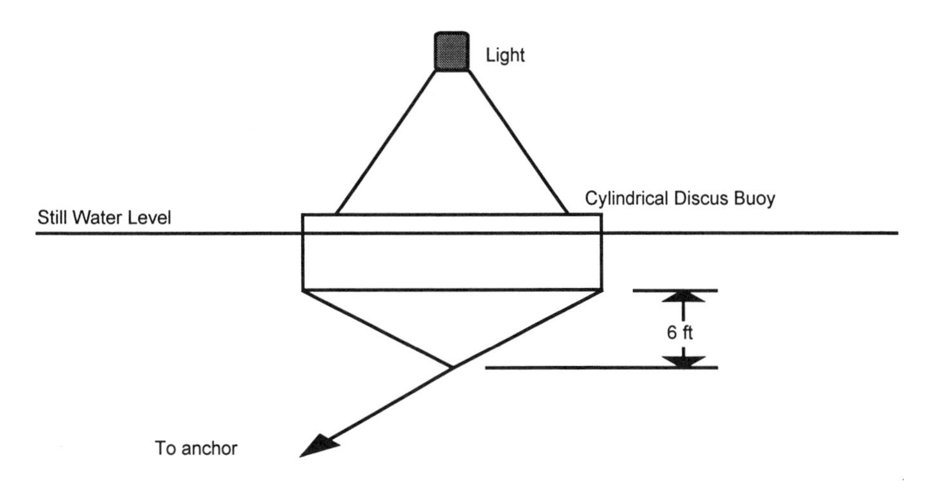

6-10. A surface buoy is located in 180 ft of water and is moored with a wire rope weighing 3 lb/ft. The drag on the buoy is 350 lb. Determine the length of wire rope necessary to maintain an angle of 25 degrees or less at the anchor. Determine the tension at the buoy and anchor. What is the distance of the buoy downstream of the anchor?

CHAPTER 7: UNDERWATER SYSTEMS

INTRODUCTION

Underwater systems provide the ability for humans to live and work beneath the sea surface. These systems can also provide the ability to view and record underwater phenomena and the operation of undersea systems remotely through the use of underwater video cameras and acoustic equipment. Many subsea systems use pressure vessels or are housed within these pressure vessels. Consequently, the elementary design principles for pressure vessels are discussed with reference to appropriate design codes. Diving and life support technology, remotely operated vehicles, submarines, underwater habitats, hyperbaric chambers, and a history of the development in some of these intriguing systems are discussed. A brief discussion addresses energy sources such as batteries, fuel cells, nuclear devices, and transmission lines that provide electrical power for the subsea systems and free the system from its dependence on surface supplied sources.

DIVING AND LIFE SUPPORT

Physiology

A basic understanding of the human circulatory system (Figure 7-1) and respiratory system (Figure 7-2) is necessary for the analysis and design of diving and underwater life support systems. Selected additional references discussing diving physiology are Miles and Mackay (1976), NOAA (1991), Nuckols et al. (1996), Schilling (1965), Schilling et al. (1976), and USN (1988). In the human circulatory system, blood is pumped by the right ventricle of the heart to the lungs through the pulmonary artery and from the lungs through the pulmonary vein to the left auricle. Blood in the pulmonary artery is deficient in oxygen and rich in carbon dioxide. The lungs provide the area for gas transfer of oxygen to the blood and removal of carbon dioxide. The blood returning to the heart is now rich in oxygen and the left ventricle pumps the oxygen rich blood to organs in the upper and lower parts of the body. Blood depleted of oxygen returns to the right atrium through veins and then is pumped to the lungs by the right ventricle completing the circuit. The distribution of blood to body organs is accomplished by the continual branching of arteries that become capillaries. Carbon dioxide and other substances are exchanged between the blood and body tissues at the thin walls of the capillaries. The blood from the capillaries flows into veins and is returned to the heart. The carbon dioxide is transported from the heart to the lungs where it is exhaled.

For the respiratory system, breathing is the result of rhythmic changes in the volume of the chest wall cavity. These changes are caused by the muscular action of the diaphragm and chest wall. This muscular action is under the control of the central nervous system that is responding to changes in blood oxygen and carbon dioxide levels. The normal respiratory rate at rest is 12 to 16 breaths per minute. The actual volume change of air involved in a given breath is called the tidal volume, which is illustrated in Figure 7-3. The respiratory minute volume is defined as the respiratory rate (breaths per minute) times the tidal volume (volume per breath).

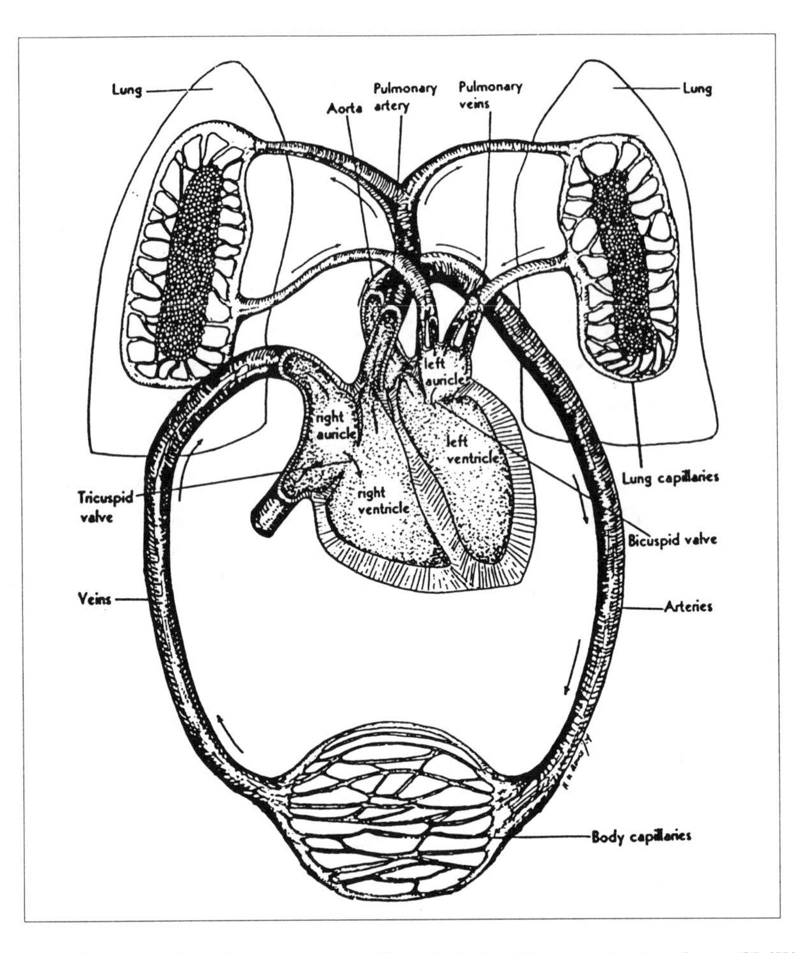

Figure 7-1. Schematic of human circulatory system. Reprinted with permission from Shilling, 1965, *The Human Machine.* (Full citing in references).

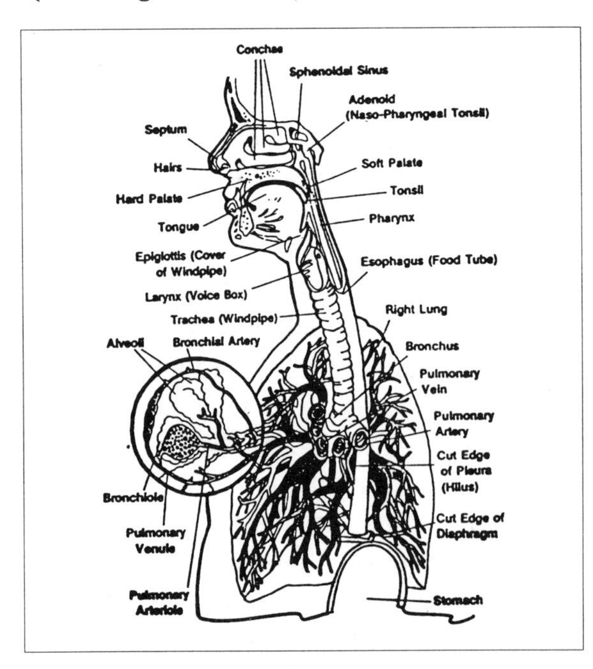

Figure 7-2. Schematic of human respiratory system. Reprinted with permission from NOAA,1991, *NOAA Diving Manual.* (Full citing in references).

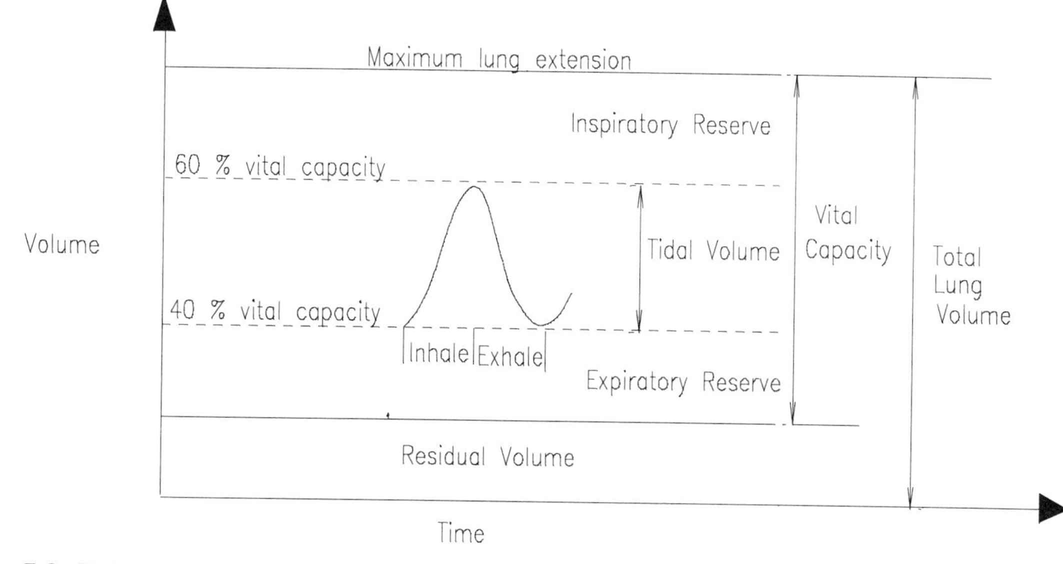

Figure 7-3. Volume relationship for a normal breath.

During respiration, air is drawn into the lung alveoli where the gas exchange takes place. The partial pressure of oxygen is less in the body tissues than in the blood because these tissues are continuously using oxygen. Consequently, oxygen is transferred from the blood to the tissues. The partial pressure of carbon dioxide is greater in the tissues because it is being produced there. Consequently, carbon dioxide is transferred from the tissue to the blood. The flexibility of the gas exchange rate is accomplished by increased movement of the chest, increased heart action, and increased differences in partial pressures.

Three main respiratory problems can occur in diving operations. The first is hypercapnia (carbon dioxide excess) that is a situation in which the tissues have an excess of carbon dioxide. This can be caused by an excess of carbon dioxide in the breathing medium, the inability to remove carbon dioxide from the breathing medium, or the inadequate removal of carbon dioxide from the tissues or blood. A relationship for the physiological effects of carbon dioxide concentration and exposure period is shown in Figure 7-4. Hypoxia (oxygen shortage) and oxygen poisoning are the other two respiratory problems. If the oxygen partial pressure drops below 0.1 ata (atmospheres absolute), then consciousness is usually lost. If oxygen is completely cut-off for 3 to 5 minutes, irreparable damage to the brain is experienced. Oxygen poisoning occurs when there is an excess of oxygen. These limits of oxygen partial pressure and concentration are illustrated in Figure 7-5.

The effects of pressure also cause some problems for divers. The most common problem is ear squeeze, but it can also cause problems in the sinuses and teeth. Gas embolism is another problem that can occur if the pressure is reduced too rapidly or the ascent from depth is to rapid. In this case dissolved gas comes out of solution, forming bubbles in the tissues and blood. These bubbles may block the circulation to the lungs, brain, heart, and spinal cord, and death can be the result. Recompression is the only treatment. This treatment requires the diver be placed in a hyperbaric chamber, and subsequently the chamber pressure is increased to the original depth of the diver. Then, a slow decompression procedure is followed to prevent the reoccurrence of the embolism.

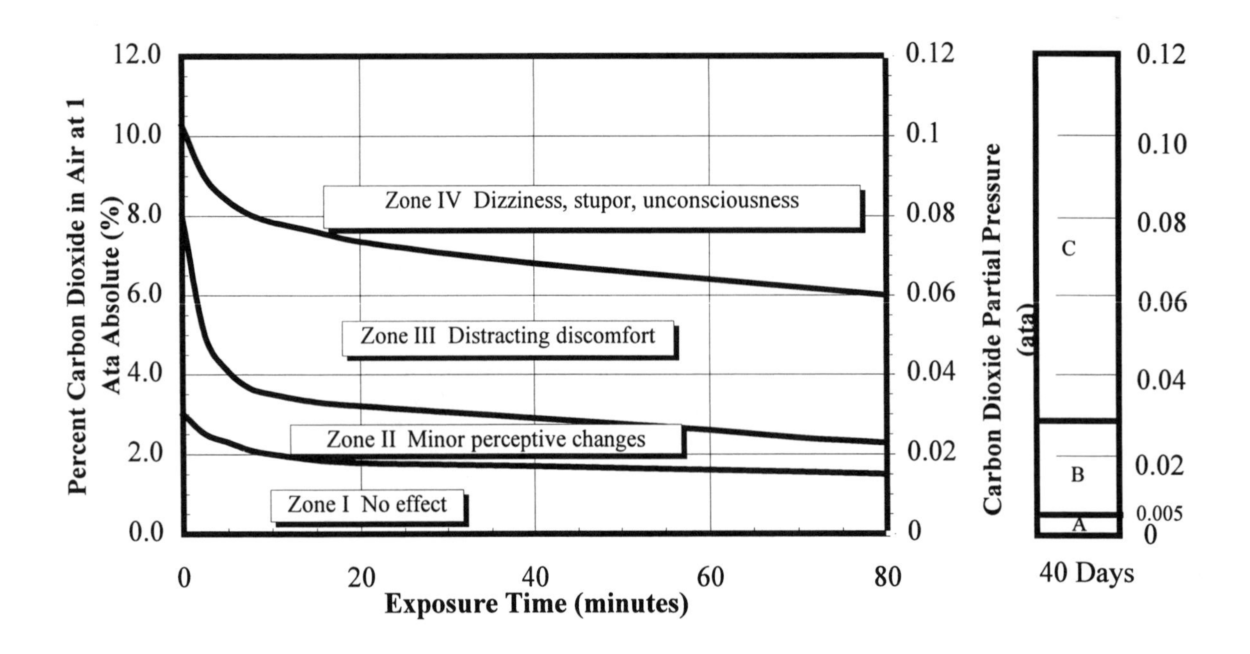

Figure 7-4. Physiological effects of carbon dioxide concentration and exposure period.

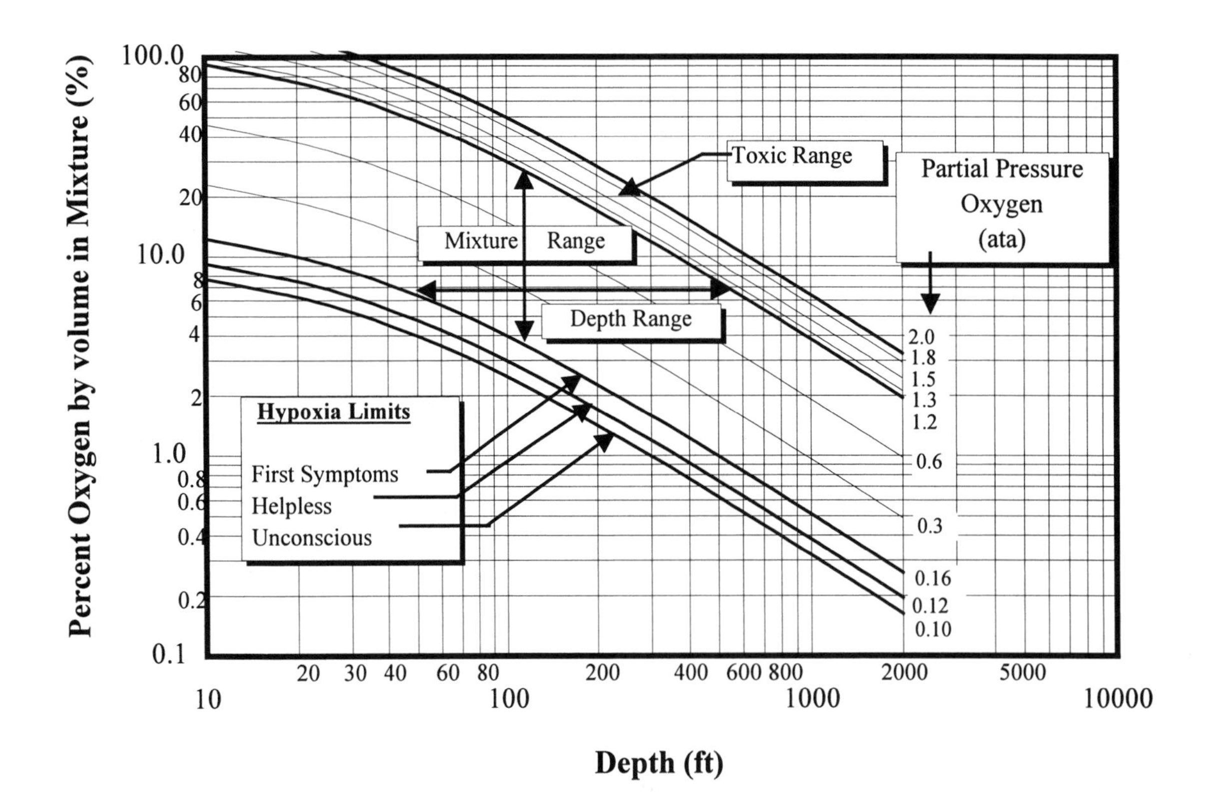

Figure 7-5. Percentage of oxygen in breathing mixture as a function of depth and oxygen partial pressure.

Two important consequences of pressure are decompression sickness (bends) and inert gas narcosis (rapture of the deep). The common inert gases (nitrogen and helium) are physiologically inert under normal pressure, but nitrogen has distinct anesthetic properties when the nitrogen partial pressure is sufficiently high. Nitrogen narcosis usually begins between the depths of 30.5 m (100 ft) and 45.7 m (150 ft). Decompression sickness occurs when the elimination of gases by the blood flowing through the lungs is slower than the rate of reduction in external pressure. The amount of super saturated inert gas in the tissues may permit it to come out of solution in the form of bubbles that cause rashes, block circulation, and distort body tissues. Again the only treatment is recompression.

An important consideration for the design of underwater life support systems is the amount of breathing gas required and oxygen consumed by the diver. Figure 7-6 shows the average required respiratory minute volume and oxygen consumption for different working conditions. The amount of carbon dioxide produced is determined from the respiratory quotient, which is the ratio of the carbon dioxide produced divided by the oxygen consumed. The typical respiratory quotient is 0.9, but it is often picked as 1.0 for convenience.

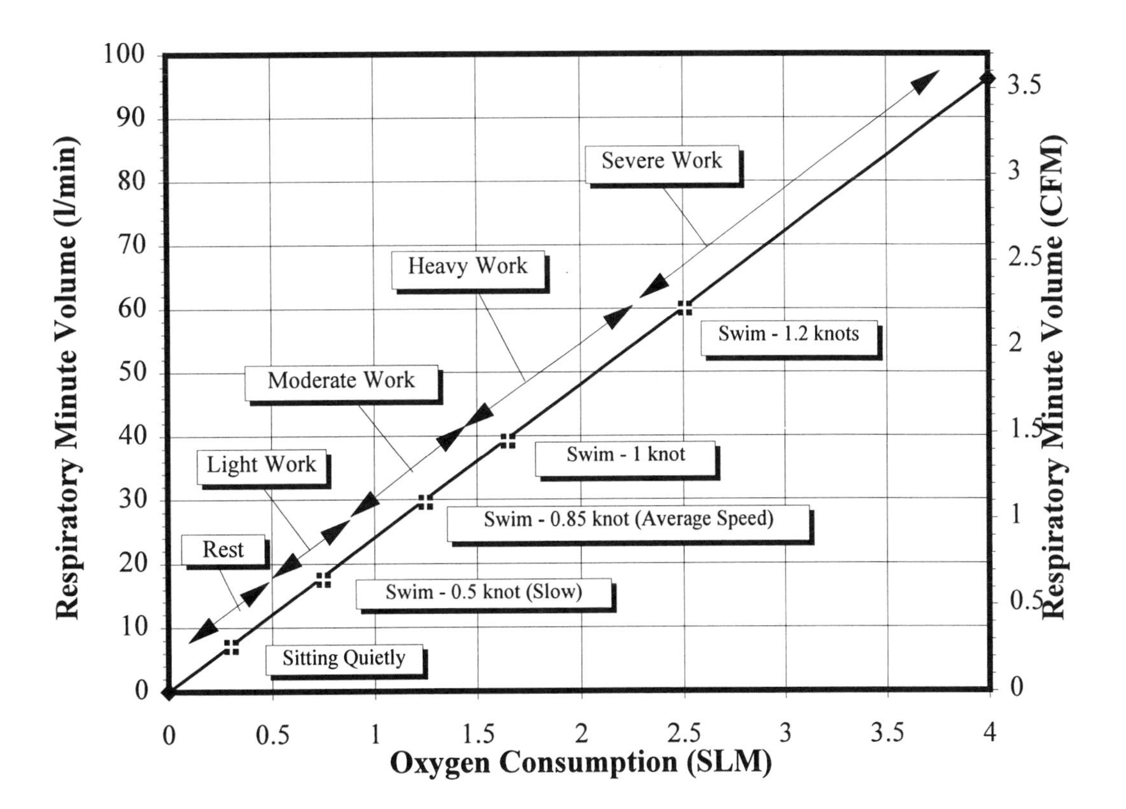

Figure 7-6. Relation of respiratory minute volume in liters per minute (l/min) or cubic feet per minute (CFM) and oxygen consumption in standard liters per minute (SLM) to type and level of exertion.

The units used to describe the respiratory minute volume are liters per minute and cubic feet per minute and both of these refer to the volumetric flowrate at the diver's depth or pressure. Standard volumetric conditions such as standard cubic feet per minute (SCFM) and standard liters per minute (SLM) refer to conditions at the surface or standard atmospheric and temperature conditions. Standard atmospheric pressure is commonly assumed as 10 m (33 ft) of sea water, 101.33 kPa (14.7 psia), 1 ata or 760 mm of Hg (mercury). Standard temperature are normally taken as 15 °C (59 °F). Therefore, flowrates in CFM or liters/min refer to diver requirements at depth and flowrates described as SCFM or SLM refer to conditions at one atmosphere (1 ata), which is the absolute pressure assumed at the ocean (water) surface.

Pressure

When pressure is measured relative to a perfect vacuum, it is called absolute pressure, and when it is measured relative to atmospheric conditions, it is called gauge pressure. These pressure relationships are illustrated in Figure 7-7. The standard atmospheric pressure is the average pressure found at sea level and is given as 10 m (33 ft) of sea water, 101.33 kPa absolute (14.7 psia) and 760 mm of Hg. Partial pressure is frequently used in diving and life support calculations. The partial pressure is the pressure a component of gas would exert if all the other gases were removed and the component gas occupied the volume alone. Dalton's Law of Partial Pressure states that the sum of the partial pressures of each component gas equals the total pressure of the gas mixture.

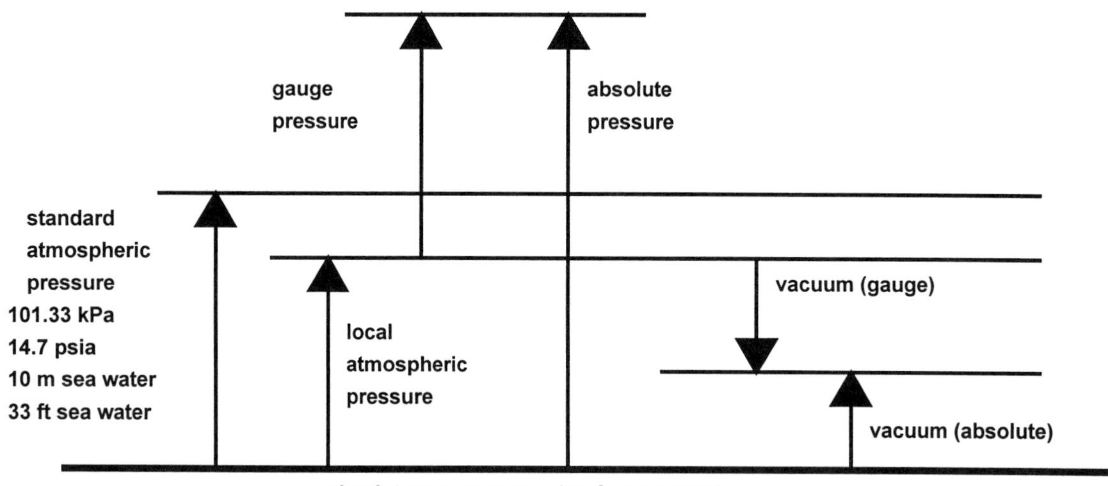

Figure 7-7. Measured pressure relationships.

Temperature

Temperature is measured as Centigrade (°C) or Fahrenheit (°F), and in absolute terms it is Kelvin (°K) or Rankine (°R), respectively. In equation form

$$^{o}K = {}^{o}C + 273 \qquad \textbf{7-1}$$

$$^{o}R = {}^{o}F + 460 \qquad \textbf{7-2}$$

The absolute temperature must be used in many thermodynamic relationships used in diver life support calculations.

Equation of State

The equation of state (perfect gas law) expresses the relationship between pressure, specific volume, and temperature for a substance. For a gas the expression is

$$pv = RT \tag{7-3}$$

where p is absolute pressure, v is specific volume, R is the gas constant and T is the absolute temperature. The gas constant is defined as

$$R = \frac{R_u}{M} \tag{7-4}$$

where R_u is the universal gas constant (1544 ft-lb/mole-^oR) and M is the molecular weight. The gas constant and molecular weight for several gases are tabulated in Table 7-1.

Table 7-1. Molecular weight and gas constant for typical breathing gas components.

Gas	Molecular Weight (M) lb_m	Gas Constant (R) ft-lb/slug-^oR	Gas Constant (R) ft-lb/lb_m-^oR
Air	29	1716.3	53.35
Carbon dioxide	44	1129.5	35.11
Helium	4	12,424.1	386.2
Hydrogen	2	24,659.3	766.53
Nitrogen	28	1774.5	55.16
Oxygen	32	1553.5	48.29

Other useful forms of the equation of state are

$$pV = mRT \tag{7-5}$$

$$p = \rho RT \tag{7-6}$$

$$pv = \frac{R_u}{M}T \tag{7-7}$$

$$pV = nR_uT \tag{7-8}$$

where V is the volume of the gas, m is the mass of the gas, ρ is the gas density, and n is the number of moles (n = m/M). Special cases of the perfect gas law are

Constant Temperature (Boyles Law) $$p_1V_1 = p_2V_2 \tag{7-9}$$

Constant Pressure (Charles Law) $$\frac{V_1}{T_1} = \frac{V_2}{T_2} \tag{7-10}$$

General Gas Law $$\frac{p_1V_1}{T_1} = \frac{p_2V_2}{T_2} \tag{7-11}$$

where the subscripts represent equilibrium state points for the gas. In some cases the perfect gas law is not satisfactory, and a real gas law must be used that considers compressibility effects. Two forms of the real gas law are

$$pv = ZRT \quad \text{or} \quad pV = ZmRT \qquad \textbf{7-12}$$

where Z is the compressibility factor which is a function of temperature and pressure.

Example Problem 7-1

A SCUBA tank is fully charged to 3015 psia at 70 °F. It contains 72 ft^3 air at standard conditions (15 psia, 70 °F). Determine the tank internal volume and assume perfect gas law applies.

Table 7-2. Results of example problem 7-1 solution.

Given	$p_1 = 3015$ psia, $V_2 = 72\ ft^3$, $p_2 = 15$ psia
Find	V_1
Solution	Assume constant temperature, $T_1 = T_2$
	Perfect gas law reduces to Boyles Law for constant temperature process
	$p_1V_1 = p_2V_2$
	$3015(144)V_1 = 15(144)72$
	$V_1 = 0.36\ ft^3$

Air Supply Calculations

The actual surface volume of air in a SCUBA tank at any pressure is given by

$$V = \frac{NV_rP_g}{P_r} \qquad \textbf{7-13}$$

where V is actual volume of air available in the tank at the surface, N is the number of tanks, V_r is the rated capacity of the tank, P_g is the gauge pressure inside the tank, and P_r is the rated tank pressure in psig. A consistent system of units must be used, SI or BG, to obtain dimensionally correct results. The effect of temperature is calculated by

$$V_t = \frac{V_dT_2}{T_1} \qquad \textbf{7-14}$$

where V_t is volume of air adjusted for temperature, V_d is the available air from previous Equation 7-13, T_2 is absolute temperature of water, and T_1 is absolute temperature of air. The duration of air supply in minutes for a SCUBA tank leaving a reserve pressure is evaluated by

$$S = \frac{NV_r(P_g - P_m)(33)T_2}{\dot{R}(D+33)P_rT_1} \qquad \textbf{7-15}$$

where S is duration in minutes, P_m is minimum reserve pressure in psig, $\dot{R}$ is breathing rate (respiratory minute volume) in cubic feet per minute, and D is depth of water in feet of sea water (Tucker 1980). The ventilation requirements for surface supplied diving rigs (helmet or mask) is

$$\dot{V} = \frac{P_{ata}\, O_{slm}\, R}{26.3\left(C_2 - C_1\, P_{ata}\right)} \qquad \textbf{7-16}$$

where $\dot{V}$ is volume rate of air required in standard cubic feet per minute (SCFM), O_{SLM} is oxygen consumption in standard liters per minute (SLM), R is respiratory quotient (volume of carbon dioxide produced / volume of oxygen consumed), C_2 is desired partial pressure of carbon dioxide in inhaled air (ata) and C_1 is partial pressure of carbon dioxide in breathing air from compressor (ata) and P_{ata} is pressure at depth in atmospheres. The constant 26.3 is a conversion factor for converting standard liters per minute to standard cubic feet per minute and includes the temperature difference between standard liters and standard cubic feet. A standard cubic foot is defined as conditions at 70 ^oF and 1.0 ata, and a standard liter is defined as the condition at 32 ^oF and 1.0 ata. If oxygen consumption is 2.6 SLM, R is 0.9, C_2 is 0.02 then

$$\dot{V} = \frac{0.0893\, P_{ata}}{0.02 - C_1 P_{ata}} \qquad \textbf{7-17}$$

and when the breathing air contains no carbon dioxide the equation reduces to

$$\dot{V} = 4.5\, P_{ata} \qquad \textbf{7-18}$$

For example, a surface supplied diver is working at 99 ft of sea water will require 18.0 SCFM of air since 99 ft is equal to 4 atmospheres pressure absolute.

Ventilation of Large Chambers

Underwater habitats are considered large chambers, and it is often necessary to evaluate the ventilation rate necessary to avoid excess carbon dioxide concentrations. If the atmosphere is assumed to be flushed continually by the incoming air, a rather simple analysis can be used. If it is assumed that pure air is entering the chamber and that complete mixing occurs, then the result is

$$P_{co_2} = \frac{\dot{m}RT}{\dot{V}_{air}}\left[1 - e^{\frac{-\dot{V}_{air}\, t}{V_t}}\right] \qquad \textbf{7-19}$$

where P_{co2} is allowable partial pressure of carbon dioxide, $\dot{m}$ is the mass flow rate of carbon dioxide, R is gas constant for carbon dioxide, T is absolute temperature, $\dot{V}_{air}$ is the volumetric flow rate of air at depth, t is time, and V_t is volume of chamber. If the time (t) is very long, then the steady state value of the carbon dioxide partial pressure is

$$P_{co_2} = \frac{\dot{m}RT}{\dot{V}_{air}} \qquad \textbf{7-20}$$

Diver Breathing Equipment

The breathing equipment used by working and recreational divers must supply the necessary respiratory minute volume containing the proper amount of oxygen. The five general types of breathing equipment are the demand regulator or self contained underwater breathing apparatus (SCUBA), semi-closed breathing apparatus, surface supplied deep-sea diving outfit, surface supplied deep-sea diving outfit with carbon dioxide absorption, and closed circuit breathing equipment.

The demand regulator breathing equipment is the primary equipment used by recreational divers. It is composed of a breathing gas supply tank, a first stage regulator that drops the tank air pressure to about 1.03 MPa (150 psi) over ambient pressure, and a second stage demand regulator that delivers air slightly above the diver's ambient pressure. The regulator is part of the diver's mouthpiece that supplies breathing gas when the diver inhales and is shut off when the diver exhales. The breathing gas is stored in a compressed gas cylinder carried on the diver's back. The rate of breathing gas usage depends upon the exertion effort of the diver and the depth of water. Dive duration with this type of equipment can vary from approximately 12 min when divers are under heavy exertion in deep water (approximately 39.6 m or 130 ft) to several hours when the diver exertion is light in shallow water.

Semi-closed breathing equipment is very efficient in the use of breathing gas. The breathing gas has oxygen supplied at a partial pressure just under the toxic limit of 1.2 ata, and it is rebreathed after being passed through a carbon dioxide absorber until the oxygen partial pressure is reduced to 0.16 ata on exhalation. Thus, a large majority of the oxygen is used as compared to about 20 percent for the open circuit systems. The semi-closed breathing apparatus is the most economical breathing gas supply when used at moderate depths. It can be supplied by compressed gas cylinders carried by the diver or through an umbilical from a compressed gas supply located in a diver lockout chamber, diving bell, or a surface vessel. The US Navy (1971) is a good source for the determination of breathing gas flow rates required for different diving operations.

Closed circuit breathing rigs have been developed such that none of the breathing gas is vented to the water. The inert gas, typically helium, is added to fill the breathing bags and adjusted as required by the diving depth. Oxygen is added to the breathing mixture at the rate of consumption, and thus all the oxygen is used by the diver. When the diver exhales the breathing mixture, the exhaled gas is passed through a carbon dioxide removal device to cleanse the gas of all the carbon dioxide, and the remaining inert gas and unused oxygen are recirculated to the inhale bag and combined with the oxygen supplied to replace the used oxygen. This system requires the automatic sensing of the partial pressure of oxygen and carbon dioxide. These systems are advantageous for deep diving operations because no inert gas is lost. Although the closed circuit systems are not well known or used by recreational divers, their application has been used for deep diving and for military or scientific applications.

Open circuit breathing equipment is used for light activity and moderate depths with air supplied from the surface by air compressors or from a bank of compressed air cylinders on a support vessel. The breathing gas is supplied to a helmet that is essentially a ventilated dead space. The supplied air flow rate must be sufficient enough to dilute the carbon dioxide exhaled by the working diver. This is a common procedure for commercial diving in shallow waters less than 30.5 m (100 ft) to 39.6 m (130 ft) of sea water. In some applications the diving helmet is equipped similar to a semi-closed breathing apparatus with the recirculation of breathing gas and carbon dioxide absorption. This is needed when the diving operation is at depths that require the use of helium and when it is uneconomical to exhale the helium to the surrounding water.

Control of Underwater Chamber Environment

The environment inside underwater habitats, diver lock-out chambers, submersible pilot chambers, personal transfer capsules, and other underwater enclosures must be controlled. In addition to the requirement for the proper breathing gas mixture and supply rate, the environment must be maintained within reasonable limits of temperature and humidity. The increased pressures and various gas mixtures require special psychrometric charts to be developed and used to evaluate the environmental conditions. These psychrometric charts are available in the US Navy Diving Gas Manual (USN 1971), Nuckols et al. (1996), and Randall (1997).

Heating and cooling systems are needed to maintain the temperature and humidity at desirable conditions. In normal air these conditions might be 23.9 ^{o}C (75 ^{o}F) and 50% relative humidity, but if the pressure requires the use of helium/oxygen mixtures, then the desirable temperature condition is more like 29.4 ^{o}C (85 ^{o}F). Heat transfer analysis of the pressure vessel system is necessary to determine the steady state and transient conditions inside the habitat, and these analyses are better discussed in more advanced literature on the subject.

PRESSURE VESSELS

Pressure vessels are used extensively in ocean engineering applications that include submarine hulls, underwater habitats, instrument housings, breathing gas storage vessels (e.g. Scuba tanks), subsurface mooring buoys, diver lockout chambers for small submersibles, buoyancy tanks for offshore structures, flotation devices, diving bells, hyperbaric chambers, underwater storage vessels, and many others. The design of pressure vessels is the subject of many handbooks and can be quite involved. The original need for pressure vessels was in the electric power industry that used pressure vessels to contain high pressure steam. Similar pressure vessels were also used in electrical power plants and also on sea going vessels. Consequently, there are many good references for designing pressure vessels such as Megyesy (1977), Chuse and Eber (1984), Roark and Young (1982), ASME (1980), and ANSI/ASME (1977). Most underwater pressure vessels are thin walled, and the basic theory for the design of this type of pressure vessel is now briefly described.

Thin Walled Cylinders and Spheres

Cylindrical pressure vessels (tanks) are used by recreational and working divers to provide breathing gas underwater. These tanks must be designed to contain the breathing gas (air or helium and oxygen mixtures) at high internal pressures such as 20.7 MPa (3000 psia) to 31.0 MPa (4500 psia) while the divers are at typical depths of up to 45.7 m (150 ft) using air and deeper depths using helium and oxygen gas mixtures and even mixtures of hydrogen and oxygen. The pressure, p, inside the tank causes tensile forces in the tank walls that resist the bursting forces as illustrated in Figure 7-8.

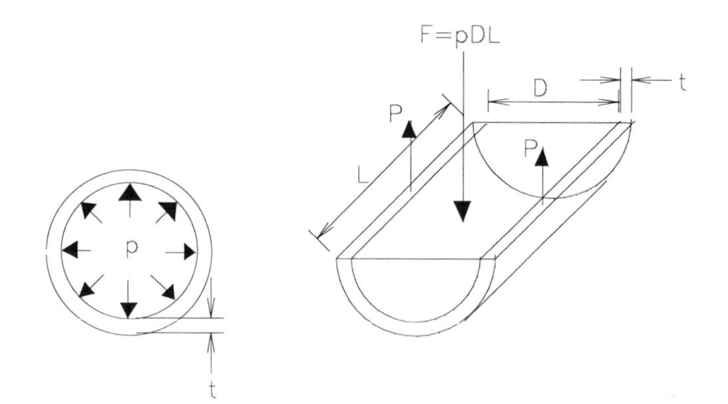

Figure 7-8. Schematic of forces acting on a thin walled cylindrical pressure vessel.

The incremental normal force acting on the cylinder located at angle ϕ from the horizontal diameter is

$$dF = pdA = pL\frac{D}{2}d\phi \qquad \textbf{7-21}$$

An equivalent force acts similarly on the other side of the vertical line, and the horizontal components of these two forces act in the opposite direction and therefore cancel. This is true for all forces so the total force due to the internal pressure is the sum of the vertical component of the elemental forces acting on the internal cylinder surface. This summation is expressed as

$$F = \int_0^{\pi}\left(pL\frac{D}{2}d\phi\right)\sin\phi = pL\frac{D}{2}\left[-\cos\phi\right]_0^{\pi} \qquad \textbf{7-22}$$

that simplifies to

$$F = pDL \qquad \textbf{7-23}$$

This total bursting force is resisted by the equal forces (P) that act on the cylinder wall surface. The stress that is resisting the bursting force is

$$S_t = \frac{F}{A} = \frac{pDL}{2tL} = \frac{pD}{2t} \qquad \textbf{7-24}$$

The stress S_t is called the hoop, circumferential, girth or tangential stress since it acts tangent to the cylinder surface. It is an average stress and is valid for cylinders with a wall thickness of less than or equal to 1/10 of the cylinder inside radius.

The bursting force acting on the ends of the cylinder are resisted by the force P acting over the transverse section shown in Figure 7-9. The area of the transverse section is the product of the thickness (t) and the mean circumference, and if the thickness is small compared to the diameter, then the mean circumference is πDt. In equation form the equivalence of the bursting force (F) and resisting force (P) is expressed as

$$P = F \qquad \textbf{7-25}$$

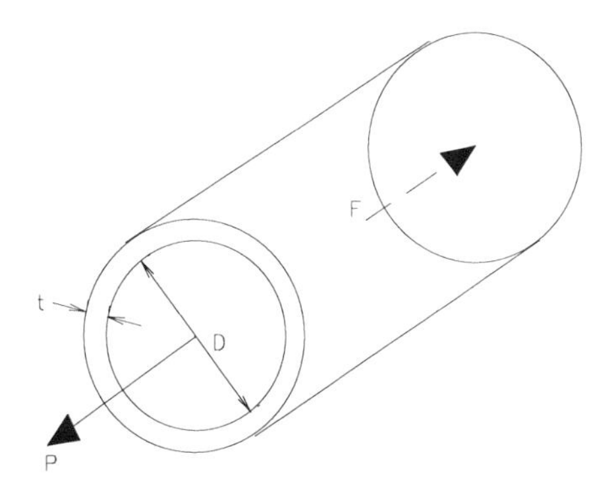

Figure 7-9. Schematic illustrating bursting force on cylinder transverse section.

Equation 7-25 can be expressed as

$$\pi DtS_l = \frac{\pi D^2}{4} p \qquad \textbf{7-26}$$

that reduces to

$$S_l = \frac{pD}{4t} \qquad \textbf{7-27}$$

where S_l is the longitudinal stress since it acts along the longitudinal axis of the cylinder. Notice the longitudinal stress is one half of the circumferential stress, and therefore, the cylindrical pressure vessel will fail much sooner along the longitudinal seam than at the transverse section. For the case of a thin walled sphere (thickness ≤ 0.1 radius) with uniform internal or external pressure, the stress in the walls is calculated using Equation 7-27.

Equations 7-24 and 7-27 also apply for external pressure acting on thin walled cylinders such as those used for underwater habitats, submarine diver lock-out chambers, and instrument housings. However, the analysis is more complicated when different end caps and welding characteristics or bolting schemes are used to construct multiple component structures. References previously mentioned Roark and Young (1982) and Megyesy (1977) should be consulted for preliminary design formulas. These pressure vessels are also reviewed by appropriate regulatory bodies such as American Bureau of Ships (ABS), det Norske Veritas (DNV), Americans Society of Mechanical Engineers (ASME) and American Petroleum Institute (API), to name a few, before the actual design and subsequent construction is approved. These handbooks and codes have more advanced design equations and procedures.

When the radius to wall thickness ratio is greater than 10, the pressure vessel is classified as a thick walled pressure vessel. These pressure vessels require a different analysis and appropriate analytical techniques are found in Roark and Young (1982) and others.

SUBMARINES

Military Submarines

Large submarines are part of the military defenses of the United States and some other nations. These submarines have the ability to remain unseen and undetected below the ocean surface and are capable of launching torpedoes, mines, and missiles to destroy enemy surface ships, submarines, and inland targets. History and information on military submarines may be found in selected references such as Sweeney (1970), Terzibashitsch (1991), Miller (1991), Burcher and Rydill (1994), Friedman (1994), Hervey (1994), and Sharp (1994).

These submarines have a unique classification indicating the propulsion and weapon system used. The designation SS means the submarine is a diesel powered submarine. The diesel engines are used to propel the submarine on the surface and to charge electrical storage batteries that are used to propel the submarine when it is submerged. These diesel submarines must return to the surface or use a snorkel to run the diesel engines to charge batteries. Typically, diesel submarines operate submerged during the day, and they surface at night to charge their batteries. The snorkel allows the submarine to raise a large tube while at periscope depth (approximately 18 m or 60 ft) that provides an air inlet for the diesel engines to charge batteries while submerged. Extended submerged operations below periscope depth can last for many days depending on the submarine speed, use of auxiliary electrical equipment, and the breathing atmosphere. A photograph of the diesel submarine USS *Grenadier* (SS-525) is shown in Figure 7-10, and she was operational in the 1960's and subsequently transferred to Venezuela in 1973. Other SS designations included SSK for killer submarines used for anti-submarine warfare and SSG for diesel submarines armed with guided missiles.

Figure 7-10. Photograph of the USS *Grenadier* (SS-525) entering the port of Malta in 1965.

Submarines that are powered by nuclear energy are designated as SSN. The nuclear submarines have almost unlimited power and need no access to surface air and are consequently true submarines. These nuclear submarines can dive to depths in excess of 457.3 m (1500 ft) and travel over 111,180 km (60,000 nautical miles) before exhausting initial nuclear fuel supplies (Hervey 1994). If these submarines are armed with cruise missiles they are designated SSGN, and with ballistic missiles they are designated SSBN. Photographs of the nuclear submarine USS *Skipjack* (SSN 585) and the ballistic submarine USS *George Washington* (SSBN-598) are illustrated in Figure 7-11.

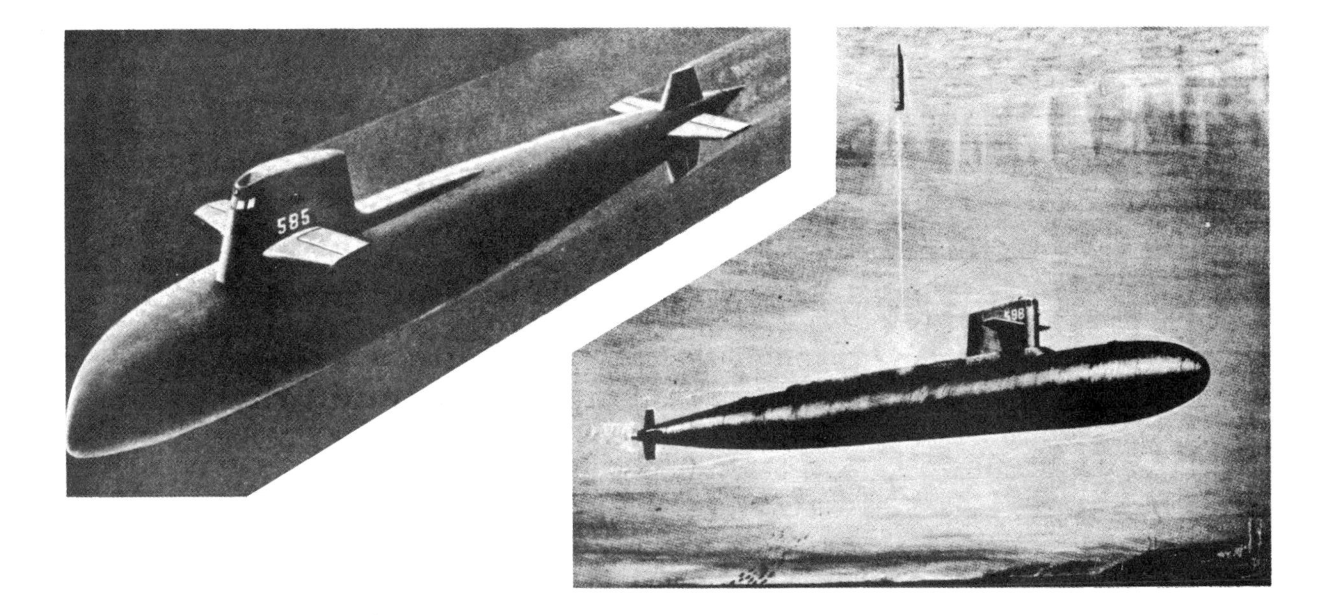

Figure 7-11. Illustration of the nuclear submarines USS *Skipjack* (SSN-585) and USS *George Washington* (SSBN-598). Reprinted with permission from Sweeney, 1970, *A Pictorial History of Oceanographic Submersibles*. (Full citing in references).

Research and Work Submersibles

The development of research and work submersibles began in the 1950's. A selected summary of manned submersibles is tabulated in Table 7-3, and a more detailed summary is found in Allmendinger (1990) and Busby (1976). In 1952 the manned submersible *Trieste* was designed and built in Italy, and this submarine was purchased by the US Navy and eventually made the historic dive in 1960 to 10,915 m (35,800 ft) in the Marina Trench near Guam in the Pacific Ocean. Also, J. Y. Cousteau of France used the French built shallow water diving saucer, *Denice*, and made hundreds of dives world wide. The US Navy modified the *Trieste* and used *Trieste II* to investigate the tragic loss the military submarine *Thresher* in 1963.

The *Alvin* that was built by General Mills and Litton industries (Figure 7-12) is one of the most frequently used US submersibles. This submarine has made over 2100 dives (Allmendinger 1990) over the first 25 years of operation and has been operated by the Woods Hole Oceanographic Institute with support from the US Navy. The original pressure hull was replaced with a titanium hull giving her a depth capability of 4000 m (13,120 ft). In the late 1980's, *Alvin* was used in conjunction with the ROV Argo-Jason to investigate the *Titanic* that rests in the deep waters of the North Atlantic.

Table 7-3. Selected research or work submersible information (Allmendinger 1990).

Name	Depth m (ft)	Dimensions, m (ft) Length	Width	Height	Displacement tons metric (short)	Crew	Builder (Country)
Aluminaut	1830 (6000)	15.5 (51)	4.7 (15.4)	5.0 (16.4)	76 (83.7)	6	General Dynamics USA
Alvin	4000 (13,120)	7.6 (24.9)	2.4 (7.9)	3.9 (12.8)	16.7 (18.4)	3	General Mills/Litton USA
Ben Franklin	610 (2000)	14.6 (47.9)	6.1 (20)	6.4 (21)	143 (157.5)	6	Grumman USA
Deep Quest	2440 (8000)	12.2 (40)	5.8 (19)	4.0 (13.1)	52 (57.3)	4	Lockheed USA
Deepstar 4000	1220 (4000)	5.5 (18)	3.3 (10.8)	2.1 (6.9)	9 (9.9)	3	Westinghouse USA
Johnson-Sea-Link I/II	800 (2620)	6.9 (22.6)	2.4 (7.9)	3.2 (10.5)	9.5 (10.5)	2 + 2 divers	Harbor Branch Ocean. Inst. USA
LR-4 lockout	457 (1500)	10.4 (34.1)	3.0 (9.8)	2.7 (8.9)	12.9 (14.2)	2 + 3 divers	Slingsby England
Mermaid IV lockout	600 (1970)	8.0 (26.2)	2.9 (9.5)	3.2 (10.5)	19.1 (21.0)	2 + 3 divers	Meerestechnik Germany
PC 12	366(1200)	6.7 (22)	2.4 (7.9)	2.4 (7.9)	7.2 (7.9)	2-3	Perry USA
PC 9	412(1350)	7.9 (25.9)	2.2 (7.2)	2.4 (7.9)	10.4 (11.5)	2-3	Perry USA
Pices 8	1000(3300)	6.2 (20.3)	3.3 (10.8)	3.2 (10.5)	10.7 (11.8)	2-3	HYCO Canada
Pisces 5	2012(6600)	6.1 (20)	3.2 (10.5)	3.3 (10.8)	10.8 (11.9)	2-3	HYCO Canada
Shinkai 2000	2000 (6560)	9.3 (30.5)	3.0 (9.8)	2.9 (9.5)	25 (27.5)	3	Mitsubishi Japan
SM360 lockout	300 (980)	10.0 (32.8)	2.8 (9.2)	2.9 (9.5)	25 (27.5)	3 + 3 divers	COMEX France
Star III	610 (2000)	7.5 (24.6)	2.1 (6.9)	2.4 (7.9)	10.5 (11.6)	2	General Dynamics USA

Figure 7-12. Photograph of the manned submersible *Alvin*. Reprinted with permission from Allmendinger,1990, *Submersible Vehicle Systems Design*. (Full citing in references).

Perry Oceanographic and HYCO have constructed many manned submersibles that were used by the offshore industry as working submersibles to complete work tasks related to the recovery of oil and gas from the Gulf of Mexico, North Sea, and other offshore locations around the world. The PC-12 and Pices 8 shown in Figure 7-13 are examples of these submarines.

PC 12 **Pices 8**

Figure 7-13. Perry *PC 12* and HYCO *Pices 8* submarines. Reprinted with permission from Allmendinger,1990, *Submersible Vehicle Systems Design*. (Full citing in references).

A manned submarine that is operated by the Harbor Branch Oceanographic Institute and used for research is the *Johnson-Sea-Link* shown in Figure 7-14. This diver lockout submarine has an acrylic pilot sphere and an aluminum lock-out chamber. This submarine has been in operation for nearly 20 years and was also used to assist in the US space shuttle disaster search in 1985.

Figure 7-14. *Johnson-Sea-Link* submersible. Reprinted with permission from Allmendinger,1990, *Submersible Vehicle Systems Design*. (Full citing in references).

Recreational and Tourist Submarines

Recreational and tourist submarines are being developed. Tourist locations such as the Caribbean have experienced some development of submarines to take passengers for underwater tours. Popular theme parks in the USA have had submarine excursion rides for many years and one of the more popular ride is the Jules Verne 20,000 leagues under the sea ride at the Disney World theme park in Orlando, Florida. These tourist submarines provide characteristic viewing windows along the port and starboard sides for observation of the underwater environment by the submarine occupants. Two person recreation submarines have also been developed for tourist locations with good underwater visibility.

Human Powered Submarines

In 1989 the first human powered submarine races were sponsored by the Perry Foundation and Florida Atlantic University, and the races took place at Riviera Beach, Florida. These races were organized to provide the opportunity for students and others interested in the advancement of underwater technology to participate in the design, construction, and racing of submarines using human power only. No electrical energy was permitted, and the submarines were free flooding which eliminated the need for the pressure hull design. The divers inside the submarine use standard SCUBA breathing apparatus.

There have been three International Submarine Races. The first race included 19 submarines built and raced by university students and the small submersible industry. The first overall performance winner was the US Naval Academy's submarine *Squid.* The 2nd International Submarine Race (ISR) was conducted at the same location in 1991 and the FAU Boat won the 100 m race with a speed of 8.7 kph (4.7 kts) and also won the 800 m underwater oval race with two submarines simultaneously on the underwater race track. The 3rd ISR was staged off Ft. Lauderdale, Florida. Nearly fifty submarines participated over a two week period with Tennessee Technical University's *Tech Torpedo II* winning the overall performance award. In 1994, the West Coast Submarine Invitational was held in the Offshore Model Basin in Escondido, California, and Florida Atlantic University's *FAU Boat* won the 100 m race with a speed of 10.94 kph (5.9 kts) that was a Guinness World Book record, and in 1996 the record was broken by the University of Quebec's *Omer* with a speed of 11.7 kph (6.3 kts). Examples of Texas A&M human powered submarines are the *Aggie Ray* and *SubMaroon* shown in Figure 7-15. Ocean Engineering students at Texas A&M University have built three human powered submarines that include *Aggie Ray* for the 1991 ISR race, *Argo* for the 1993 ISR, and *SubMaroon* for the 1996 World Submarine Invitational (WSI).

REMOTELY OPERATED VEHICLES

History

Remotely operated vehicles (ROV) are unmanned underwater vehicles controlled from a remote location such as a ship, fixed offshore platform, floating platform, or other above water structure. The development of unmanned underwater vehicles began in the 1950's, but reliable waterproof electrical connectors held back their early development. Shell Oil Company developed the first oil wellhead ROV called Mobot (Shatto 1991). World publicity for ROVs was gained for the recovery of the H-bomb in 868.9 m (2850 ft) of water near Palomares, Spain in 1966 by the US Navy's remotely operated vehicle CURV, which is illustrated in Figure 7-16. CURV II and III followed, and it was CURV III that was used to rescue the submersible, Pisces II, that sank in 477.1 m (1565 ft) of water off Ireland. These operations spurred the development of remotely operated vehicles especially in the offshore industry. One of the very popular early ROVs was built by Hydroproducts (now Honeywell AMSO) and named the RCV 225, which was dubbed the flying eyeball. It was very maneuverable and contained an underwater video camera that allowed surface personnel the ability to see underwater. Figure 7-17 shows the RCV 225 with its handling system and exiting the launcher, or protective cage.

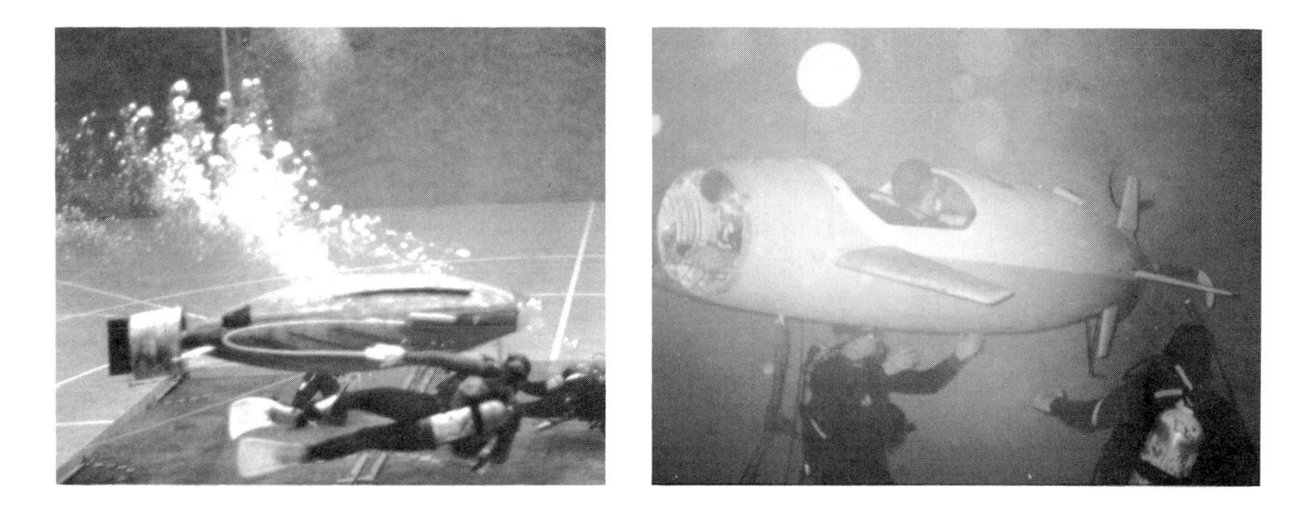

Aggie Ray **SubMaroon**

Figure 7-15. Human powered submarines *Aggie Ray* and *SubMaroon* being tested in the Offshore Technology Research Center's wave basin.

One of the first ROV's that was capable of reliably doing underwater work was the ROV Scorpio, Figure 7-18, built by Ametek Straza. It was a very popular vehicle and nearly sixty of these vehicles were built and used in the offshore industry. Scorpio did not use a cage and was found to be very effective for observation and medium to light work. The first one was built in 1977, but it is no longer in production. Another vehicle that experienced considerable use was the Triton, Figure 7-19, built by Perry Offshore. Approximately forty Tritons were built and equipped for cleaning offshore structures with high pressure jets and suction cups for attaching itself to the structure. A limited summary of ROV development is tabulated in Table 7-4.

Figure 7-16. Illustration of the remotely operated vehicle CURV. Reprinted with permission from Shatto, 1991, *Handbook of Coastal and Ocean Engineering*, Ch. 6. (Full citing in references).

Figure 7-17. Photograph of RCV 225 system. Reprinted with permission from Miller and Koblich, 1984, *Living and Working in the Sea*. (Full citing in references).

Figure 7-18. Photograph of Scorpio built by Ametek Straza. Reprinted with permission from Shatto, 1991, *Handbook of Coastal and Ocean Engineering*, Ch. 6. (Full citing in references).(Shatto 1991).

Figure 7-19. Photograph of Triton built by Perry. Reprinted with permission from Shatto, 1991, *Handbook of Coastal and Ocean Engineering*, Ch. 6. (Full citing in references).

Classification and Applications

Remotely operated vehicles are generally classified (MTS 1984) into six types and the primary difference is the means of propulsion. Table 7-5 lists the six categories of remotely operated vehicles as tethered, towed, bottom reliant, structure reliant, autonomous, and hybrid. Some typical characteristics and applications are also shown. The ROV is usually near neutral buoyant and thus can hover for inspection or observation purposes using video, sonar, or still camera. It can also push or pull, lift, and connect or disconnect systems using only small forces on the order of 444.8 N (100 lb) to 889.6 N (200 lb). Vehicle shapes are typically block shape when forward speeds greater than 3.71 kph (2 kts) are not needed, and torpedo shaped when greater speeds are desirable. The response of the vehicle is largely determined by the comparison of the available thrust to the weight of the vehicle. This thrust to weight ratio defines the quickness of response or how fast the vehicle can accelerate to its steady state speed.

Table 7-4. Summary information for selected remotely operated vehicles.

Name	Date Built	Application	Depth ft	Weight lb	Cage	Major Equipment	Builder/User
Mobot	1960	wellhead maintenance	na	na	none		Hughes/Shell
Unumo	1962	offshore	na	na	none	Two manipulators	Hughes/Shell
Curv I, II, & III	1965 1967 1971	military			none	manipulator, sonar, camera	US Navy
RCV-225	1978	offshore	6600	180	yes	video camera	Hydroproducts
TROV	1976	offshore	3000	1600	none	video camera, manipulator	ISE
TREC	1977	offshore	1200	400	none	video camera, echo sounder	ISE
Hydra	1983	offshore					ISE/Oceaneering
Deep Drone	1976	military	2000	1600	none	video camera, sonar	Ametek Straza/USN
Scarab II	1979	cable inspection/repair	6000	5000	none	video camera, manipulator, sonar	Ametek Straza/
Scorpio	1977	offshore	3000	1500	none	video camera, sonar, manipulator	Ametek Straza/ Solus Ocean Systems
Recon V	1977	offshore	1200	848	yes	video camera, manipulator	Perry
Pioneer	1983	offshore					Subsea International
Triton XL	1994	offshore	4920	7700	none	manipulators, video camera, hydraulic tool interface, sonar	Perry Tritech
Deep Tow	1985	scientific research	20,000	2000	none	video camera, magnetometer, plankton sampler	Scripps Institute of Oceanography
Flexjet	1992	Pipeline trencher	1640	26,400	none	jetting system, pipeline navigator system,	Perry Tritech

Table 7-5. Types and applications for remotely operated vehicles.

Classification	Characteristics	Applications
Tethered, free swimming	Large majority of ROV's; Closed circuit video; Cable connected to surface; Positively buoyant	Observation; Diver assistance
Towed	Propelled by surface vessel	Broad area reconnaissance; Mineral and geological surveys
Bottom reliant	Power and control from surface vessel; Closed circuit video; Propulsion by wheels or tracks	Pipeline/cable trenching; Bottom excavation and observation; Geotechnical investigations
Structurally reliant	Power and control from surface; Closed circuit video; Propulsion from wheels or tracks in contact with structure	Subsea production system maintenance and inspection; Tank soundings; Hull cleaning
Untethered (Autonomous)	Self powered and 3-D maneuverability; No physical connection to surface	Undersea surveillance
Hybrid	Combination of above types	Pipe trenching and pipe laying

Vehicle Velocity and Thrust Calculations

An estimate of the speed the vehicle can move through the water or the amount of current in which the vehicle can hold its position is made by equating the available thrust in a particular direction to the drag force given by

$$F_d = \rho C_d A_d \frac{V_c^2}{2} \tag{7-28}$$

where ρ is the mass density of water, C_d is the drag coefficient that is a function of the Reynolds number and may vary from 0.2 to 2.0, A_d is the projected area normal to the flow direction, and V_c is the current velocity or speed of the vehicle.

The available thrust for a propeller is usually determined from the standard bollard pull tests and this information is sometimes available from manufacturers. It can also be estimated from the available power by assuming that a thruster develops about 111.2 N (25 lb) to 133.4 N (30 lb) of thrust per horsepower (Shatto 1991). Also, the available thrust can be theoretically determined by computing the change in momentum of the water that is accelerated through the thruster using

$$F_T = \rho A_T V_T^2 \tag{7-29}$$

where F_T is thruster force, ρ is the mass density of water and A_T is the cross-sectional area of the thruster's stream of water being accelerated from zero velocity to the average velocity (V_T) through the thruster. The velocity of the water through the thruster is determined from

$$V_T = \frac{NPE}{60} \tag{7-30}$$

where N is the revolutions per minute (RPM) of the propeller, E is the propeller efficiency, and P is the propeller pitch. Thrusters usually have a large hub, and its cross sectional area is subtracted when calculating A_T. Thus,

$$A_T = \frac{\pi}{4}\left(d_p^2 - d_h^2\right) \tag{7-31}$$

where d_p and d_h are the propeller and hub diameters. The thruster intake is usually not in still water because the vehicle is moving through the water or holding against a current. When the thruster has a fixed pitch and a fixed maximum speed, the thruster force (F_T) is reduced approximately by the ratio of V_C/V_T and is expressed as

$$F_T = \rho A_T V_T^2 \left[1 - \frac{V_C}{V_T}\right] \tag{7-32}$$

Calculation of vehicle performance requires knowledge of the vehicle drag coefficient C_d and the propeller efficiency E. These parameters are usually determined from empirical results. The drag force of the tether should also be considered, and the drag on a long tether can exceed that of the vehicle.

The vertical thrust developed by the vehicle affects the response in the vertical direction, and it also determines the amount of weight that can be lifted or carried by the vehicle. The amount of weight the vehicle can lift is typically called the dead lift weight. The amount of thrust available for vertical acceleration is determined by subtracting the dead lift weight. Vertical thrust is also used to control the vehicle attitude and to adjust the pitch angle of the vehicle when the manipulator lifts or drops a weight. This is only needed when the loads to be lifted cause excessive pitch or roll. The ROV usually has a relatively large BG that makes them very stable. For a typical work vehicle, it might have a BG of 27.9 cm (11 in), a weight and buoyancy of 3500 lb and a stiffness of 56 ft-lb/degree at zero pitch and roll. The angle of pitch when picking up a load can be determined by

$$\theta = \sin^{-1}\left[\frac{HW_L}{\left(\frac{W_a}{BG}\right)}\right] \tag{7-33}$$

where H is the horizontal moment arm from the load centerline to the ROV's vertical center of thrust, W_L is in-water weight of load, W_a is the in-air weight or buoyancy of the ROV, and BG is the distance between center of buoyancy and center of gravity. For the vehicle described above, the ROV could lift a weight of 100 lb at a radius of 2.1 m (7 ft) resulting in a pitch angle of 12.6^o.

ONE ATMOSPHERE DIVING SYSTEMS

One atmosphere diving systems were developed to overcome the difficulties associated with decompression sickness and high pressure nervous syndrome. The diver in these systems remains at one atmosphere pressure and can work for relatively long periods without having to spend many hours or days in decompression. The JIM suit (Figure 7-20) was initially developed in England and has undergone design improvements to attain greater depth and flexibility capabilities. Greater bottom times, security, and protection from cold are additional capabilities provided by the one atmosphere diving systems. In 1976, the system was used for a dive in 275.8 m (905 ft) of sea water in the Canadian Arctic through a 4.9 m (16 ft) diameter hole in the ice, and the diver worked for a 6 hr period. Using saturation diving for the same operation would have required more than eight days of decompression. Newer JIM systems have been constructed with magnesium alloy and carbon steel fiber. The record dive for the JIM system is 548 m (1780 ft) in the Gulf of Mexico. An adaptation of this one atmosphere diving system that allows a diver to operate at midwater locations is called the WASP (Figure 7-20) where the legs have been removed and movement in the water is accomplished by small propulsors similar to those found on small ROVs. Advances in manipulator technology are expected to further improve the performance of divers working in these one atmosphere systems.

UNDERWATER HABITATS AND HYPERBARIC CHAMBERS

The use of underwater habitats began in the early 1960's and reached its peak in the late 1960's. The use of these habitats diminished during the 1970's, and in the 1990's there are very few in operation. NOAA's National Undersea Research Program (NURP) currently operates the habitat Aquarius near the Florida Keys. The habitat La Chalupa, now called the Jules Verne Lodge, is operated by private enterprise near Key Largo, Florida. An excellent description of nearly sixty different habitats is described by Miller and Koblick (1984). A historical listing and basic characteristics of selected underwater habitats are tabulated in Table 7-6.

The early habitats were designed to determine their engineering feasibility and to evaluate man's ability to survive undersea living. After the success of the initial habitats such as Man in the Sea, Conshelf, and SeaLab, the habitat construction increased, and they were used for observation stations, seafloor laboratories for scientists, and operational bases for working divers.

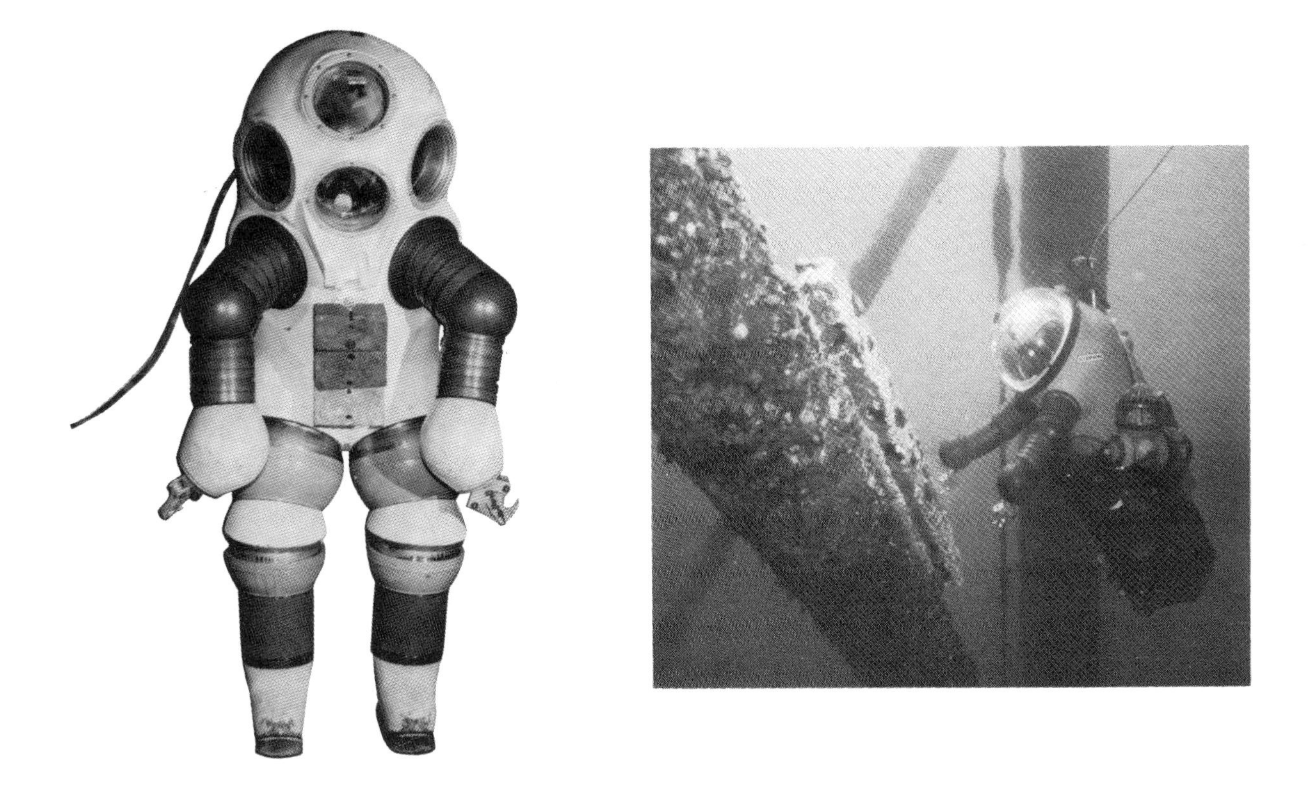

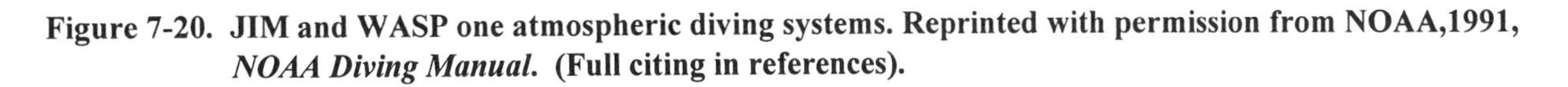

Figure 7-20. JIM and WASP one atmospheric diving systems. Reprinted with permission from NOAA,1991, *NOAA Diving Manual.* (Full citing in references).

Underwater habitats are designed to allow diver scientists, engineers, and technicians easy access to the ocean environment that enables them to make observations, conduct experiments, and test new diving systems. Since the habitats are normally open to the ambient pressure at the habitat depth, the aquanauts, or divers, are saturated at the depth with the breathing gas mixture designed for the habitat or breathing apparatus used for the habitat excursions. Consequently, decompression is required at the completion of the habitat stay unless the depth is less than 10 m (33 ft) of sea water.

Habitats are normally pressure vessels that can withstand external and internal pressures during installation and decompression. The shape of the habitat is usually a cylinder or sphere because of their ability to best resist the pressure with minimum thickness of material and providing the most volume for space. Combinations of spheres, cylinders, hemispheres are often used to increase the space available for the divers.

Table 7-6. Characteristics of selected underwater habitats (Miller and Koblich, 1984).

Name	Country	Date	Location	Depth (m)	Crew	Duration (Days)	Size (m)	Weight (tons)	Breathing Gas	Surface Support	Mobility	Deco mpre ssion (hr)	Comment
Aegir	USA	1969-71	Hawaii	24-157	4-6	14	2 cyl 27x4.6 plus 3m sph	200	N2/O2 He/O2	Ship	Towable		Internal control of ascent/ descent
Chernomor-1	USSR	1968	Black Sea	5-14	4-5	4-6	L=7.9 D=2.9	62 Displ	Air	Ship	Towable		
Conshelf I	France	1962	Med. Sea	100	2	7	L=5.2 D=2.4		Air	Ship Shore	Readily movable	3	
Conshelf-III	France	1965	Med. Sea	100	6	22	D=5.5 sphere	130+	2.5% O2 97.5% He	Ship	Towable	84	
Edalhab	USA	1968 1972	Alton's Bay, NH Miami, FL	12.2 13.7	3	3-5	L=3.6 D=2.4	14 25	Air	Shore ship	Readily movable	19	
Helgoland II	Germany	1971 1977	North Sea USA	22-31	4	4-14	L=13.8 W=6.0	102	N_2/O_2	Buoy	Towable	Varied	
Hydrolab	USA	66-70 70-74 75-84	Florida Bahamas Virgin I.	12-18	2-4	7-14	L=4.9 W=2.4	40	Air	Buoy	Towable	13-20	Most utilized habitat
La Chalupa	USA	1971-74	Puerto Rico	15-30	4-5	14	2 cyl 2.4x6.01 rm 3x6	150	N_2/O_2	Buoy	Readily movable	36-48	
Lakelab	USA	1972	Grand Traverse Bay, MI	15.2	2	2	D=3.0 H=2.1	25 Bal.	Air	Shore	Readily movable	NA	
Man-in-Sea I	USA	1962	Med. Sea	61	1	1	L=3.2 D=0.9	2.1	3% O_2 97%He	Ship	Readily movable	65.5	First open sea saturation
Meduza II	Poland	1968	Gdansk, Baltic Sea	26	3	7	L=3.6 W=2.2 H=2.1	3	37%O_2 63%N_2	Shore	Readily movable	22	
Portalab	USA	1972	Narragansett Bay, RI	11.3	2	<1	L=2.4 W=1.8 H=2.1	7.2 Bal	Air	Shore	Readily movable	0	
Sealab-II	USA	1969	LaJolla, CA	62.5	10	15-30	L=17.5 D=3.6 H=3.6	200	4%O_2 25%N_2 17%He	Ship	Movable	30	
SPID (Man-in-Sea II)	USA	1964 1974	Bahamas Arctic	131.7 4.3	2	1	L=2.4 W=1.2	Inflatable	3.6%O_2 5.6%N_2 90.8% He	Ship shore	Readily movable	92 0	Inflatable habitat
Sub-Igloo	Canada	1972-75	Cornwallis I.	12.2	2-4	1	D=2.5 sphere	8 bal.	Air	Shore	Readily movable	NA	Under ice
Suny-lab	USA	1976	New York	12.2	2-3	1	1.5	6	Air	Ship	Readily movable		
Tektite I-II	USA	1969-70	Virgin I.	13.1	4-5	6-59	D=3.8 H=5.5	79	92%N_2 8%O_2	Ship shore	Fixed	19.5	Longest open-sea saturation

When designing habitats, various technical, logistical, and habitability considerations are investigated. Some of these considerations include comfort, simplicity, and functionality. Desired features for an underwater habitat were determined in a University of New Hampshire study and are tabulated in Table 7-7. The desirable features include an overall size of 2.4 m (8 ft) (diameter, height, width) and 11.6 m (38 ft) in length with separate wet room and living room. The interior should have temperature and humidity control, and the divers need communications between themselves and the habitat. External survival shelter and on-bottom and surface decompression capability are desirable.

More than 72 underwater habitats have been constructed around the world since 1962. These habitats range from simple shelters to sophisticated large facilities used for extended stays on the ocean bottom. The most used habitats include Tektite, Hydrolab, and La Chalupa in the United States, Chernomor in Russia, and Helgoland in West Germany. Several universities in the United States also ventured into underwater habitat design and operation. These habitats included Edalhab at the University of New Hampshire, Portalab at the University of Rhode Island, and Suny-Lab at the State University of New York.

The Hydrolab underwater habitat is shown in Figure 7-21 and is briefly described herein because of its simplicity, representative design, relatively low operating cost, and its significant use over a nearly 20 year period. It was decommissioned by NOAA in 1985 and can be observed at the Smithsonian Museum of Natural History in Washington, D.C. The habitat consisted of a 2.4 m (8 ft) diameter and 3.7 m (16 ft) long steel cylinder with a hemispherical viewing port at one end. It was supported by four short legs providing a 0.91 m (3 ft) clearance above the concrete support base. The habitat could be towed for short distances and was submerged by venting ballast tanks on the side of the main structure and several tanks in its base structure. Hydrolab was designed for depths 30.5 m (100 ft) or less. An entrance hatch was located at the bottom and near one end of the cylinder, and it also was used as a lock when there was a pressure difference between the inside and ambient water pressure.

Table 7-7. Desirable design features for underwater habitats (NOAA 1991).

GENERAL	SEPARATE WET ROOM	LIVING ROOM
Size: 8 ft (D, H, W) by 38 ft (L) 2.4 m by 11.6 m	Large entry trunk	Bunks
Hemispheric windows	Wet suit rack	Microwave oven
Temperature and humidity control	Hot shower	Food freezer and refrigerator
Separate double chambers	Hookah and built-in breathing system	Water heater
Bottom and surface decompression capability	Scuba charging	Toilet
Suitable entry height off the bottom	Wet lab workbench	Individual desk and storage
Submersible decompression chamber for emergency escape	Freezer for samples	Laboratory workbench
External survival shelter	Clothes dryer	Trash compactor
External lights at trunk and viewports	Diving equipment storage	Library
External breathing gas cylinder storage and charging	Rebreathers	Tapes, TV, radio, CD player
Habitat to diver and diver to diver communications		Emergency breathing system
Adjustable support legs, mobility		Computer terminal
External or protected internal chemical hood		

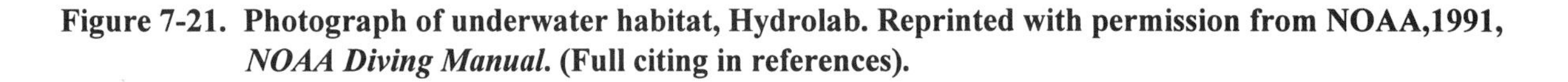

Figure 7-21. Photograph of underwater habitat, Hydrolab. Reprinted with permission from NOAA,1991, *NOAA Diving Manual.* (Full citing in references).

The habitat one room interior was furnished with three bunk beds, dehumidifier, air conditioner, folding chairs, sink, electric hot plate, and table surface. The close quarters are illustrated in Figure 7-22. A shower hose was located in the entrance trunk, and a 7 m (23 ft) long life support surface vessel was moored to the top of the habitat. This boat shaped vessel supplies electrical power, high and low pressure air and water, and communications through an umbilical. The habitat was used for many years by NOAA's Manned Undersea Research Program and based in St Croix, Virgin Islands.

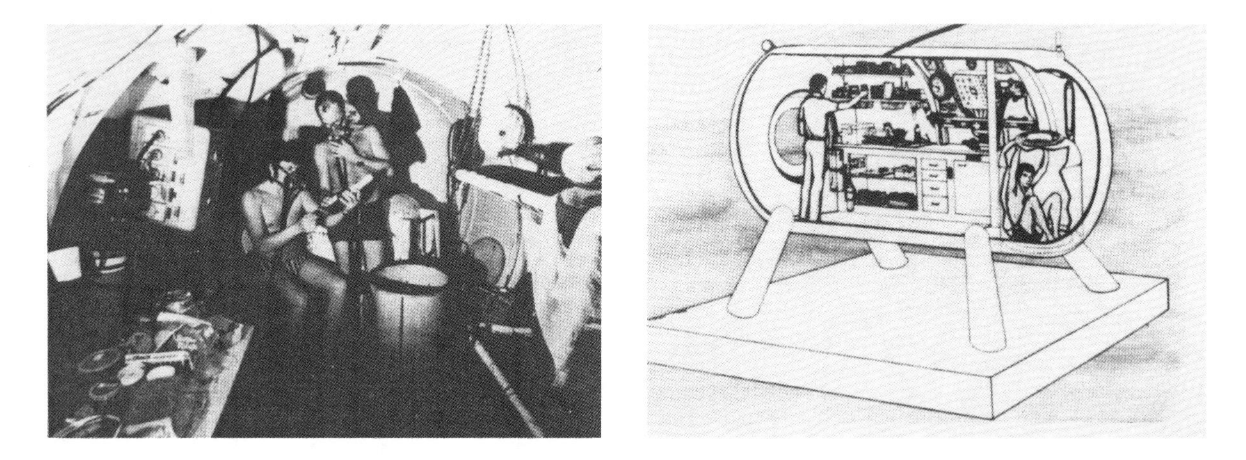

Figure 7-22. Interior view of Hydrolab. Reprinted with permission from Miller and Koblich, 1984, *Living and Working in the Sea.* (Full citing in references).

The habitat, Tektite, shown in Figure 7-23 was a four person habitat consisting of two vertical cylinders connected by a horizontal cylinder and mounted on a rectangular support base. The General Electric company sponsored the construction and operation of Tektite. The vertical cylinders were divided into two compartments containing an equipment room, wet room, living quarters, and control room (dry scientific laboratory). The living quarters contained four bunk beds, small galley, storage, and entertainment equipment. Environmental control, food freezer and toilet facilities were contained in the equipment room. Umbilicals from the shore provided air, water, electricity, and communications to the habitat. Several hemispherical windows and a cupola at the top of one cylinder provided scientists a view of the surrounding area. A personnel transfer capsule (PTC) was lowered to the bottom for decompression, and the divers entered the PTC and were transported to a surface where they entered a deck decompression chamber. Tektite I and II missions were conducted in 1969-70 in the Virgin Islands and were the worlds longest open sea saturation operation (59 days).

A new habitat, Aquarius, was constructed by NOAA for their Manned Undersea Research Program in the late 1980's. It is shown in Figure 7-24 and has been used at various sites throughout the Caribbean Sea. Aquarius was designed to operate at depths of 36.6 m (120 ft) or less and accommodate six scientists.

The use of underwater habitats has also been considered at subsea bases for the offshore oil wellheads. The concept was to bring divers down by submersible to the underwater base that was kept at one atmosphere pressure inside. The submersible is mated to the hatch on the habitat and the divers are transferred to the habitat to conduct their maintenance work. A similar scenario is used to rescue personnel from large submarines.

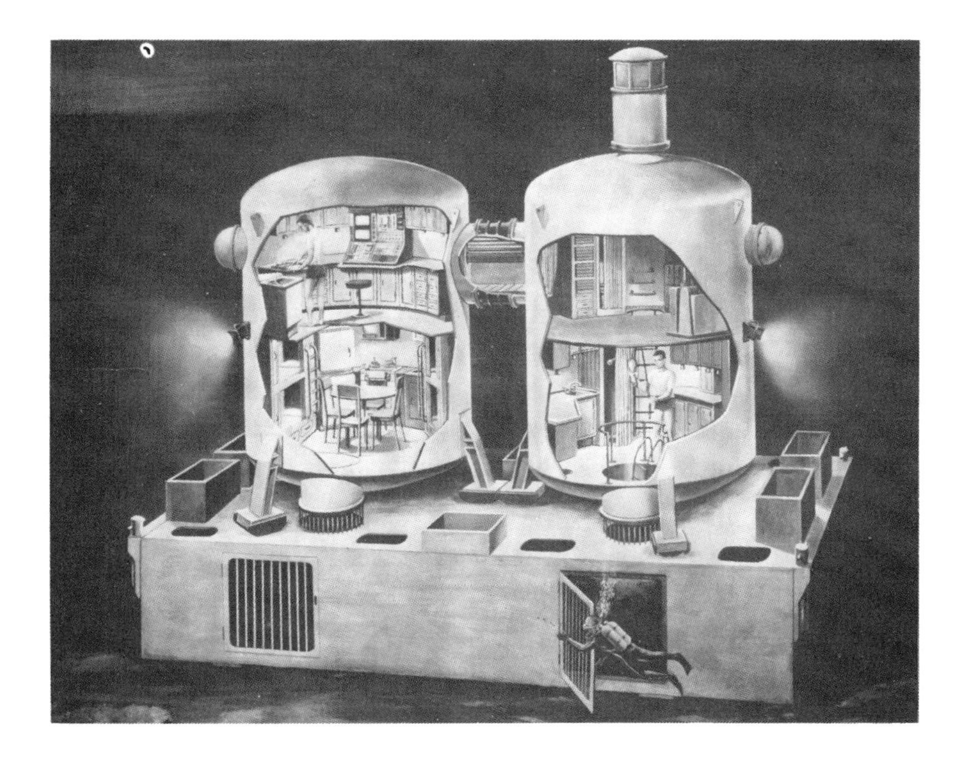

Figure 7-23. Conceptual view of the Tektite habitat. Reprinted with permission from NOAA,1991, *NOAA Diving Manual.* (Full citing in references).

Figure 7-24. NOAA underwater habitat, Aquarius. Reprinted with permission from NOAA,1991, *NOAA Diving Manual.* (Full citing in references).

The use of habitats for recreational facilities such as underwater hotels and restaurants has received some interest. Currently, there is a small habitat in the Florida Keys that can accommodate divers for about one week. It is located in approximately 9.1 m (30 ft) of water off Key Largo. Ocean engineering students at Texas A&M University have conducted preliminary designs for an underwater hotel and restaurant, and their Underwater Wotel received some interest but was never constructed. Several conferences have been devoted to the concepts of underwater cities. These areas for underwater facilities may expand in the years ahead to accommodate expanding populations in land starved areas and to provide recreational facilities.

Hyperbaric chambers are facilities that are normally used to treat divers for decompression sickness, conduct controlled diving experiments, and provide a means for divers to decompress after saturation diving. These chambers are found at medical hospitals, diving surface support vessels, US Navy submarine rescue vessels, and operational diving locations where decompression is a possibility. Most hyperbaric chambers are constructed with a double lock capability that allows medical doctors and technicians to enter the chamber to treat and observe patients and then leave. Most hyperbaric chambers are used primarily to treat non-diving medical diseases.

ENERGY SYSTEMS FOR UNDERWATER APPLICATIONS

The power sources used for underwater applications include diesel generators, transmission lines from local commercial power utilities, closed cycle diesel systems, electrical storage batteries, fuel cells and small nuclear plants. Underwater habitats have commonly used diesel powered generators located on board a mother ship, nearby buoy, or at a shore station with emergency facilities available when needed. In 1969, the habitat Hydrolab used a fuel cell for its total electrical power (Miller and Koblick 1984). The fuel cell was called Powercel 8 and built by Pratt and Whitney. This fuel cell provided power for interior and exterior lighting, CO_2 scrubbers, instruments, data recording, communication equipment, compressor, and a motor generator. The powercel was located just outside one of the viewing ports so that the divers could monitor the generator status. After 1970, the Hydrolab received its power from a surface buoy that was the shape of a small 7 m (23 ft) boat hull and was moored above the habitat with cables and hoses supplying air, water, electricity, and communications. A 12.3 kw (16.5 hp) diesel engine powering a 7.5 kw generator was located in the surface buoy along with a low and high pressure compressor. Storage facilities on the buoy included 946.3 l (250 gal) of freshwater and 946.3 l (250 gal) of diesel fuel. These facilities provided approximately a one month supply before refueling. The German habitat, Helgoland, used a similar buoy support system. Electrical power requirements for underwater habitats range from 10 to 100 kw (UNH 1972).

For remotely operated vehicles and submersibles, energy sources are typically batteries, fuel cells, and closed cycle engines. Batteries used include the lead acid, nickel cadmium, and silver zinc type, and their energy densities are tabulated in Table 7-8. Lead acid batteries are heavy and significantly increase the weight, and they produce oxygen and hydrogen gases during the charging process that must be carefully vented. The silver zinc batteries have been shown to be very attractive for submersibles because of their high energy density. Fuel cells have been tried with some success, but they are expensive and have some safety concerns. Closed cycle engines have also been tried, but the combustion by-products are a danger.

Table 7-8. Energy densities for submersible batteries (Allmendinger 1990).

Battery Type	Theoretical watt-hours/ lb	Actual	
		watt-hours/lb	watt-hours/in^3
Lead acid	80-115	10-20	0.8-1.2
Nickel cadmium	105	12-19	0.7-1.25
Silver zinc	205	40-60	2.4-3.8

Battery energy sources are placed either externally or internally. The external storage requires the battery container be pressure compensated. Oil compensated batteries have been used for many submersibles, but they require maintenance and are less reliable than the dry storage containers. However, a majority of the submersibles have used properly designed oil compensated batteries for electrical energy. The internal containers require a pressure vessel, and the greatest danger is associated with handling the gases that occur during the charging process. Dry cells or rechargeable batteries are commonly used for emergency energy sources for communication equipment, carbon dioxide scrubbers, and emergency lighting.

REFERENCES

Allmendinger, E. E., Editor. *Submersible Vehicle Systems Design.* Jersey City: Society of Naval Architects and Marine Engineers, 1990. Figures 7-12, 7-13, and 7-14 reprinted with permission: "Source: Allmendinger, *Submersible Vehicle Systems Design.* Copyright Society of Naval Architects and Marine Engineers, 1990."

American National Steel Institute/American Society of Mechanical Engineers (ANSI/ASME). "Safety Standard for Pressure Vessels for Human Occupancy." New York: American Society of Mechanical Engineers, 1977.

American Society of Mechanical Engineers (ASME). "Boiler and Pressure Vessel Code." New York: American Society of Mechanical Engineers, 1980.

Burcher, R., and L. Rydill. *Concepts in Submarine Design.* Cambridge: Cambridge University Press, 1994.

Busby, R. *Manned Submersibles*. Alexandria: Office of the Oceanographer of the Navy. 1976.

Chuse, R., and S. M. Eber. *Pressure Vessels: The ASME Code Simplified*, Sixth Edition. New York: McGraw-Hill, 1984.

Friedman, N. *U. S. Submarines Since 1945: An Illustrated Design History*. Annapolis: Naval Institute Press, 1994.

Hervey, J. B. "Submarines," Vol. 7, *Brassey's Sea Power: Naval Vessels, Weapons Systems and Technology Series*. London: Brassey's (UK), 1994.

Marine Technology Society (MTS). *Operational Guidelines for Remotely Operated Vehicles*, Washington: Marine Technology Society, 1984.

Megyesy, E. F. *Pressure Vessel Handbook.* Tulsa: Pressure Vessel Handbook Publishing Co., 1977.

Miles, S., and D. E. Mackay, *Underwater Medicine*, Fourth Edition. Philadelphia: Lippincott Co., 1976.

Miller, D. *Submarines of the World.* New York: Orion Books, 1991.

Miller, J. W., and I. G. Koblick. *Living and Working in the Sea.* New York: Van Nostrand Reinhold Company, 1984. Figures 7-17 and 7-22 reprinted with permission: "Source: Miller and Koblick, *Living and Working in the Sea,* CopyrightVan Nostrand Reinhold Company, 1984."

National Oceanic and Atmospheric Administration (NOAA). *NOAA Diving Manual.* Washington: Government Printing Office, 1991. Figures 7-2, 7-20, 7-21, 7-23, and 7-24 reprinted with permission: "Source: *NOAA Diving Manual.* Copyright, Government Printing Office, 1991."

Nuckols, M. L., W. C. Tucker, and A. J. Sarich. *Life Support Systems: Diving and Hyperbaric Applications*. Needham Heights: Simon and Schuster Custom Publishing, 1996.

Randall, R. E. "Introduction to Diving and Life Support Technology." Unpublished Class Notes, Ocean Engineering Program, Texas A&M University, College Station, 1997.

Roark, R. J., and W. C. Young. *Formulas for Stress and Strain*, Fifth Edition. New York: McGraw-Hill, 1982.

Schilling, C. W. *The Human Machine*. Annapolis: US Naval Institute, 1965. Figure 7-1 reprinted with permission: "Source: Schilling, *The Human Machine*. Copyright US Naval Institute, 1965."

Schilling, C. W., M. F. Werts and N. R. Schandelmeier, Editors. *The Underwater Handbook: A Guide to Physiology and Performance for the Engineer*. New York: Plenum Press, 1976.

Sharp, F., Editor. *Janes Fighting Ships*. London: Butler and Tanner Limited, 1994.

Shatto, H. L. "Remotely Operated Vehicles." *Handbook of Coastal and Ocean Engineering*, J. B. Herbich, Editor, Ch. 6, Vol. 2. Houston: Gulf Publishing Co., 1991. Figures 7-16, 7-18, and 7-19 with permission: "Source: Shatto, *Handbook of Coastal and Ocean Engineering*, Ch. 6, Vol. 2, Copyright Gulf Publishing Co., 1991."

Sweeney, J. B. *A Pictorial History of Oceanographic Submersibles.* New York: Crown Publishers, 1970. Figure 7-11 reprinted with permission: "Source: Sweeney, *A Pictorial History of Oceanographic Submersibles,* Copyright Crown Publishers, 1970."

Terzibaschitsch, S. *Submarines of the US Navy.* New York: Arms and Armour Press, 1991.

Tucker, W. *Underwater Diving Calculations.* Centreville: Cornell Maritime, 1980.

University of New Hampshire (UNH). The Impact of the Requirements of the United States Scientific Diving Community on the Systems' Design, Operation and Management of Underwater Manned Platforms." Tech. Rep. No. 111, Manned Undersea Science and Technology Office, National Oceanic and Atmospheric Administration, Rockville, 1972.

US Navy (USN). *US Navy Diving Gas Manual.* Washington: Government Printing Office, 1971.

US Navy (USN). *US Navy Diving Manual*, Vol. 1, NAVSEA 0994-LP-001-0910, Revision 2. Washington: Government Printing Office, 1988.

PROBLEMS

7-1. A dive is planned for two divers at a depth of 85 ft of seawater. The SCUBA tanks to be used are rated at 72 standard cubic feet at 2514.7 psia. How many bottles are required for a one hour dive. Assume the perfect gas law applies and that each diver requires 1.5 cubic feet per minute of air at depth. Assume the temperature is 60 degrees F and that one atmosphere equals 14.7 psia.

7-2. Two divers are staying in a 750 cubic feet chamber for one day. The partial pressure of carbon dioxide is to be maintained at 0.005 ata, and the temperature in the chamber is 75 degrees F ($T = 75\ ^oF$). Each diver has a carbon dioxide production rate of 0.8 standard liters per minute. The depth of the chamber is 60 ft. Determine the air flow rate to maintain the partial pressure of carbon dioxide at 0.005 ata.

7-3. A diver is inside a chamber that has been pressurized with normal air to depth of 150 ft of sea water where the temperature is 75 ^oF. Determine the partial pressure of oxygen in the chamber. If the chamber is pressurized with normal air (O_2 is 21% by volume), determine the pressure at which oxygen poisoning will occur. Assume oxygen poisoning occurs when the partial pressure of oxygen is 1.6 ata.

7-4. A surface supplied breathing apparatus is to supply air to a diver at a depth of 120 ft of sea water. Assuming the breathing air has no initial concentration of carbon dioxide, estimate the air flow rate in standard cubic feet per minute (SCFM).

7-5. If the allowable stress for the steel used in a 1.0 ft internal diameter pressure vessel is 33,500 psi, evaluate the thickness of the cylinder walls for an internal pressure of 3000 psi.

7-6. Three divers are doing light work in a 120 ft^3 habitat that is located at a depth of 50 ft. Determine the partial pressure of carbon dioxide (P_{co2}) after a period of three hours. Assume that carbon dioxide production equals oxygen consumption. Will the divers experience any physiological effects at the end of the three hours?

7-7. Two divers are resting quietly in a diver lock-out chamber when the ventilation system fails. How long will it take for the divers to immediately experience minor perceptive changes due to excess carbon dioxide?

7-8. A habitat is designed to withstand an external pressure equal to 80 ft of sea water with a safety factor of 4. The inside diameter of the spherical habitat is 8 ft, and the thickness of steel is 0.25 inches. Determine the allowable stress for the steel using a safety factor of 4.

7-9. A dive is planned for three divers at a depth of 66 ft of seawater. The SCUBA tanks to be used are rated at 80 standard cubic feet at 3014.7 psia and 70 ^{o}F. How many bottles are required for a one hour dive. Assume each diver is doing moderate work during the entire dive and the temperature of the water is constant at 50 degrees F. The tank reserve pressure is 500 psig.

7-10. If divers are going to stay in a habitat for 4 days, what should be the maximum allowable carbon dioxide concentration? What is the permissible range of oxygen concentrations and partial pressures for a depth of 60 ft of sea water?

CHAPTER 8: UNDERWATER ACOUSTICS

INTRODUCTION

Like many physical science areas, underwater acoustics had its beginnings many centuries ago. In the 15th century, Leonardo da Vinci found that when a ship was stopped and one opening of a long tube was placed in the water and the other end was placed next to one's ear, other ships could be heard from great distances (MacCurdy 1942). Sir Isaac Newton provided the first mathematical analysis of sound and related the propagation of sound in fluids to physical quantities such as density and elasticity in his study of theory of sound in 1687. Colladon and Sturm measured the speed of sound in water in 1827 using a light flash and the sounding of an underwater bell in Lake Geneva, Switzerland to obtain a sound speed of 4707 ft/s when the lake water temperature was 8 ^{o}C. In 1840, Joule quantified and described the magnetostriction effect which is the change in the dimensions of a magnetic material caused by the presence of a magnetic field. Rayleigh published the book *Theory of Sound* in 1877, and this publication provided the fundamental theories for describing the generation, propagation and reception of sound. Curie later discovered in 1880 the piezoelectric effect in which a mechanical strain applied to certain crystal types (e.g. quartz) causes an electrical charge on the crystal faces (Albers 1965). The piezoelectric and magnetostrictive effects are used in the production of transducers to generate and receive underwater sound.

Fessenden developed the first electrically driven underwater sound source (known as the Fessenden oscillator) in 1914 (Urick 1983), and it operated as an underwater transmitter and receiver in the frequency range of 500-1000 Hz, where 1 Hz is a cycle per second. Fessenden also developed the first commercial application of an underwater acoustic device for ships to determine their range to a lightship by simultaneously sounding a foghorn and underwater bell that were both located on the lightship. The difference in the time of arrival of the sound signal from the foghorn (airborne) and underwater bell were used to evaluate the distance from shore.

After the start of World War I (WWI) in 1915 acoustical sensing devices were acknowledged as the only practical devices for detecting submerged submarines used by the Germans to attack shipping. The United States developed the SC tube, Figure 8-1, that was a biaural air tube listening device that used stethoscope ear pieces. The SC tube was effective around 500 Hz and was successfully used to locate both submarines and surface ships. Subsequent devices, called MB, MV and U-3 were developed to improve sensitivity, allow higher ship speeds, and reduce effects of self-noise (Burdic 1991). During WWI(1914-18) a system for underwater echo-ranging was investigated under the acronym ASDIC (Anti Submarine Division Investigation Committee). The principle of echo-ranging (echo location) was that a pulse of sound was transmitted into the water, and any reflection (echo) from a target (e.g. submarine) was received by a hydrophone, which is the underwater equivalent of a microphone. The received signal was heard on headphones, and the time delay between transmission and reception was used as a measure of the range to the submarine. The war ended before underwater echo ranging could make any contribution.

Following WWI, advances in underwater acoustics continued with the development of depth sounders, echo ranging transducers, and the bathythermograph as well as a better understanding of the physics of underwater sound propagation. When a sound pulse is directed vertically downwards, an echo is received from the sea bottom, and the depth of water is subsequently determined (depth sounder). This particular instrument has continually been used as a navigational instrument. In the U. S., Hayes pioneered the field of passive sonar arrays which were listening devices used to detect sounds in the ocean such as submarines. The term SONAR was coined which is an acronym for sound navigation and ranging. Spilhaus invented the bathythermograph in 1937, and it measures the temperature versus depth of water. This instrument was installed on submarines to measure the temperature profile of the ocean to assist in the determination of characteristics of sound propagation and sonar detection. Ultrasonic frequencies were used to obtain greater directionality with hydrophones of modest size and transducer arrays were housed inside streamlined domes to minimize interfering noise effects. The standard SONAR equipment for surface ships was the QC equipment, and JP sonars were installed on submarines. Operators on surface ships searched in bearing (different directions) and listened for submarine echo returns with headphones. Similarly, sonar operators on submarines searched in bearing while listening for radiated noise of surface ships and other submarines. A great amount of knowledge was attained on the propagation of sound in the sea and published in a series of reports that are collectively titled Physics of Sound in the Sea (NDRC 1969).

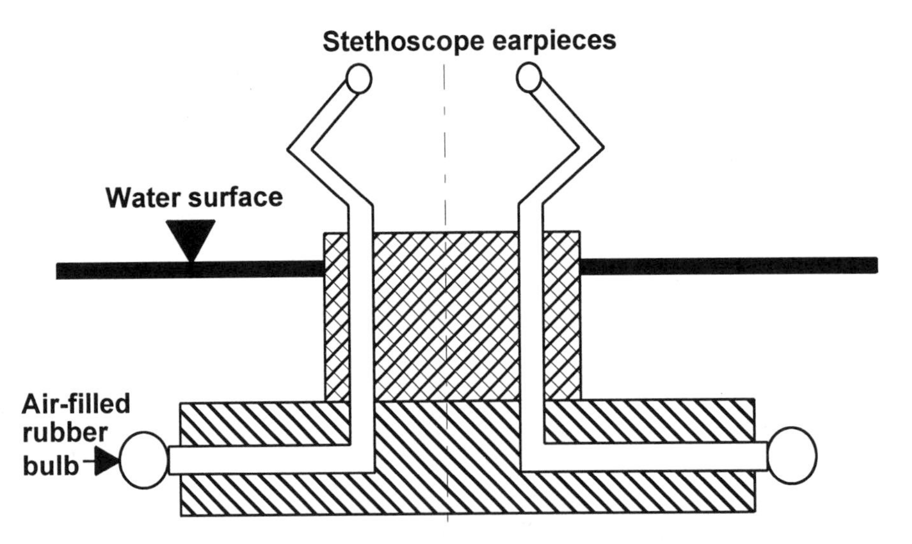

Figure 8-1. The SC binaural air tube listening equipment used during World War I.

After WWII, the advances in electronic equipment, signal processing techniques, computer equipment, understanding of ambient noise, underwater sound propagation models, and causes of absorption fueled the further development of underwater acoustic equipment. Requirements in private industry have also increased the use of underwater sound devices. For example, the search for natural resources beneath the seafloor has advanced the development of seismic exploration and subbottom profiling equipment and techniques. The development of remotely operated vehicles uses sonars for searching, tracking, communications, and navigation. Seafloor mapping and searching have propelled the use of side scan sonars. Oceanographic instrumentation has utilized underwater sound principles for the development of acoustic release mechanisms, pingers and transponders, inverted echo sounders, and acoustic devices for measuring ocean currents now called acoustic Doppler current profilers (ADCP). Acoustic velocimeters are also finding use in laboratory studies and in measuring flow velocities in pipes. Oceanographers are using acoustic tomography to examine ocean seafloor sediments.

Current underwater acoustic applications include fishing aids, hydrographic surveying, oceanographic and ocean engineering instrumentation, geophysical research, underwater communications, navigation, underwater search, and sea bottom surveying. Four types of underwater acoustic systems and examples are described in Table 8-1. Natural sonar systems are found in marine mammals such as dolphins, porpoise, and whales that use sound for their navigation. Bats also use sound for navigation in air.

One reason acoustic waves are used in underwater detection and communications is that electromagnetic waves hardly propagate in water except a very large wavelengths. At these large wavelengths, the waves are not useful. There are many excellent references that treat the subject of underwater acoustics in greater depth, and the reader is referred to these texts for additional reading. Some of these references include those authored by Albers (1965), Burdic (1991), Clay and Medwin (1977), Coates (1989), Horton ((1959), Kinsler et al. (1982), Officer (1958), Ross (1976), Strutt (1945), Tolstoy and Clay (1966), Tucker and Gazey (1966) and Urick (1983). Underwater acoustics research is reported in journals such as Journal of the Acoustical Society of America, Journal of Sound and Vibration, Journal of Geophysical Research, Journal of Ocean Engineering, and Ultrasonics.

UNDERWATER SOUND FUNDAMENTALS

Sound is defined as the periodic variation in pressure, particle displacement, and particle velocity in an elastic medium. Water is the elastic medium for underwater acoustics, and the water may be either fresh or saline. Sound waves are produced by mechanical vibration, and the energy from the vibrating source is normally transmitted as a longitudinal wave similar to electromagnetic waves used in radios and radars. Sound waves are longitudinal waves because the molecules transmitting the wave move back and forth in the direction of wave propagation, producing alternate regions of compression and rarefaction. The transmission of sound waves is very complicated. Thus, plane waves of sound are studied initially because they are the simplest type of wave motion propagating through a fluid medium. Plane waves are easily produced in a rigid pipe with a vibrating piston. For a plane wave the acoustic pressures, particle displacement, density changes, etc. have common phases and amplitudes at all points on any given plane perpendicular to the direction of wave propagation.

In a homogeneous medium, plane wave characteristics are attained at large distances from a vibrating source. The term particle of the medium refers to a volume element that is large enough to contain millions of molecules such that it may be considered a continuous fluid, yet small enough so that the acoustic properties of pressure, density, and velocity are considered as constants throughout the volume element. For the case of a plane wave of sound, the acoustic pressure (p) is related to the particle velocity (u) by

$$p = \rho c u \qquad \textbf{8-1}$$

where p is pressure, ρ is density, c is the propagation velocity of the plane wave, and u is the particle velocity. The term ρc is called the specific acoustic resistance, and its value for seawater is 1.5 x 10^5 g/cm^{2}s and for air is 42 g/cm^{2}s. Equation 8-1 is also known as Ohm's Law for acoustics where the acoustic pressure (p) is analogous to voltage, specific acoustic resistance (ρc) is analogous to electrical resistance, and particle velocity (u) is analogous to electrical current.

Table 8-1. Examples of four types of underwater acoustic systems.

Equipment System	Brief Description
Active Sonar Systems	
Active echo ranging sonar	Used by ships to locate submarine targets. In these sonars, a short pulse of sound is generated uniformly by a transmitting transducer. For reception, the same unit is used as a hydrophone.
Torpedoes	Use moderately high frequencies to echo range on targets and then steer on reflected signals.
Depth sounders	Equipment sends short pulses downward and the time of the bottom return is used to indicate depth of water.
Side-scan sonars	Maps ocean terrain at right angles to a ship's track.
Fish finding aids	Forward looking active sonars for spotting fish schools.
Diver hand held sonars	Used by divers to locate underwater objects.
Position marking beacons	Transmits sound signal continuously.
Position marking transponders	Transmits sound only when interrogated.
Sub-bottom profilers	Uses high low and high frequency sound signals to delineate the upper layers of the seafloor structure.
Flow and wave height sensors	Used to measure flow rate and wave height respectively.
Sonobuoy	Link between an aircraft and underwater sound source.
Seismic Systems	
Seismic Reflection Systems	Used to map the structure of the ocean floor. The acoustic pulses are basically unidirectional pressure pulses which are generated by underwater sound sources such as explosive charges, underwater arc (sparker), electromagnetic (thumper) devices, and air guns.
Seismic Refraction System	Similar to seismic reflection systems, but it measures different seafloor properties by using wide angle (or oblique paths.
Underwater Communication, Telemetry, and Navigation Systems	
Underwater telephone	Device used to communicate between a surface ship and a submarine or between two submarines.
Diver communications	Diver has a full face mask which allows him to speak normally underwater and a throat microphone to obtain speech signals. A transducer is used to transmit and receive signals that are subsequently passed to the diver via an ear piece.
Telemetry systems	Transmit data from a submerged instrument to the surface through the water column without the aid of cables.
Doppler navigation	Uses pairs of transducers pointing obliquely downward to obtain speed over the bottom from the Doppler shift of the bottom returns.
Passive Systems	
Passive ship sonar	Detects acoustic radiation from another vessel or object.
Acoustic mines	Explode when acoustic radiation reaches a certain value.
Torpedoes	Home on acoustic radiation of submarine or ship.

The energy involved in propagating acoustic waves through a fluid medium is due to kinetic energy (particle motion) and potential energy (stress in elastic medium). For a plane wave, the acoustic intensity (I) of a sound wave is the average rate of flow of energy through a unit area normal to the direction of wave propagation. The instantaneous acoustic intensity (I) is

$$I = \frac{\overline{p^2}}{\rho c} \qquad \textbf{8-2}$$

where p^2 is the time average of the instantaneous acoustic pressure squared. The intensity is also the acoustic power (P) per unit area (A) and the units are often watts/cm^2.

In practical engineering and experimental work, it is customary to describe sound intensities and sound pressures with logarithmic scales known as sound levels. The most generally used logarithmic scale for describing sound levels is the decibel scale. The intensity level (IL) of a sound of intensity I_1 is

$$IL = 10 \log \frac{I_1}{I_2} \qquad \textbf{8-3}$$

where I_2 is a reference intensity. By substituting Equation 8-2 the sound pressure level (SPL) is

$$SPL = 20 \log \frac{p_1}{p_2} \qquad \textbf{8-4}$$

where p_2 is a reference pressure. The units of the intensity and sound pressure levels are decibels (dB).

The reference level must be known to insure proper interpretation of the dB value. The current reference pressure level used in underwater acoustics is 1 micropascal (1 μPa). Some old reference levels are 1 dyne/cm^2 and 0.0002 dyne/cm^2 (threshold of hearing). Note that 1 μPa equals 10^{-5} dyne/cm^2 and 1.45 x 10^{-9} psi. It is often necessary to convert from one reference pressure (p_2) to another (p_3), and this conversion is accomplished using

$$N_{p_3} = N_{p_2} + 20 \log \frac{p_2}{p_3} \qquad \textbf{8-5}$$

where N_{p_3} is the level at reference p_3 and N_{p_2} is level at reference p_2.

Example Problem 8-1

Express 125 dB relative to 0.0002 dyne/cm^2 in dB relative to 1 dyne/cm^2 as tabulated in Table 8-2.

Table 8-2. Results of example problem 8-1.

Given	N_{p_2} = 125 dB, p_2 = 0.0002 dyne/cm^2; p_3 = 1 dyne/cm^2
Find	N_{p_3}
Solution	$N_{p_3} = 125 + 20 \log (0.0002/1)$
	$N_{p_3} = 125 - 74 = 51$ dB

The level of a sound wave (N) is the number of decibels by which its intensity differs from the intensity of the reference sound wave. In the case of a sound wave with an intensity I_1 and a reference intensity I_2, the level of the sound wave is

$$N = 10 \log \frac{I_1}{I_2} \qquad \textbf{8-6}$$

For clarity the level should be written as N dB re (the intensity of a plane wave of pressure equal to) 1 μPa where the words in parentheses are normally omitted. If a sound wave has an intensity 500 times that of a plane wave of root mean squared (rms) pressure 1μPa, then the level N is

$$N = 10 \log \frac{500}{1} = 27 \text{ db re } 1 \mu\text{Pa} \qquad \textbf{8-7}$$

SONAR EQUATIONS

The sonar equations are a means for determining the effects of the medium (ocean), target, and sonar equipment on sound propagation. These equations are part of the design and prediction tools available to the ocean engineer for underwater sound applications. Practical functions of the sonar equations are the prediction of performance of sonar equipment of known or existing design and the design of new sonar systems.

The total acoustic field at a receiver may be defined as a desired portion (signal) and an undesired portion (background). The background is the noise or reverberation due to scattering of the output signal back toward the source. The objective of the design engineer is to discover a method for increasing the overall response of the sonar system to the signal and for decreasing the response of the system to the background. A sonar system is just accomplishing its purpose when the signal level equals the background masking level. Masking means that the background interferes with a portion of the signal.

Sonar parameters are defined in relation to the equipment, medium, and target and are tabulated in Table 8-3. All parameters are levels in decibels relative to the standard reference intensity of a 1 μPa plane wave. In order to describe the meaning of the sonar parameters, let us consider the example of an active sonar. When the sound source and receiver are in the same location as shown above in Figure 8-2, the sound source produces an intensity level of SL dB at a distance of 1 m (1 yd) along its acoustic axis from the acoustic center of the source. The radiated sound is attenuated by the transmission loss before it reaches the target, and the level at the target is then (SL-TL). As a result of reflection and scattering by the target whose target strength is TS, the reflected or backscattered level is SL-TL+TS at a distance 1 m (1 yd) from the acoustic center of the target in the direction back toward the source. In traveling back, the level is again attenuated due to transmission loss. Therefore, the intensity level of the reflected signal when it reaches the receiver is SL-2TL+TS. Assuming the background is isotropic ambient noise, then the background noise level is NL. Next the background level is reduced by the directivity index DI, so the relative noise level is NL-DI. Therefore, the echo (SL-2TL+TS) to noise (NL-DI) ratio is

$$SL - 2TL + TS = NL - DI \qquad \textbf{8-8}$$

Since detection is normally the purpose of the sonar, then some means of determining when the target is present is desired. Thus, a detection threshold (DT) is established. When the signal to noise ratio equals the DT then a decision is made that a target is present. This may be accomplished by a person using headphones or by electronic means. Therefore, the active sonar equation is

$$SL - 2TL + TS - (NL - DI) = DT \quad \textbf{8-9}$$

Table 8-3. Definition of sonar equation parameters.

Parameter	Symbol	Reference	Definition
Equipment Related Parameters			
Projector Source Level	SL	1 yd from source acoustic center along acoustic axis	$10 \log \frac{\text{source intensity}}{\text{reference intensity}^{1}}$
Noise Level	NL	At hydrophone location	$10 \log \frac{\text{noise intensity}}{\text{reference intensity}^{1}}$
Receiving Directivity Index	DI	At hydrophone terminals	$10 \log \frac{\text{noise power generated by equivalent nondirectional hydrophone}}{\text{noise power generated by actual hydrophone}}$
Detection Threshold	DT	At hydrophone terminals	$10 \log \frac{\text{Signal power to make decision on presence of target}}{\text{noise power at hydrophone terminals}^{1}}$
Medium Related Parameters			
Transmission Loss	TL	1 yd from source acoustic center and at target or receiver	$10 \log \frac{\text{signal intensity at 1 yd}}{\text{signal intensity at target or receiver}}$
Reverberation level	RL	At hydrophone terminals	$10 \log \frac{\text{reverberation power at hydrophone terminals}}{\text{power generated by signal of reference intensity}^{1}}$
Ambient Noise Level	NL	At hydrophone terminals	$10 \log \frac{\text{noise intensity}}{\text{reference intensity}^{1}}$
Target Related Parameters			
Target Strength	TS	1 yd from acoustic center of target	$10 \log \frac{\text{reflected signal (echo) intensity at 1 yd from acoustic center of target}}{\text{incident intensity}}$
Target Source Level	SL	1 yd from acoustic center of target	$10 \log \frac{\text{source intensity at 1 yd from acoustic center of target}}{\text{reference intensity}^{1}}$

[1] Reference intensity is that of a plane wave that has an rms pressure of 1 μPa

Another common form for the active sonar equation is

$$SL - 2TL + TS = NL - DI + DT \quad \textbf{8-10}$$

where the left side of the equation is the signal level and the right side is the noise masking level. In the monostatic case, the source and receiver are coincident, and the target return is back toward the source. The bistatic case has the source and receiver separated. Therefore, transmission loss path lengths are not always the same, and Equation 8-10 must account for the different transmission loss paths.

A modification of the active sonar equation is necessary when the background is reverberation instead of noise. In this case DI is not correct as defined for noise. For a reverberation background, the NL-DI is replaced by an equivalent reverberation level RL, and the active sonar equation becomes

$$SL - 2TL + TS = RL + DT \quad \textbf{8-11}$$

In the case of a passive sonar system, the target itself produces the signal or sound source (SL) by which it is detected. Therefore, the source level parameter refers to the level of the radiated signal of the target at the distance of 1 m (1 yd) from its acoustic center. Also, the TS parameter is no longer meaningful, and the transmission loss occurs only from the target to the receiver. Thus, the passive sonar equation is

$$SL - TL = NL - DI + DT \quad \textbf{8-12}$$

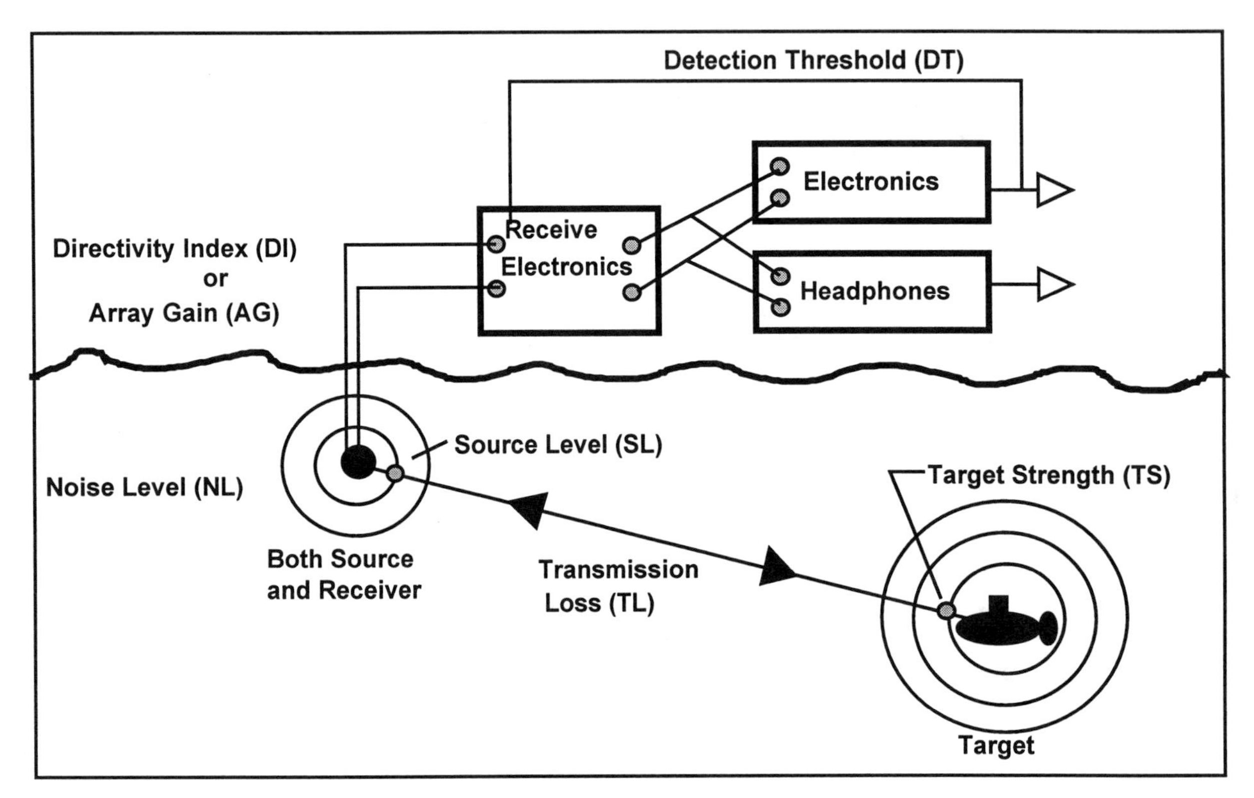

Figure 8-2. Schematic of active sonar system.

The sonar performance figure (PF) is the difference between the source level and the noise level (SL-(NL-DI) measured at the hydrophone. The Figure of Merit (SL-(NL-DI+DT)) is the maximum allowable transmission loss in passive sonars, or the maximum allowable two way loss for active sonars when the TS is 0 dB. An example of the use of the passive sonar equation is illustrated in Table 8-4.

Example Problem 8-2

A passive sonar system is being used to detect an object that has a source level of 20 dB re 1 dyne/cm^2 and a directivity index of 12 dB. If the detection threshold is 15 dB 1 μPa and the transmission loss is 70 dB, determine the noise level that will just permit detection of the target.

Table 8-4. Results of example problem 8-2.

Given	SL = 20 dB re 1 dyne/cm^2, DI = 12 dB, DT = 15 dB, TL = 70 dB
Find	NL
Solution	Passive Sonar Equation: SL - TL = NL - DI + DT
	$N_{p_3} = N_{p_2} + 20\log\frac{p_2}{p_3}$
	$N_{1\mu Pa} = N_{1 dyne/cm^2} + 20\log\frac{p_2}{p_3} = 20 + 20\log\frac{1 dynes/cm^2}{10^{-5} dyne/cm^2}$
	$N_{1\mu Pa} = 20 + 20(+5) = 120$ db re 1μPa; $\therefore$SL=120 db re 1μ Pa
	SL - TL = NL - DI + DT
	NL = SL-TL+DI-DT =120 - 70 +12 -15 = 47 dB re 1μPa

The sonar equations are a statement of the equality between the desired part of the acoustic field (signal - echo or noise from target) and the undesired part (background of noise or reverberation). This equality holds at only one range in most cases. Echo and reverberation signals decrease with range, but the background noise is relatively constant. In conclusion, the sonar parameters fluctuate randomly with time, and there are unknown changes in equipment and platform conditions. Therefore, the solution of the sonar equations yields a time averaged result of a stochastic problem.

TRANSDUCERS AND BEAM PATTERNS

Underwater sound equipment provides the means for an observer to detect the presence of an underwater sound wave. This equipment consists of a hydrophone or hydrophone array that transduces, or converts, acoustic energy into electrical energy, and this energy is delivered to electronic signal processing equipment for display in an acceptable manner. Transducers are devices that convert sound and electrical energy into each other. In the most general sense, transducers are used to convert one form of energy into another. In underwater acoustics a hydrophone is a transducer that converts sound into electrical energy, and a projector is a transducer that converts electrical energy into sound energy.

The conversion of sound into electrical energy and vice versa is accomplished using special materials that have certain specific properties. These properties are:

1. Piezoelectric - Material such as quartz, ammonium dihydrogen phosphate (ADP), and Rochelle salt acquire a charge between crystal surfaces when placed under pressure, and conversely it acquires a stress when a voltage is placed across the surfaces. Subsequently, the electrical potential between the crystal surfaces can be varied periodically at the frequency of the desired sound signal, and thus the material vibrates at the desired frequency and generates an acoustic signal.

2. Electrostrictive - Material such as barium titanate and lead zirconate have the same effect as piezoelectric materials. However, these materials are ceramics that have been properly polarized.

3. Magnetostrictive - Ferrous material that changes its dimensions when it is subjected to a magnetic field, and conversely its magnetic field is changed when it is placed under stress.

The design of transducers is still an art and not an exact science, and therefore transducer design is a special technology in its own right. In most underwater acoustic applications, hydrophone arrays consist of a number of elements spaced in a particular way so as to generate the most desirable beam pattern for a particular application are used. The advantages of arrays over a single element hydrophone are increased sensitivity, better directional capabilities, and greater signal to noise ratio.

The response of a transducer array varies with direction relative to the array. This very desirable property of array directionality provides for the determination of the direction of signal arrival and also facilitates the resolution of closely adjacent signals. In addition, directionality reduces noise relative to the signal by discriminating against noise arriving from directions other than the signal direction. The response of an array varies with direction in a manner specified by the beam pattern of the array, and the two dimensional beam patterns for a line and circular plane array are illustrated in Figure 8-3. Further information on the theory and design of acoustic transducers is discussed by Wilson (1988).

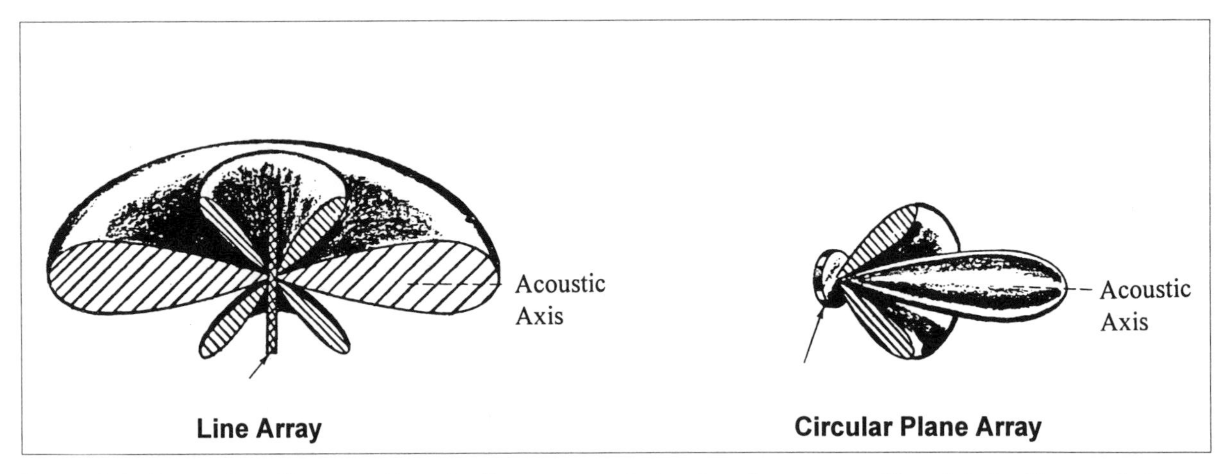

Figure 8-3. Examples of beam pattern for line array and circular plane array . Reprinted with permission from Urick, 1983, *Principles of Underwater Sound.* (Full citing in references).

UNDERWATER SOUND GENERATION

Active sonars use a projector to generate acoustic energy, and this projector normally consists of an array of individual elements that produce a directional beam pattern. The projector source level (SL) designates the amount of sound radiated by the projector. The directionality of the array is described by the transmitting directivity index (DI_T) that is the difference (measured at a point on the acoustic axis of the beam pattern) between the intensity level of the sound generated by the projector (I_D) and the intensity level (I_{ND}) that would be produced by a nondirectional projector radiating the same total amount of acoustic power. The transmitting directivity index is

$$DI_T = 10 \log \frac{I_D}{I_{ND}} \qquad \textbf{8-13}$$

The projector source level can be expressed as a function of the total radiated acoustic power (P),

$$SL = 171.5 + 10 \log P \qquad \textbf{8-14}$$

If the projector is directional, then

$$SL = 171.5 + 10 \log P + DI_T \qquad \textbf{8-15}$$

In the case of an electroacoustic projector, the radiated power (P) is less than the electric power (P_e) input to the projector. The ratio of the total radiated power to the input electric power (P_e) is the projector efficiency (E),

$$E = \frac{P}{P_e} \qquad \textbf{8-16}$$

and the projector source level can be expressed as

$$SL = 171.5 + 10 \log P_e + 10 \log E + DI_T \qquad \textbf{8-17}$$

The typical range of total radiated power (P), transmitting directivity index (DI_T), source level (SL) and projector efficiency (E) for ship sonars are 0.3 - 50 kW, 10 - 30 dB, 210 - 240 dB re 1 μPa, and 20 - 70 %, respectively.

Limitations of sonar power are mainly due to cavitation and interaction between array elements. Cavitation bubbles form on the face of the projector when power is increased to a certain value. Interaction between sonar array elements can also reduce sonar power as a result of one element absorbing power from another.

UNDERWATER SOUND PROPAGATION

The flow of acoustic energy from a source to a receiver is described in terms of its intensity at a distance 1 m or 1 yd from the source and the reduction in intensity between this point and the receiver. The transmission (propagation) loss (TL) is the reduction in intensity between the reference point and the receiver as defined by

$$TL = 10 \log \frac{I_1}{I_2} \qquad \textbf{8-18}$$

where I_1 is the source intensity referenced to 1 m (1 yd) and I_2 is the intensity at a distant point. Transmission (propagation) loss is affected by two primary factors, called spreading and attenuation. In the case of spreading, acoustic energy becomes diluted as it spreads over a larger area and thus intensity is reduced. Near the source, spreading is spherical, and the loss is proportional to the inverse square of the distance. At larger distances the spreading is affected by refraction, which is the bending of rays along the paths that the waves travel. For the case of attenuation, the loss of energy from the wave is a result of absorption and scattering. Absorption results from the conversion of acoustic energy into heat (frictional effects), and scattering is the process whereby objects in the medium cause some of the energy to be deflected in various directions.

Spreading Laws

The two main spreading laws are spherical and cylindrical spreading. Spherical spreading considers a source located in a homogeneous, unbounded, and lossless medium. The power generated by the source is radiated equally in all directions so as to be distributed over the entire surface of a sphere surrounding the source. The transmission loss due to spherical spreading is

$$TL = 10 \log \frac{I_1}{I_2} = 10 \log r_2^2 = 20 \log r_2 \quad \textbf{(Spherical Spreading Law)} \qquad \textbf{8-19}$$

where I_1 is the intensity 1 m (1 yd) from the acoustic center and I_2 is the intensity at a range r_2. The intensity decreases as the square of the range, and the TL increases as the square of the range. The expression for cylindrical spreading is

$$TL = 10 \log \frac{I_1}{I_2} = 10 \log r_2 \qquad \textbf{(Cylindrical Spreading)} \qquad \textbf{8-20}$$

where I_1 is again the intensity at 1 m (1 yd) from the acoustic center of the source and I_2 is the intensity at the range or distance r_2. Cylindrical spreading occurs at moderate and long ranges whenever sound is trapped in a sound channel as illustrated in Figure 8-4.

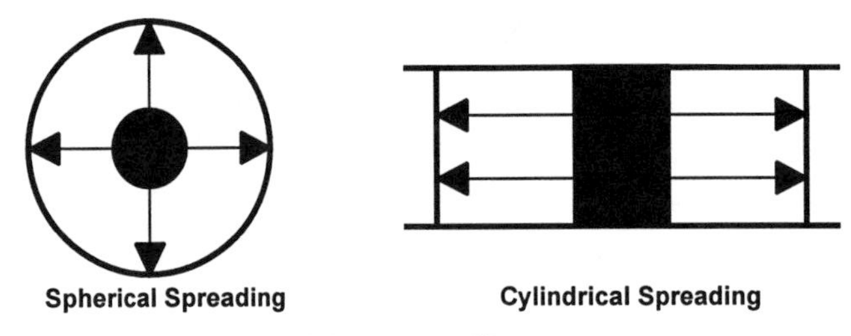

Figure 8-4. Illustration of spherical and cylindrical spreading.

Absorption

The absorption of sound is due to viscosity of pure water and the presence of dissolved salts in the water whose effect is dominant in seawater when the frequency is below 100 kHz. Francois and Garrison (1982) presented a relationship for evaluating absorption that considers the sum of the effects of boric acid, magnesium sulfate, and pure water. The expression for sound absorption (α) is

$$\alpha = \frac{B_1 D_1 f_1 f^2}{f^2 + f_1^2} + \frac{B_2 D_2 f_2 f^2}{f^2 + f_2^2} + B_3 D_3 f^2 \qquad \textbf{21}$$

where f is frequency in kHz, f_1 and f_2 are relaxation frequencies in kHz, and α is the absorption coefficient in dB/km (multiply by 1.0936 to get dB/kyd). The first term of Equation 21 represents the effect of boric acid in seawater, and the equations for evaluating the coefficients B_1, D_1, and f_1 are

$$B_1 = \frac{8.86 \times 10^{(0.78\,pH - 5)}}{c}, \quad D_1 = 1, \quad f_1 = 2.8\left(\frac{S}{35}\right)^{0.5} 10^{\left(4 - \frac{1245}{273 + T}\right)}, \text{ and } c = 1412 + 3.21T + 1.19S + 0.0167d \qquad \textbf{22}$$

where c is sound speed in m/s, T is temperature in °C, S is salinity in parts per thousand (o/oo), and d is depth in m. The second term in Equation 21 accounts for the effects of magnesium sulfate, $MgSO_4$, in seawater, and the coefficients B_2, D_2, and f_2 are determined from

$$B_2 = \frac{21.44\,S}{c}(1+0.025\,T)\ ,\ D_2 = 1-1.37\text{x}10^{-4}\ d+6.2\text{x}10^{-9}\ d^2\ ,\ \text{and}\ f_2 = \frac{8.17\text{x}10^{\left(8-\frac{1990}{273+T}\right)}}{1+0.0018(S-35)} \qquad \mathbf{23}$$

The third term represents the contribution of pure water to absorption, and the coefficients B_3 and D_3 are evaluated using

$$D_3 = 1-3.83\text{x}10^{-5}\ d+4.9\text{x}10^{-10}\ d^2,$$
$$\text{for } T \le 20^\circ C; \quad B_3 = 4.937\text{x}10^{-4} - 2.59\text{x}10^{-5}T + 9.11\text{x}10^{-7}T^2 - 1.50\text{x}10^{-8}T^3 \qquad \mathbf{24}$$
$$\text{for } T \rangle 20^\circ C; \quad B_3 = 3.964\text{x}10^{-4} - 1.146\text{x}10^{-5}T + 1.45\text{x}10^{-7}T^2 - 6.5\text{x}10^{-10}T^3$$

Figure 8-5, developed by Francois and Garrison (1982), shows the variation of the absorption coefficient (α) as a function of frequency from 0.1 to 1000 kHz at zero depth (surface) for a salinity of 35 o/oo and pH of 8.0. The accuracy of the predicted absorption coefficients is estimated as ± 5% for the ranges of 0.4 to 1000 kHz, -1.8 to 30 °C, and 30 to 35 o/oo.

Spreading and Absorption Loss

Propagation measurements made in the ocean indicate that spherical spreading together with absorption yields a reasonable approximation to measured data for a wide variety of conditions. Therefore, transmission loss may be expressed as

$$TL = 20 \log r + \alpha r \text{ x } 10^{-3} \qquad \mathbf{8\text{-}25}$$

where r is the range in yards and α is the absorption coefficient in dB/kyd. This is a rough approximation, but a good rule of thumb.

Example Problem 8-3

An active sound source operates at a frequency of 80 kHz and is located at a depth of 30 ft where the temperature is 10°C and the salinity is 35 o/oo. It is desired to detect a target at a range of at least 3000 yds. If the major causes of transmission loss are spherical spreading and absorption, predict the magnitude of the transmission loss. The solution to this example problem is shown in Table 8-5 and the predicted transmission loss is 154 dB.

Table 8-5. Results of example problem 8-3.

Given	f = 80 kHz, r = 5,000 yds, d = 30 ft, T = 10°C, pH = 8, and S = 35 o/oo
Find	TL
Solution	TL = 20 log r + α r x 10^{-3}
	Figure 8-5 shows α = 28 dB/kyd when d = 0 ft, S = 35 o/oo, pH = 8, T = 4°C. The depth of 30 ft is assumed to have negligible effect. Equations 8-21 through 8-24 can be solved to get the exact value for α.
	TL = 20 log (3000) +28(3) = 69.5 + 84 = 154 dB

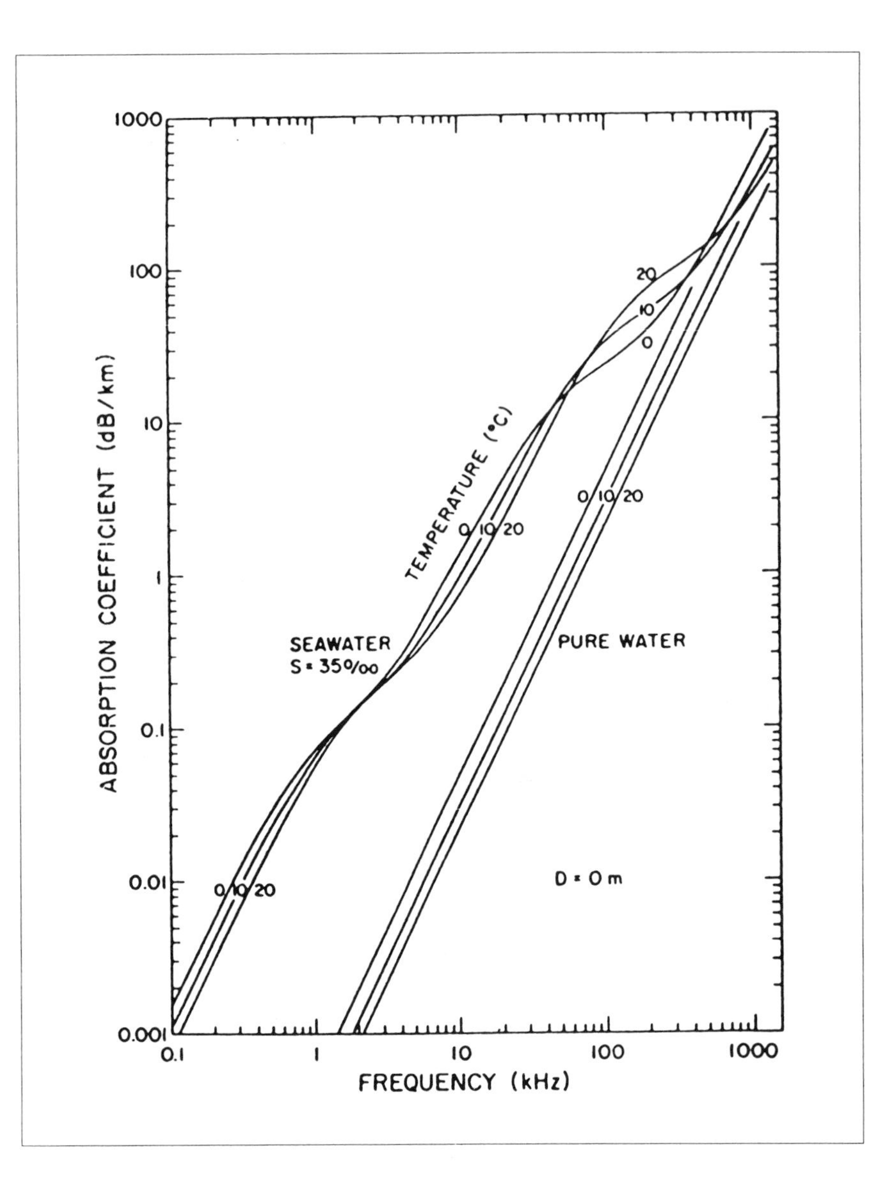

Figure 8-5. Absorption coefficient and correction for depth over useful sonar frequency range for salinity of 35 o/oo and pH of 8. Reprinted with permission from Francois and Garrison, 1982, "Sound Absorption Based on Ocean Measurements. Part II: Boric Acid Contribution and Equation for Total Absorption." (Full citing in references).

Sound Velocity Variation with Depth

The velocity profile or variation of sound velocity with depth is illustrated in Figure 8-6. The main features of the sound velocity profile are the surface layer, seasonal thermocline, main thermocline, and deep isothermal layer. The surface layer is where the sound velocity is subject to daily and local changes in heating, cooling, and wind action. The seasonal thermocline is the negative thermal, or velocity, gradient that varies with the season. In the summer and fall, the near surface waters are warm, and the seasonal thermocline is well defined. However, in the winter and spring it tends to merge and be indistinguishable from the surface layer. The main thermocline is affected only slightly by seasonal changes, and it is in this layer that the major decrease in temperature occurs. The deep isothermal layer is where a nearly constant temperature of 39 °F occurs, and consequently, the sound velocity increases due to the effect of pressure with depth. The sound velocity profile also varies with latitude, season, and time of day.

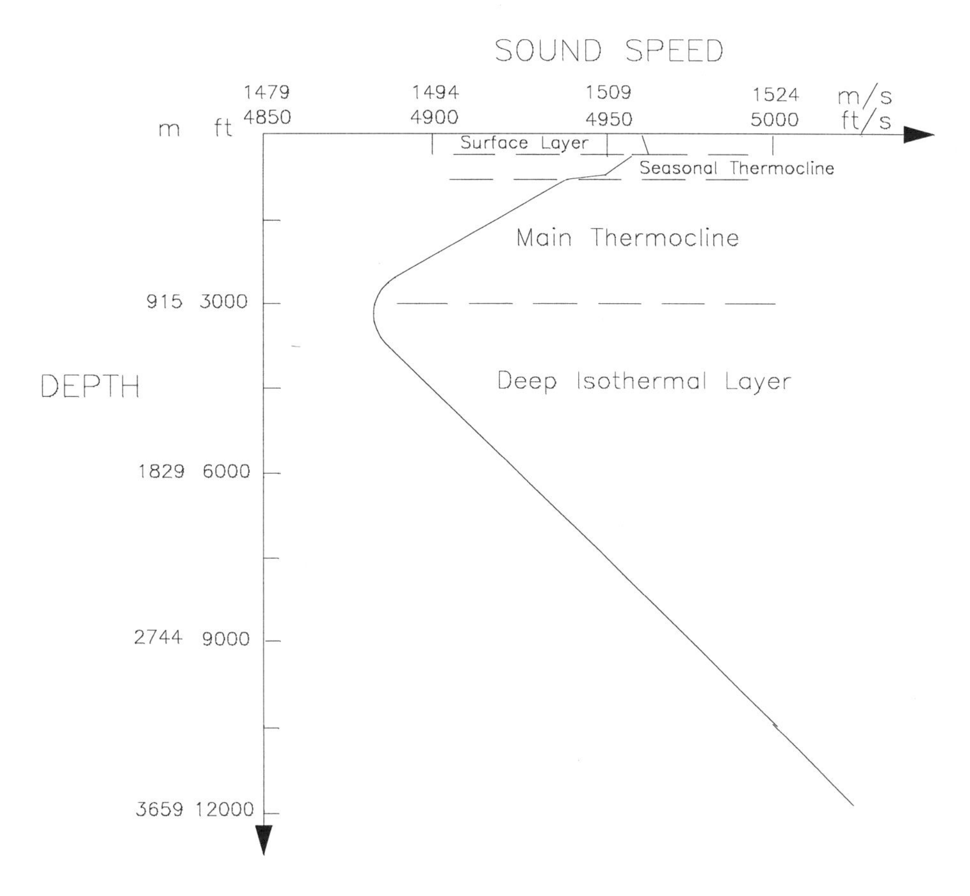

Figure 8-6. Typical deep water sound velocity profile.

The speed of sound in water has been determined both theoretically and experimentally. One equation is that due to Mackensie (1981) that is given as

$$\begin{aligned} c = {} & 1448.96 + 4.591T - 5.304\text{x}10^{-2}T^2 + 2.374\text{x}10^{-4}T^3 + 1.340(S-35) + 1.630\text{x}10^{-2}d + 1.675\text{x}10^{-7}d^2 \\ & - 1.025\text{x}10^{-2}T(S-35) - 7.139\text{x}10^{-13}Td^3 \end{aligned} \qquad \textbf{8-26}$$

where c is sound speed (m/s), T is temperature (°C) at the depth, S is salinity (o/oo), and d is depth (m). The range of validity for this equation is : $0\ ^\circ C \le T \le 30\ ^\circ C$, 30 o/oo $\le S \le 40$ o/oo, and $0\text{ m} \le d \le 8000\text{ m}$. The expression is good for practical work, and it shows that sound speed increases with temperature, salinity, and depth.

Instruments that are used to measure sound speed either directly or indirectly are the bathythermograph, sound velocimeter, and the expendable bathythermograph. The bathythermograph measures temperature as a function of depth as it is lowered into the water. The sound velocimeter measures sound velocity in terms of the travel time of the sound over a fixed path. The expendable bathythermograph (XBT) measures temperature versus depth without having to retrieve sensing unit.

Snell's Law

To illustrate Snell's Law, two sound rays of a plane wave are shown propagating from a slow sound velocity medium to a faster sound velocity medium (Figure 8-7).

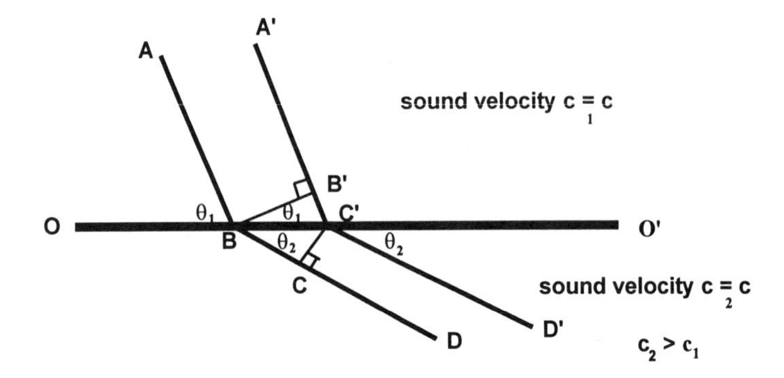

Figure 8-7. Schematic of plane sound wave propagation.

A plane wave is traveling downward in the first medium with rays (normal to the wave front) at an angle θ_1 with the boundary. The wave front at the instant when the ray AB reaches the boundary is shown as BB' which is perpendicular to A'B'. As the wave crosses the boundary the speed of propagation is suddenly changed from c_1 to c_2. While ray A'B'C'D' is traversing the distance B'C', the ray ABCD traverses a different distance BC. When ray A'B'C'D' reaches the point C', the wave front lies along the line CC'. To locate the point C, swing an arc of radius BC about the point B and then draw a tangent line from C' to the arc. The time to travel from B'C' and BC is a constant, and the magnitude of BC equals c_2t. The tangent line determines the direction θ_2 in which the ray BCD travels. The relationship between BC, B'C' and the sound speeds is expressed as

$$\frac{BC}{c_2} = \frac{B'C'}{c_1} \tag{8-27}$$

$$BC = BC' \cos \theta_2 \; ; \; B'C' = BC' \cos \theta_1 \tag{8-28}$$

Therefore

$$\frac{BC' \cos \theta_2}{c_2} = \frac{BC' \cos \theta_1}{c_1} \tag{8-29}$$

$$\frac{\cos \theta_2}{c_2} = \frac{\cos \theta_1}{c_1} \tag{8-30}$$

For many layers, Snell's Law is expressed as

$$\frac{c_1}{\cos \theta_1} = \frac{c_2}{\cos \theta_2} = \frac{c_3}{\cos \theta_3} = \ldots\ldots = \text{constant} = c_V \tag{8-31}$$

Approximating the ocean with layers in which the sound speed is constant, letting the number of layers approach infinity, and letting the thickness of the layer go to zero, then

$$c = c_v \cos\theta \tag{8-32}$$

The limitations of Snell's Law are that it is valid only when the speed of sound is a one dimensional space function and the constant c_v applies only to a particular ray. The constant c_v is the speed of propagation at the depth at which the ray is horizontal. The critical angle (θ_c) is defined as

$$\cos \theta_c = \frac{c_1}{c_2} \quad (c_1 < c_2) \tag{8-33}$$

and is illustrated in Figure 8-8.

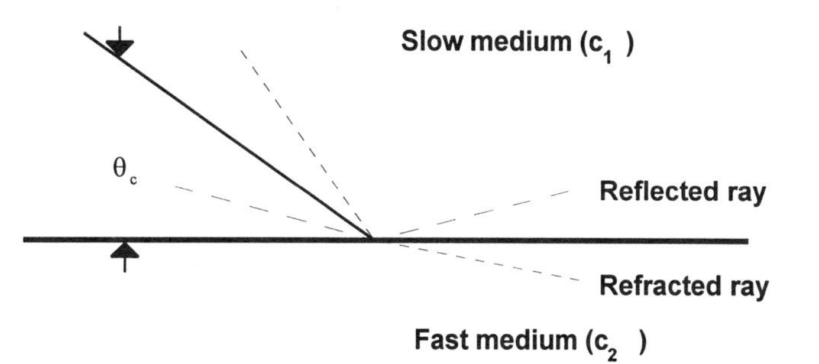

Figure 8-8. Critical angle definition.

When a ray in the slower medium is incident upon the layer boundary at an angle $\theta_1 > \theta_c$ with the horizontal, the ray enters the faster medium and is bent toward the horizontal. At the critical angle θ_c, the refracted ray travels along the interface ($\theta_2 = 0$). When the incident ray is more nearly horizontal than the critical ray ($\theta_1 < \theta_c$), the ray does not enter the faster medium but is totally reflected. When a ray is incident from the faster medium, there is no critical angle, and refraction occurs for all angles of incidence. Also, the angle between the refracted ray and the horizontal is never less than θ_c.

APPLICATIONS USING UNDERWATER ACOUSTIC PRINCIPLES

There are numerous applications of underwater acoustic principles used by ocean engineers and other engineers and scientists in exploring ocean resources and working in coastal and deep oceanic waters. Some of these applications include seismic exploration, side scan sonars, subbottom profilers, depth sounders, acoustic Doppler current meters, sonic flowmeters underwater telephones, sound velocimeters, ultrasonic thickness gauges, ship sonars, submarine sonars and weapon systems, marker beacons and transponders, data telemetry, and acoustic tracking systems. A brief description of a few of these applications are discussed to give a brief overview of the use of underwater acoustics and to demonstrate the importance of acoustics in ocean engineering and other disciplines involved in working in the oceans, lakes, and other water bodies.

The purpose of underwater seismic exploration is to locate possible oil and mineral deposits located deep beneath the ocean floor. Interpretation of the seismic data reveals the layered structure below the sea bed, and certain structural formation lend themselves to having oil and gas trapped that can be extracted using offshore drilling and production techniques. Some of the various names for sea bed structural formations where oil and gas are often trapped are anticlines, salt domes, fault traps, pinchouts, and limestone reefs. An anticline is the most common type of trap, and it occurs when an impermeable bed is overlaying a permeable bed. Salt domes are formed when a mass of salt flows upward and causes a petroleum trap to form around the sides of the dome. Fault traps occur when an impermeable bed overlays a permeable bed and is faulted by an impermeable bed. Pinchouts occur when a reservoir bed gradually thins and eventually pinches out. When limestone reefs are covered by deposition of impermeable material, the reef material is usually porous and acts as a trap for petroleum.

There are two types of seismic profiling procedures used for oil prospecting, and these are seismic refraction which is used for engineering analysis of sediment properties and seismic reflection which is used for oil prospecting or exploration. In seismic reflection the velocity of a sound wave passing through a stratum of rock depends on the elastic property of rock. When an interface is encountered between two rock layers of different sound velocities, part of sound wave is reflected according to Snell's Law. The seismic survey method, illustrated in Figure 8-9, uses an explosion that is produced just below the water surface at a known time. Sound waves radiate downward through the water and penetrate deep into the sea floor. When there is a major discontinuity between one type of rock and another, part of the signal is reflected back to the surface. Then, by measuring the time taken for this signal to reach and return from each stratum, an estimate of the depth of the stratum below the surface can be determined. Hydrophones are strung out in a straight line at known intervals along an underwater streamer to record characteristics of signals and times of their arrivals. Records are stored and played back for interpretation. Grid lines are laid out for tracts that the seismic vessels used for navigation and are subsequently used in characterizing the seafloor structure. Coffeen (1986) is suggested as an additional reading source on seismic reflection analyses and procedures.

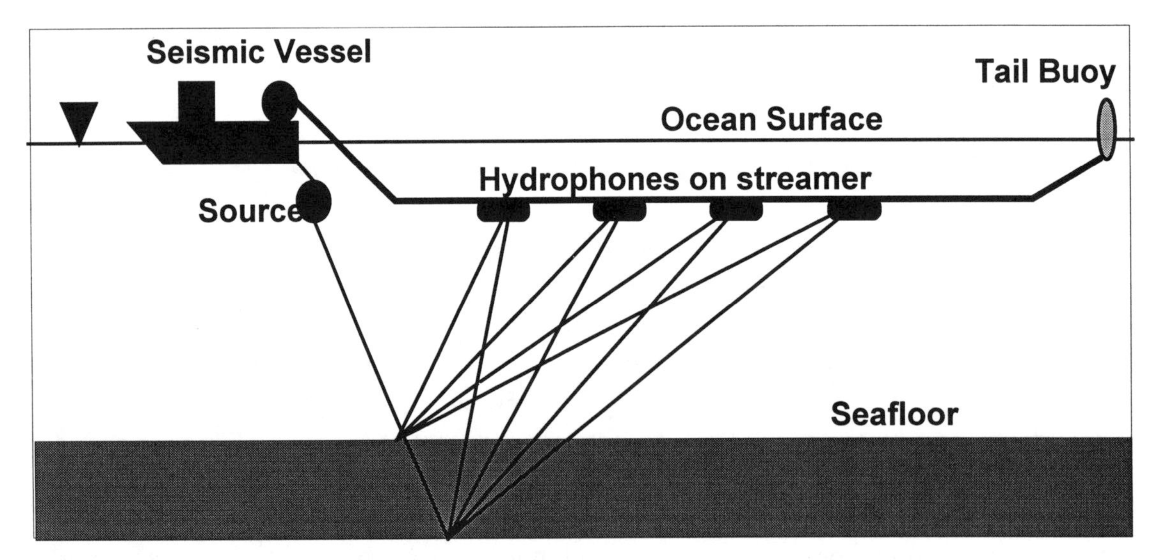

Figure 8-9. Schematic of seismic reflection system.

A dynamic positioning system (Morgan 1980) is composed of several basic elements as shown in Figure 8-10, and it is used to maintain position of drilling rigs over wellheads in deep water when conventional mooring systems can not be used. The purpose of the position and heading sensor is to gather information with sufficient speed and accuracy for the controller to calculate the thruster commands so that the vessel performs the desired task. Information required is vessel position, heading, and wind speed and direction. The offshore position reference needs to include not only navigation position but also an accurate, repeatable, local position reference. The navigation position is a location on earth, and the local position, needed for thruster control, is the location relative to point of interest on seafloor.

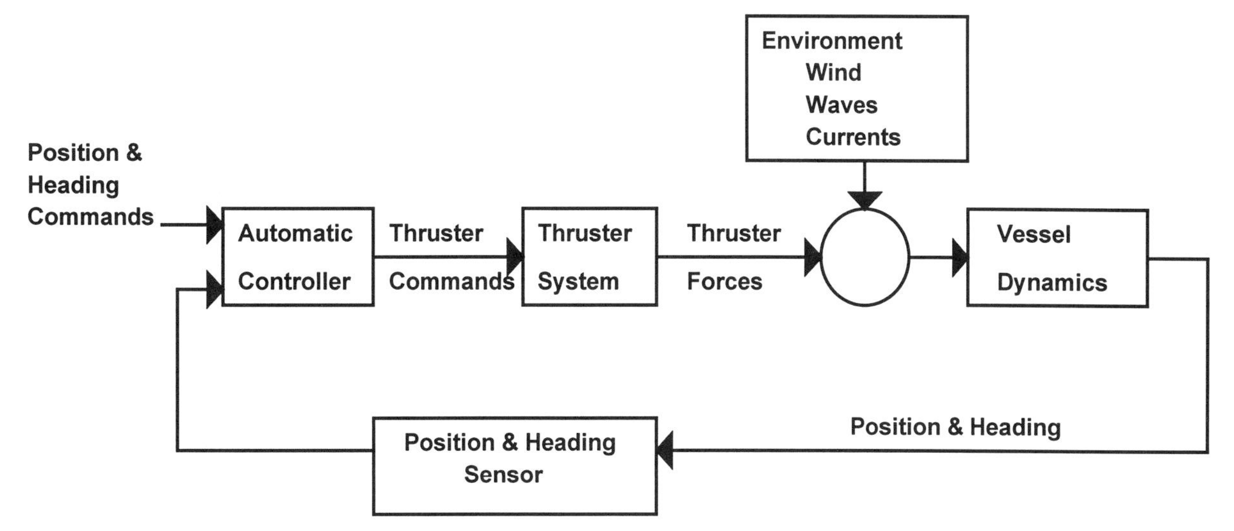

Figure 8-10. Schematic of dynamic positioning system.

One of the most used local position sensors is an acoustic position reference system. It is restricted to a small coverage area. Types of acoustic position reference systems are short baseline and long baseline. These systems use time of arrival, phase comparison, pingers, and transponder acoustic systems. All systems depend on propagation of acoustic signals from one point to another through the ocean medium. Therefore, the propagation characteristics of acoustic energy in water affect the performance of the system. A schematic of a basic acoustic position reference system is shown in Figure 8-11.

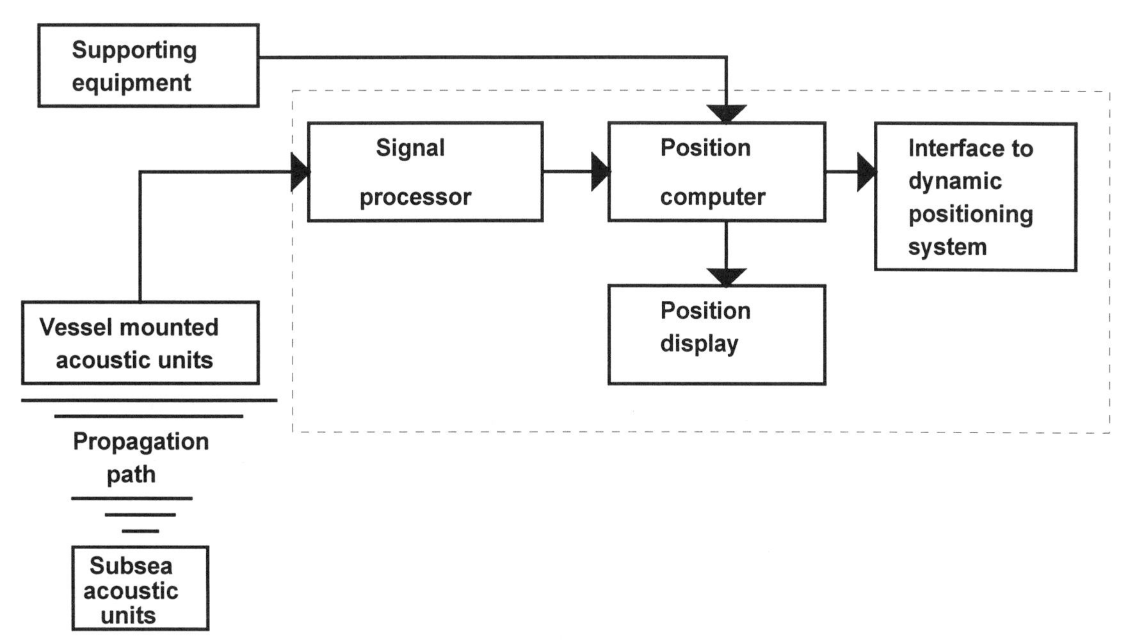

Figure 8-11. Schematic of basic acoustic position reference system.

Acoustic systems operate by projecting acoustic energy into the ocean medium. In simple systems the acoustic energy travels only from the subsea beacon to receivers on the vessel. In more complicated acoustic systems the acoustic energy is transmitted from the vessel to a subsea transponder, and then the transponder transmits an acoustic pulse back to the vessel. The transmitted acoustic signal is affected by the medium through which it must travel, and the acoustic signal experiences transmission loss. Sound is refracted and rays follow curved paths that affect accuracy and range. Noise is also a problem that must be considered.

The components of the side scan or subbottom sonar system are the transducer, data processor, display, and electromechanical cable as shown in Figure 8-12. For the side scan sonar, the acoustic beam is directed perpendicular to the direction of travel. The ship moves at a constant velocity, and each sonar ping insonifies a slightly different wedge of the bottom. The backscattered sound is recorded on a graphic recorder. The depth becomes slant range, and the display is a rectangular coordinate map of slant range and distances along the track as shown in Figure 8-13. The subbottom profiler system directs a sound signal directly below the towfish or survey vessel, and the display shows the depth of the seafloor and the near surface structure of the sea bed. Additional descriptions and example graphical results are discussed in Milne (1980) and Klein (1985).

Figure 8-12. Side scan sonar equipment. (Courtesy of Klein Associates Inc.)

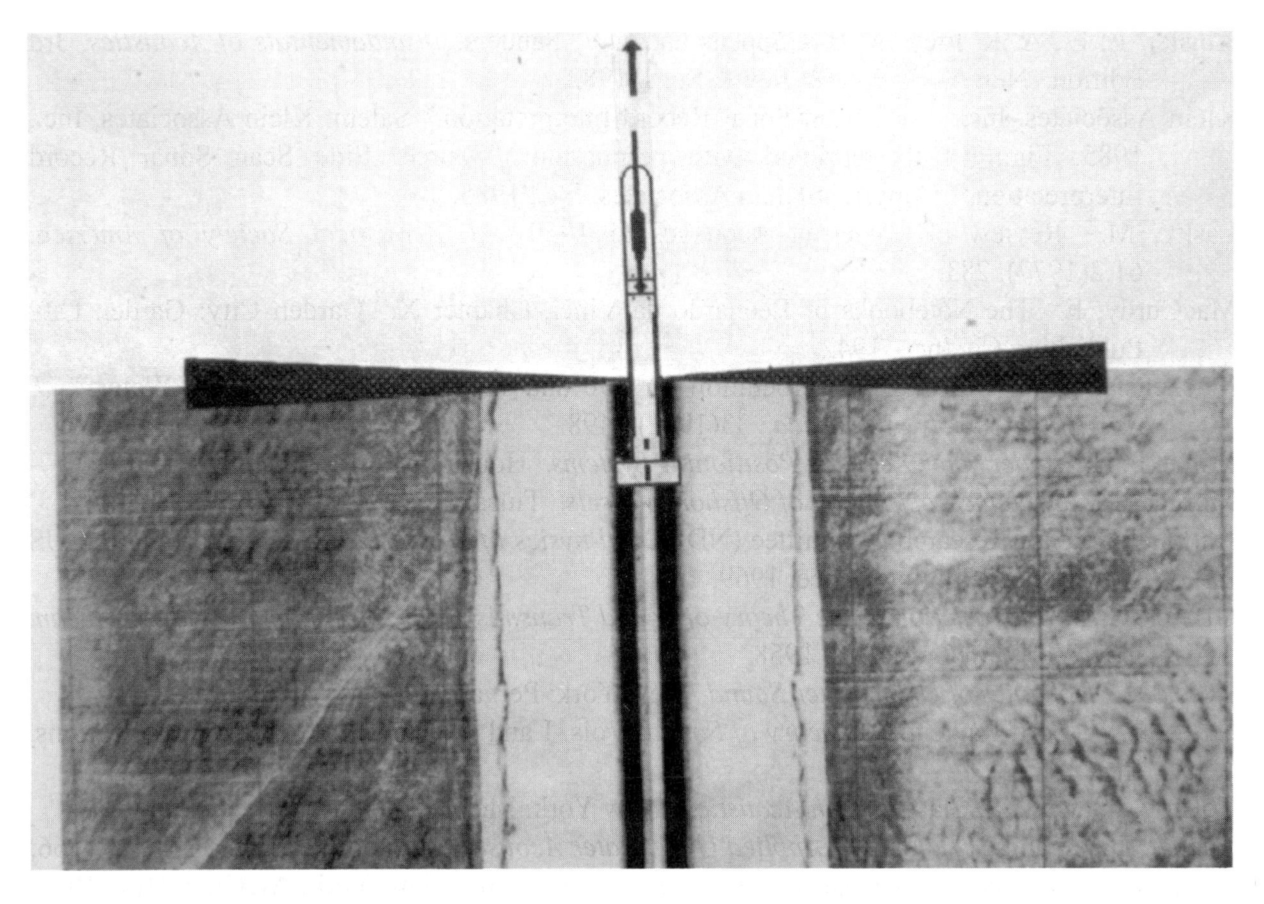

Figure 8-13. Typical side scan sonar output. Reprinted with permission from Klein Associates, Inc., 1985, "Side Scan Sonar record Interpretation." (Full citing in references).

REFERENCES

Albers, V. M. *Underwater Acoustics Handbook II.* University Park: Penn State University Press, 1965.

Burdic, W. S. *Underwater Acoustic System Analysis*, 2nd Edition. Englewood Cliffs: Prentice Hall, 1991.

Clay, C. S. and H. Medwin. *Acoustical Oceanography: Principles and Applications*. New York: John Wiley & Sons, 1977.

Coates, R. F. W. *Underwater Acoustic Systems*. New York: John Wiley & Sons, 1989.

Coffeen, J. A. *Seismic Exploration Fundamentals*, 2nd Edition. Tulsa: PennWell Publishing Company, 1986.

Francois, R. E., and G. R. Garrison. "Sound Absorption Based on Ocean Measurements. Part II: Boric Acid Contribution and Equation for Total Absorption." *J. Acoustical Society of America*, 72.6:(1982). Figure 8-5 reprinted with permission: "Source: Francois and Garrison. "Sound Absorption Based on Ocean Measurements. Part II: Boric Acid Contribution and Equation for Total Absorption." Copyright *J. Acoustical Society of America*, 1982."

Horton, J. W. *Fundamentals of Sonar*, 2nd Edition. US Naval Institute, 1959.

Kinsler, L. E., A. R. Frey, A. B. Coppens and J. V. Sanders. *Fundamentals of Acoustics*, 3rd Edition. New York: John Wiley & Sons, 1982.

Klein Associates, Inc. "Side Scan Sonar Record Interpretation." Salem: Klein Associates, Inc., 1985. Figure 8-13 reprinted with permission: "Source: Side Scan Sonar Record Interpretation." Copyright Klein Associates, Inc., 1985."

Lasky, M. Review of Undersea Acoustics to 1950. *J. Acoustical Society of America.* 61.2(1977): 283.

MacCurdy, E. The Notebooks of Leonardo da Vinci, Chapter X. Garden City: Garden City Publishing Co. Inc., 1942.

Mackensie, K. V. "Bottom Reverberation for 530 and 1030 cps Sound in Deep Water." *J. Acoustical Society of America.* 33(1981):1498.

Milne, P. H. *Underwater Acoustic Positioning Systems.* Houston: Gulf Publishing Co., 1980.

Morgan, M. *Dynamic Positioning of Offshore Vessels.* Tulsa: PennWell, 1980.

National Defense Research Committee (NDRC). *Physics of Sound in the Sea.* Washington: US Government Printing Office, 1969.

Officer, C. B. *Introduction to the Theory of Sound Transmission with Application to the Ocean.* New York: McGraw-Hill, 1958.

Ross, D. *Mechanics of Underwater Sound.* New York: Pergamon Press, 1976.

Strutt, J. W.(Lord Rayleigh). *Theory of Sound,* Vols. I and II. New York: Dover Publications, Inc., 1945.

Tolstoy, I., and C. S. Clay. *Ocean Acoustics.* New York: McGraw-Hill, 1966.

Tucker, D. G., and B. K. Gazey. *Applied Underwater Acoustics.* London: Pergamon Press, 1966.

Urick, R. J. *Principles of Underwater Sound,* 3rd Edition. New York: McGraw-Hill, 1983. Figure 8-3 reprinted with permission: "Source: Urick, R. J. *Principles of Underwater Sound,* 3rd Edition. Copyright McGraw-Hill, 1983."

Wilson, O. B. *An Introduction to the Theory and Design of Underwater Transducers.* Los Altos: Peninsula Publishing, 1988.

PROBLEMS

8-1. The intensity level of a sound wave is 75 dB relative to 1 μPa. What is the intensity level relative 1 microbar?

8-2. An acoustic plane wave in seawater has a sound pressure level of 160 dB re 1 μPa. Find its intensity in watts/m^2.

8-3. If a sound wave has an intensity of 50 dB re 0.0002 dynes/cm^2, then determine its intensity relative to 1 μPa

8-4. Determine the transmission loss for a sound source at a frequency of 80 kHz and at a depth of 3000 m. The salinity and temperature of the sea water at this depth are 35 o/oo and 4 $^{\circ}$C.

8-5. A sound projector is to be used in an active sonar system and it has a transmitting directivity index of 12 dB. The total acoustic power radiated by the projector is 300 W. If the expected target has a target strength of 15 dB, detection threshold of 10 dB and a transmitting and receiving directivity index of 10 dB, then determine the maximum allowable transmission loss when the noise level is 45 dB re 1 μPa.

8-6. An active sonar must be designed to detect a submarine with a target strength of 5 dB re 1 μPa at a range (r) of 20,000 yds. The detection threshold for the sonar is 10 dB re 1 μPa and the directivity index is 25 dB re 1 μPa. The noise level is set at 60 dB re 1 μPa. Assuming spherical spreading (TL = 20 log r) can be used to evaluate the transmission loss, what must be the design source level of the sonar system?

8-7. A passive sonar just detects a target which has a source level of 115 dB re 1 μPa. The directivity index and detection threshold are 15 dB re 1 μPa. The noise level in the area was 45 dB re 1 μPa due to the high sea state. What is the range to the target if the spherical spreading law applies?

8-8. At a location in the ocean the temperature, salinity and depth are 10 ^{o}C, 34 o/oo, and 2500 m. Determine the speed of sound at this location.

8-9. Consider a layered ocean where the sound velocity in the top layer is 1475 m/s and in the bottom layer it is 1500 m/s, evaluate the critical angle according to Snell's Law.

8-10. Determine the speed of sound in the ocean at a depth of 8000 ft, 40 ^{o}F, and 30 o/oo.

CHAPTER 9 : INSTRUMENTATION FOR OCEAN APPLICATIONS

INTRODUCTION

A unique set of instrumentation is used by ocean engineers and oceanographers to measure characteristics of the ocean which are crucial for the design of ocean systems. Current meters, anemometers, tide and wave gauges, multiparameter probes, strain gauges, load cells, accelerometers, laser systems, water samplers, thermistors, acoustic systems, and others are used to evaluate physical characteristics necessary for design of ocean systems and to model these systems in the laboratory. Environmental protection of the ocean requires monitoring of the ocean environment using many of the same instruments and other sophisticated laboratory instruments to determine small but possibly toxic concentrations of contaminants in the ocean waters and sediments. Ocean engineers use instrument systems to measure characteristics of oceans, lakes, estuaries, and rivers. In laboratory facilities such as wave basins, open channel flumes, towing basins, recirculating flumes, circulation water tunnels, and wave tanks, similar instruments are employed to measure the physical behavior of model ocean systems.

An instrument system normally consists of a transducer or sensor, a signal conditioner, and a readout or display device. There are a number of desirable attributes for instruments described by Williams (1973) as accurate, sensitive, rugged, durable, convenient, simple, cost effective, and understandable. All instruments can not meet all of these requirements, but it should meet as many as possible. The accuracy of an instrument is usually defined by several parameters such as uncertainty (error), sensitivity, and repeatability. Uncertainty or error is the difference between the measured value and the true value of the parameter. Sensitivity relates how much of a change in the measured parameter results in a change in the output of the instrument. Repeatability is a measure of the instruments ability to produce the same result for a measurement of the parameter after successive measurements. Errors can be systematic or random, and the systematic errors can be corrected through calibration of the instrument. Calibration is the comparison of the instrument measurements to a known standard and is a critical process for any measurement program.

TEMPERATURE

Temperature is usually defined as the thermal state of a body and is related to the Centigrade or Fahrenheit temperature scales. For bodies of water such as the oceans, estuaries, lakes and rivers, the temperature is a commonly measured property, and the temperature range of most interest is -2 to 40 ^{o}C. The accuracy required for many deep ocean measurements is typically $\pm$ 0.02 ^{o}C, and in estuaries and in coastal zones, a typical accuracy requirement is $\pm$0.2 ^{o}C. For meteorological measurements, an accuracy of near $\pm$ 0.5 ^{o}C is commonly attained.

Thermometers and pyrometers are used to measure water temperature. Liquid-in-glass thermometers such as mercury filled thermometers have been in use for oceanic measurements for nearly 100 years. Another thermometer consists of a bulb or capillary tube filled with a liquid whose dimensions change as a result of the temperature variation. The end of the bulb or tube is connected to a Bourdon tube such that the expansion and contraction of the liquid inside the Bourdon tube produces a mechanical motion moving an indicator device.

The bathythermograph was developed in the 1930's to measure temperature as a function of depth in the upper layer of the ocean, Figure 9-1. The liquid filled tube is wrapped around the tail fins and is attached to a stylus that moves as the temperature changes. A pressure bellows moves a slide in relation to the pressure change and the motion is perpendicular to the temperature stylus movement. Therefore, the slide moves in one direction with a change in depth, while the temperature stylus moves in a perpendicular direction with the temperature variation.

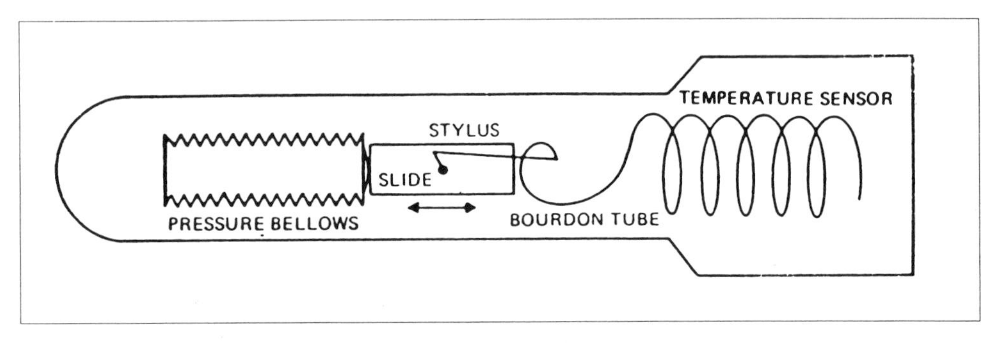

Figure 9-1. Sketch of a mechanical bathythermograph. Reprinted with permission from Williams, 1973, *Oceanographic Instrumentation*. (Full citing in references).

Resistance thermometers are also commonly used. There are four types: platinum, copper, nickel and semi-conductor thermistors. The thermistor type is currently the more commonly used temperature measuring sensor. The expendable bathythermograph uses the thermistor as the temperature sensor as shown in Figure 9-2. Temperature distribution is determined by the thermistor, and time is used to determine the depth. The device is designed to descend through the water column at a constant rate. A quartz crystal thermometer is another type of thermometer in which the resonant frequency of the crystal varies with temperature. This type of sensor is very stable, accurate, and readily input to computers.

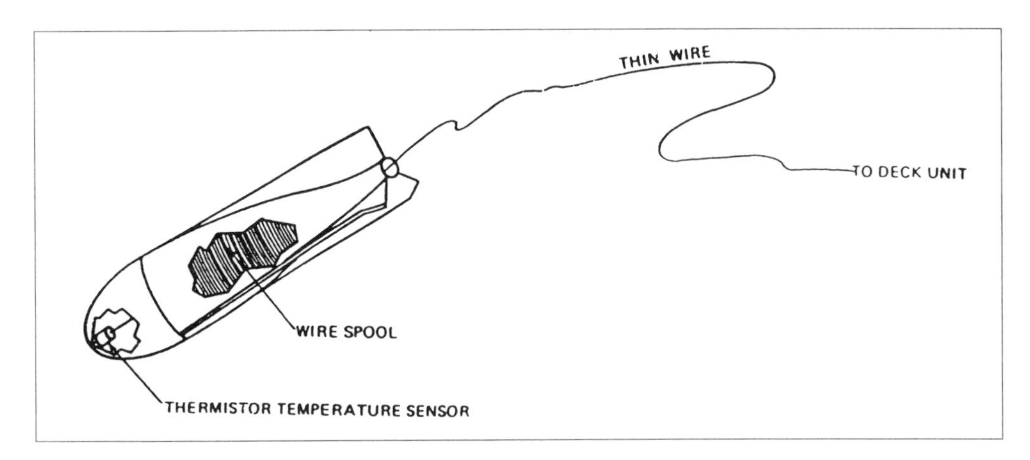

Figure 9-2. Schematic of expendable bathythermograph. Reprinted with permission from Williams, 1973, *Oceanographic Instrumentation*. (Full citing in references).

Pyrometers are capable of measuring temperature over a large area and give good synoptic pictures. These instruments use infrared thermometers that can be used from aircraft and satellites. The accuracy of these devices are not as good as thermometers, but they have the advantage of covering large areas in a relatively short period of time.

DEPTH

The most common method for determining the depth of water is to use an acoustic device commonly known as a depth sounder. This device transmits a sound signal from a transducer, and the signal is reflected by the sea bottom and returns to the transducer. The depth can be determined if the speed of sound in water is known. Thus, the knowledge of sound velocity, which is a function of the salinity and temperature of the water column, is required. The exact location of the transducer must also be known. These transducers are typically located on the bottom of ships and other vessels or instruments. A schematic of a ship depth sounder is shown in Figure 9-3.

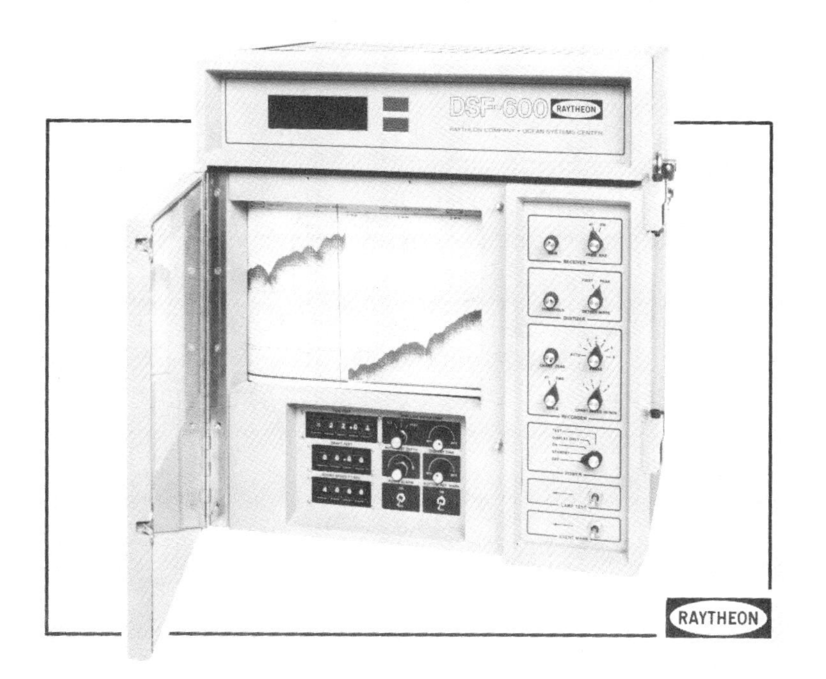

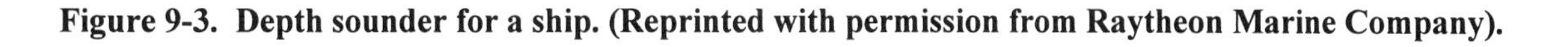

Figure 9-3. Depth sounder for a ship. (Reprinted with permission from Raytheon Marine Company).

Another method for determining the depth of water is to measure the length of cable and the angle of cable supporting a device or instrument at the end of the cable (Figure 9-4). However, currents and motion of support vessels tend to cause the cable to become curved instead of remaining straight and considerable errors are experienced. None the less, this procedure is a good estimate of depth of a sensor at the end of a cable and an approximate check for other pressure measuring devices.

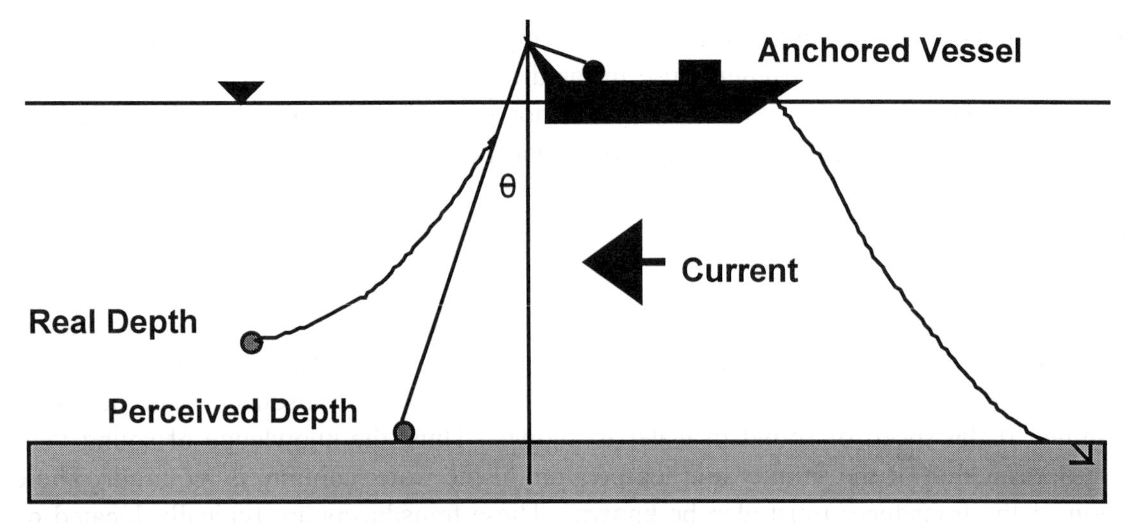

Figure 9-4. Cable length and angle used to measure depth of sensor.

The hydrostatic equation is a relationship between the pressure (p) at a point and the specific weight (γ) and height of a column of fluid (h) above the point.

$$p = \gamma h \qquad \textbf{9-1}$$

An aneroid element, or spring bellows, consists of a chamber which compresses a spring in the form of a bellows (Figure 9-5). As the pressure increases, the bellows is compressed and the movement is used to drive an electrical transducer. A Bourdon tube that is a fluid filled curved tube is another common pressure sensor. The curvature of the fluid filled tube tends to straighten as the pressure changes. The change in curvature can be attached to a mechanical or electrical device to indicate the pressure. Through the use of the hydrostatic equation, the depth can be evaluated. Semi-conductor material undergoes a change in resistance when the pressure applied to the material is changed, and consequently this material is often used as a pressure sensing device.

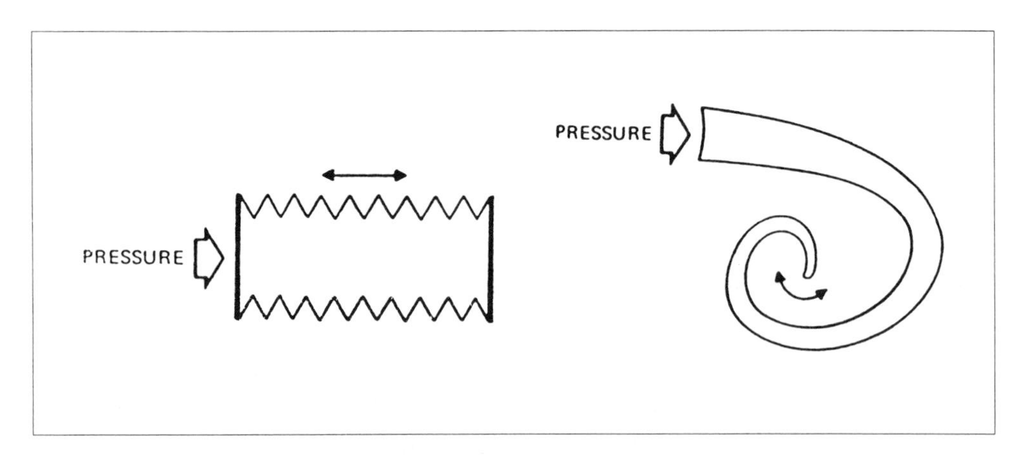

Figure 9-5. Example of bellows and Bourdon tube pressure sensors.

The output signal from the pressure sensors is conditioned using several types of signal conditioners such as strain gauges, linear differential transformers, variable frequency oscillators, potentiometers, and photocells. Strain gauges are often used with aneroid elements and Bourdon tubes. Linear differential transformers connect to an aneroid element and convert the mechanical motion to electrical output. The potentiometer is probably the most common device for converting the mechanical motion of the pressure sensor to an electrical signal. One of the most difficult problems of designing depth or pressure sensors is the large range in pressure the instruments must measure. In order to attain the desired accuracy, the pressure sensors are frequently limited to a specific pressure or depth range.

SALINITY

Seawater is a solution that contains the majority of the known elements, and the more abundant components are chloride (55%), sulfate (7.7%), sodium (30.6%), magnesium (3.7%), and potassium (1.1%). The total amount of dissolved material in seawater is called salinity, and it is defined as the total amount of solid materials in grams contained in one kilogram of seawater (Pickard 1975). Salinity units are parts per thousand (ppt, ‰). The salinity in the ocean varies typically between 33 and 37‰ in 99% of oceanic water with the average being 35‰. In coastal regions salinity is commonly less than 35‰ as a result of river runoff and in tropical areas it is larger due to evaporation. In deep ocean areas the change in salinity over the depth of water is usually less than 1‰, and therefore very accurate measurements of salinity (i.e. ± 0.02‰) are required. In coastal regions, the salinity can vary 5 to 10‰ over a tidal cycle and in these coastal areas the accuracy of salinity measurements of ±0.1‰ is satisfactory.

The relationship between salinity and chlorinity is

$$S‰ = 1.80665 \; Cl \; ‰ \qquad \textbf{9-2}$$

and the relationship between salinity and electrical conductivity (Williams 1973) is

$$S‰ = -0.08996 + 28.2972 R_{15} + 12.80832 R_{15}^2 = 5.98624 R_{15}^4 - 1.32311 R_{15}^5 \qquad \textbf{9-3}$$

where

$$R_{15} = \frac{\text{conductivity of sample at 15 } ^{o}\text{C and 1 atmosphere}}{\text{conductivity of water at 15 } ^{o}\text{C and 1 atmosphere with salinity of 35 ‰}} \qquad \textbf{9-4}$$

In 1978 the practical salinity scale was established that defines salinity in terms of the ratio K_{15} of the electrical conductivity of seawater at a temperature of 15 °C and standard atmospheric pressure to that of potassium chloride solution at the same temperature and pressure (Pickard and Emery 1993. A detailed explanation of the definition of salinity and the practical salinity scale (PSS 78) are described by Lewis (1980).

Laboratory determination of salinity from water samples collected from offshore and coastal environments is determined by either chemical titration or more commonly by the use of laboratory salinometers. Refractometers are also used based upon the fact that the refractive index of water is a function of salinity. Refractometers have achieved an accuracy of 0.05‰. Most laboratory salinometers measure the conductivity and then convert to salinity using the above equation. These laboratory salinometers (Figure 9-6) attain an accuracy of ±0.003 ‰.

Figure 9-6. Laboratory salinometer. (Reprinted with permission from Guildline Instruments Inc.).

The electrical conductivity of seawater is a function of temperature and dissolved solids. Any instrument that uses conductivity as a measure of salinity must take into account the temperature effect through a temperature compensation circuit or must include the measurement of temperature of the sample in the salinity evaluation process. Electrodes and induction cells are the two types of sensors commonly used to measure conductivity. The electrode type requires that the bare metal electrodes be immersed in the water sample to determine the electrical resistance between the electrodes. Fouling is a problem for long term immersion such as days or months, but this type of sensor is less expensive and simpler to use. Although normally less accurate, it is found to be acceptable for coastal environments and brackish water. The induction cell is electrodeless, and consequently it is less susceptible to corrosion and fouling and has better accuracy. However, it is more complicated and more expensive. It remains the common choice for oceanographic measurements. Examples of two electrode type conductivity and temperature instruments are illustrated in Figure 9-7.

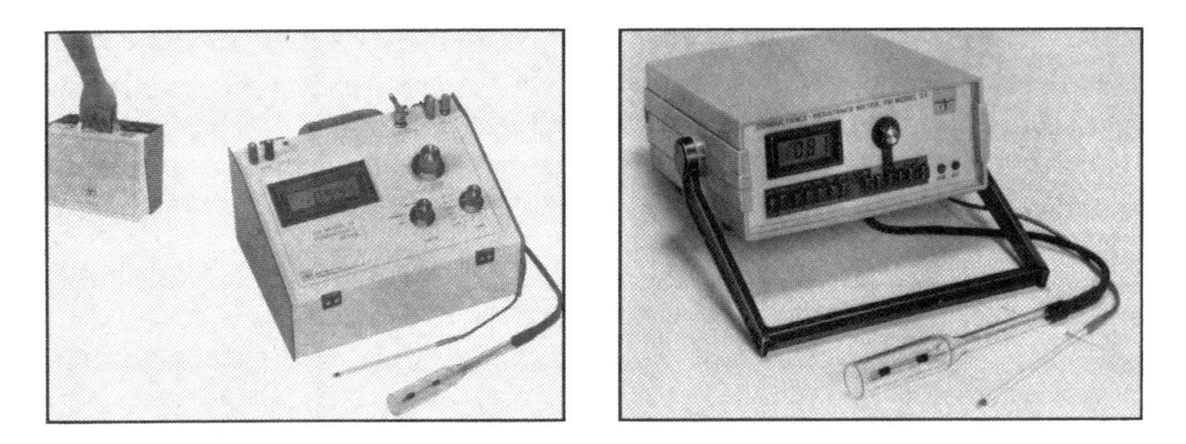

Figure 9-7. Examples of electrode conductivity and temperature instruments used in the field and laboratory. (Reprinted with permission from ENDECO/YSI Inc.).

MULTIPARAMETER INSTRUMENTS

For coastal and ocean measurements, the instrument systems commonly combine the measurement of conductivity (salinity), temperature, and depth into a single instrument called a multiparameter probe (CTD or STD). In many cases other environmental sensors such as dissolved oxygen, pH, turbidity, reduced oxygen potential (REDOX), water current, and sound velocity are added to the probe system, and it is capable of being raised and lowered through the water column on a cable attached to an oceanographic winch. Other systems are placed in-situ on a moored system or attached to a fixed pile, and the data are stored in computer memory or on magnetic tape. Examples of these multiparameter probes are shown in Figure 9-8.

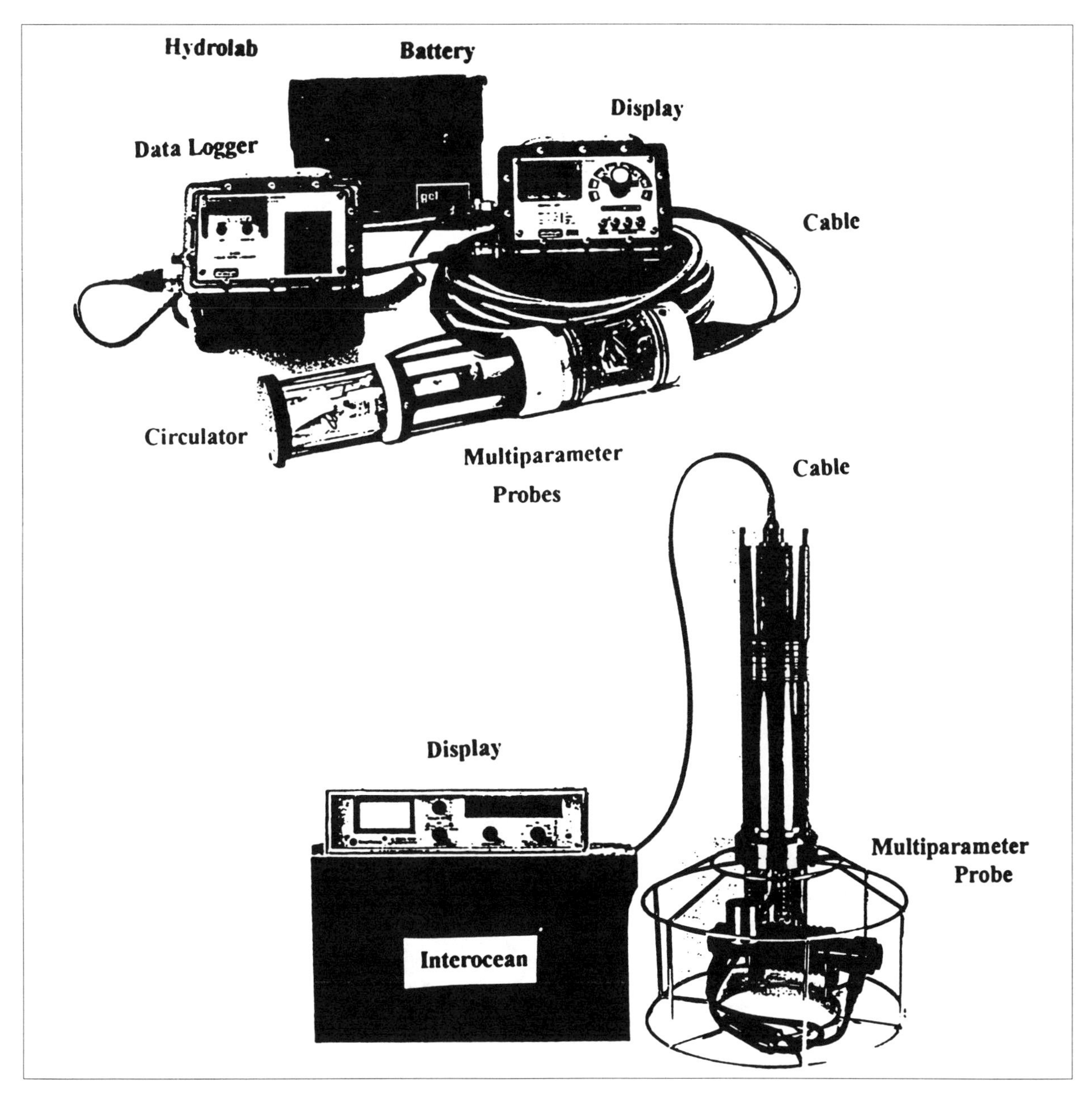

Figure 9-8. Examples of field multiparameter probes. (Reprinted with permission from Hydrolab Corporation and InterOcean Systems Inc.).

The sensors that are included on the multiparameter probe systems depend upon measurements required and manufacturer preferences. The conductivity sensor may be an electrode or inductive type. The salinity is then computed from the conductivity and temperature measurements. The temperature sensor is commonly a thermistor, and the depth sensor is normally a semiconductor strain gauge transducer. Sound velocity sensors usually send a sound signal over a known path, and the time to travel the known path distance is used to determine the sound velocity.

Dissolved oxygen in sea water is necessary to support marine life, and therefore is a frequently measured parameter. The sensor commonly used is a polargraphic membrane type. The membrane separates the polargraphic cell from the water, and the membrane is permeable to dissolved oxygen. Some sensors require flow of the water past the membrane and uses some type of stirrer to move the water sample past the sensor. Temperature and salinity affect the solubility of dissolved oxygen in water, and the measurements from the respective sensors are used to correct for these effects in the dissolved oxygen measurement.

The pH sensor combines a reference and measuring electrode into a single sensor and measures the pH on a scale from 1 to 14. Ocean water pH is typically around 8.5. The turbidity sensor is commonly a transmissometer type that uses a light source and a calibrated photosensor spaced a given distance apart. The attenuation of the known light source by the turbid water is measured.

WATER CURRENT METERS

Ocean and coastal waters are in continuous motion and in most cases are affected by turbulence. Currents, the general movement of the water, are generated by the mechanical action of wind and the thermal activity of uneven heating. Other forces such as Coriolis and centrifugal also influence the currents. Gravitational forces cause tidal currents. Surface currents are typically near 1 knot or less but may reach values in excess of 5 knots. The higher values are usually tidal in nature and found near the shore in most cases. Non-tidal currents that sometimes occur with speeds as high as 6 knots are usually located near the surface and are caused by storms or hurricanes. Exceptions occur such as found in the Gulf Stream and equatorial undercurrents. Deep ocean currents are small in magnitude (i.e. 0.02 to 0.2 knots, 1 to 10 cm/s). Vertical currents are usually 1 to 3 orders of magnitude smaller than the horizontal current values. In some coastal and estuarine locations, the vertical currents are less than an order of magnitude different and can not be neglected. Many current meters measure only horizontal currents, but some of the newer instruments measure three components of current (velocity).

Currents are measured directly or indirectly. Direct methods measure the flow itself, and indirect methods measure some property of the medium to infer the flow. An indirect instrument that is in common use is the electromagnetic current meter where the gradient of electrical potential produced by a moving conductor (seawater) within the earth's magnetic field is measured. Two electrodes are placed within the volume of the water and the potential difference is measured. Examples of an electromagnetic current meter are shown in Figure 9-9.

Direct methods of current measurement are usually characterized as Lagrangian or Eulerian. Lagrangian methods use drifting objects to determine the current magnitude and direction by measuring the sequential position of the object as a function of time. The Eulerian method places an instrument in a fixed location and measures the current moving past it. Drift devices are the oldest form of measuring system for determining currents. The familiar stoppered bottle with a note inside requesting the finder to communicate the time and place that the bottle was found is one of the oldest techniques. Drogues are used in the same way and can be positive (surface currents) and negatively buoyant (bottom currents). Another form of Lagrangian current measurement is the use of radioactive tracers and fluorescent dyes. Submerged drifters are often tracked by sonic devices. Ships without propulsion can also be used to infer surface currents using their accurate navigation position systems to determine their motion. Some examples of drifter devices are illustrated in Figure 9-10.

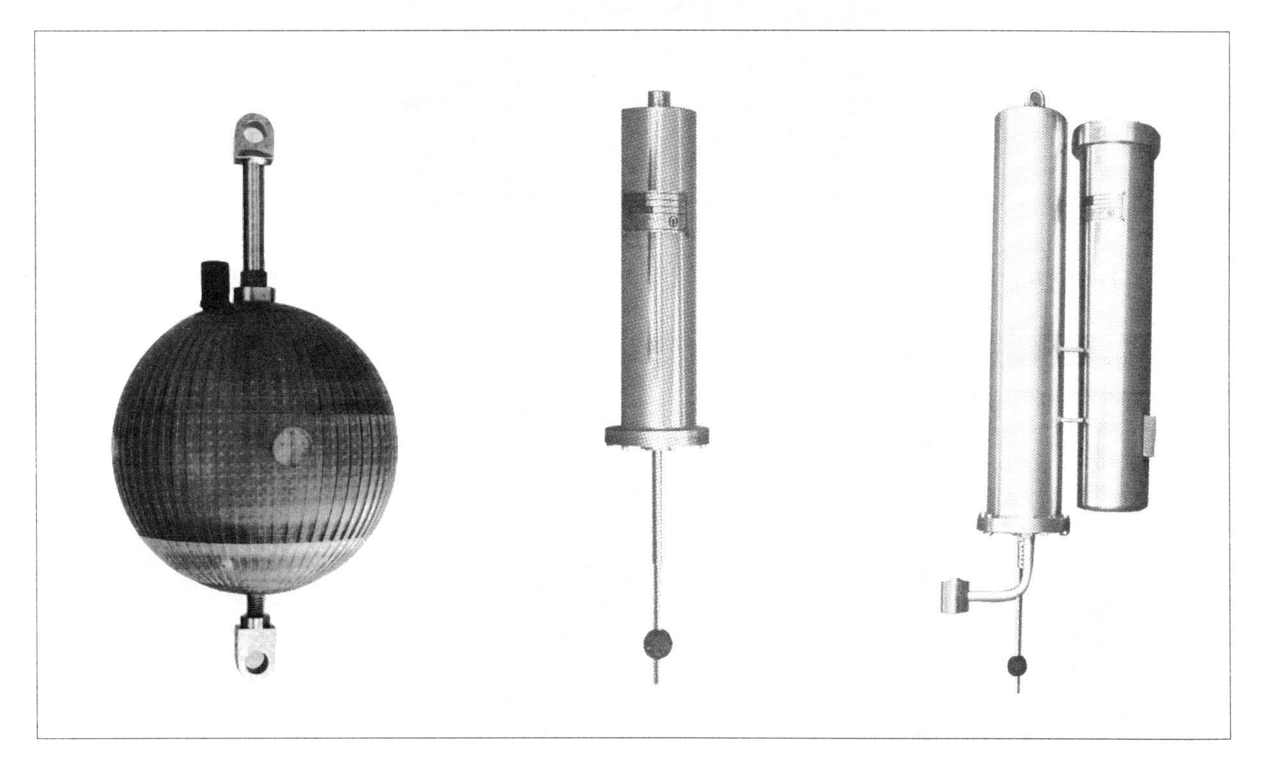

Figure 9-9. Example electromagnetic current meter. (Reprinted with permission from InterOcean Systems Inc.).

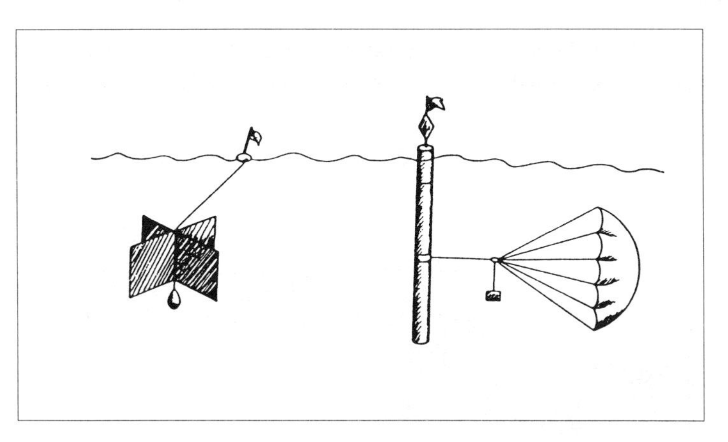

Figure 9-10. Drifter current measuring devices.

Eulerian direct methods include open propellers (Figure 9-11), savonius rotors, and ducted propellers (Figure 9-12), and acoustic Doppler current meters (Figure 9-13) for open water measurements. Laboratory current measuring devices include hot wire anemometers (Figure 9-14), laser Doppler anemometers (Figure 9-15), and acoustic Doppler velocimeters (Figure 9-16).

Figure 9-11. Open propeller small current meter. (Reprinted with permission from Ott Company).

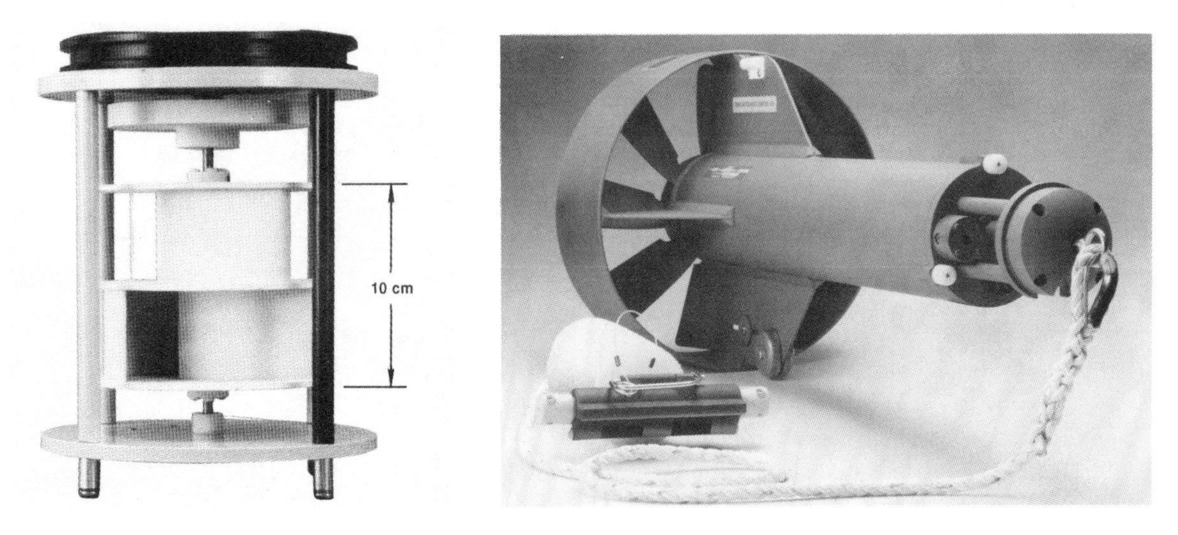

Savonius rotor ducted propeller

Figure 9-12. Savonius rotor (Reprinted with permission from InterOcean Systems Inc.) and ducted propeller current meters. (Reprinted with permission from ENDECO/YSI Inc.).

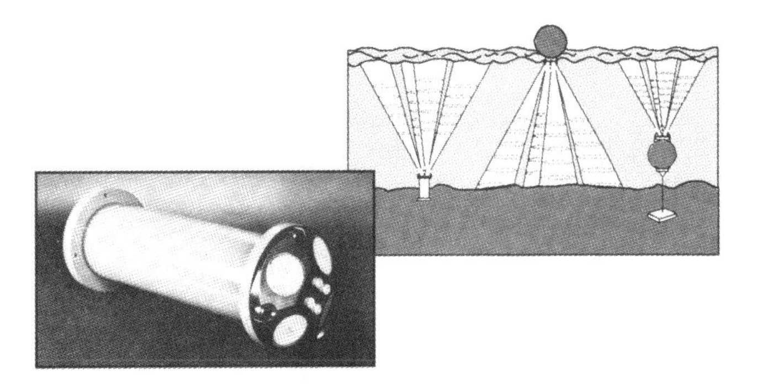

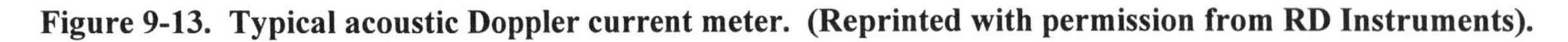

Figure 9-13. Typical acoustic Doppler current meter. (Reprinted with permission from RD Instruments).

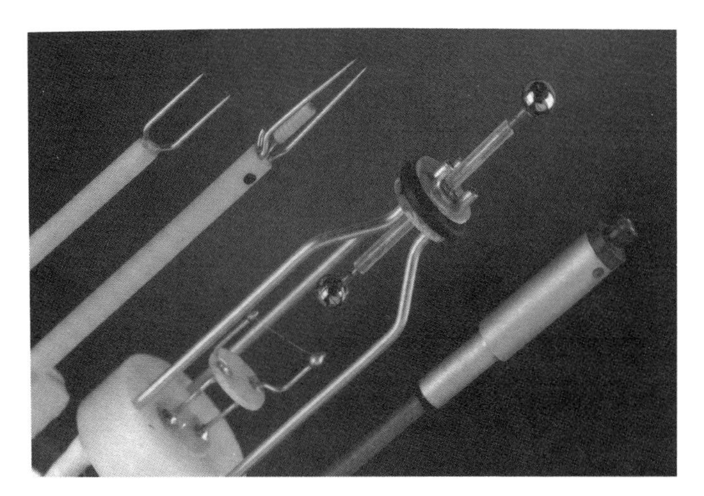

Figure 9-14. Hot wire anemometer probes. (Reprinted with permission from DANTEC).

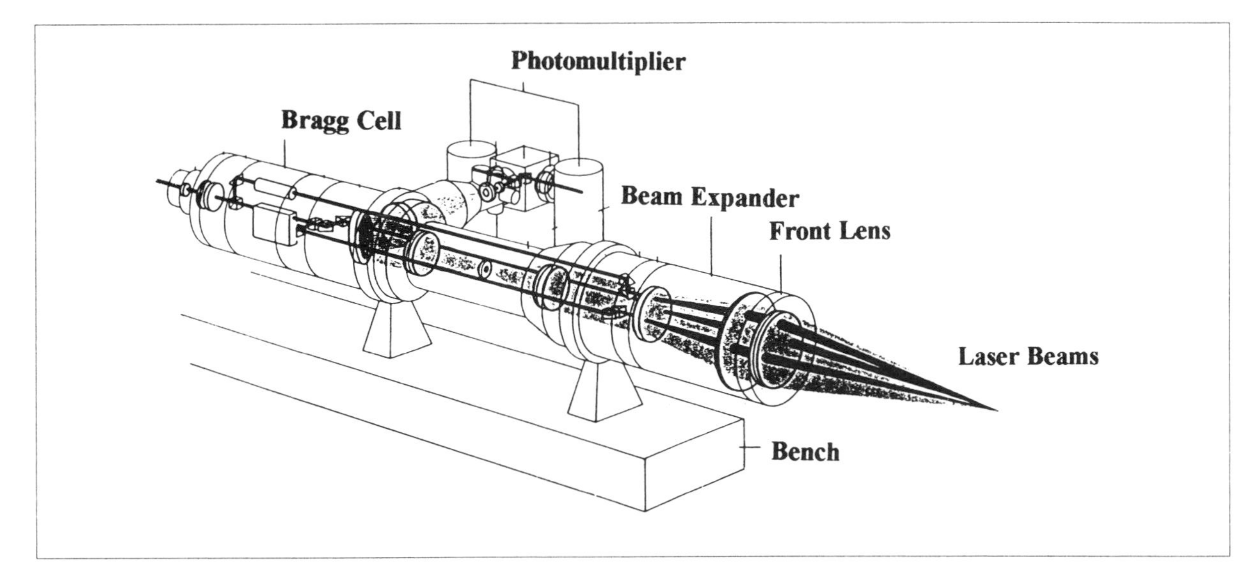

Figure 9-15. Example three beam laser Doppler anemometer. (Reprinted with permission of DANTEC).

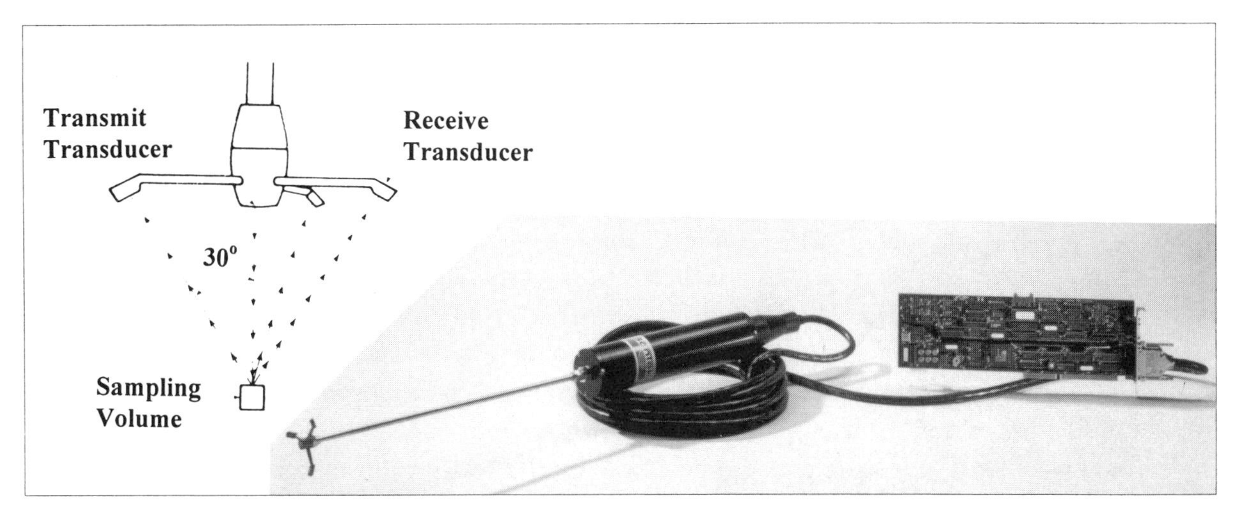

Figure 9-16. Acoustic Doppler velocimeter for laboratory. (Reprinted with permission from SonTek).

METEOROLOGICAL

Instrumentation for measuring meteorological data include sensors that measure wind speed and direction, barometric pressure, air temperature, and humidity. Figure 9-17 shows a wind sensor mounted on the tail of a directional vane that is free to rotate about its vertical axis. The wind direction is determined by an integral flux-gate compass that senses the North direction. The wind speed is measured using a vortex counting technique. The wind speed is usually given in miles per hour (mph), nautical miles per hour (kts), and kilometers per hour (kph). The standard height of the wind sensor is 10 m (33 ft) above sea level. The wind direction is shown in degrees. The temperature sensor is a hermetically sealed platinum resistance thermometer while the humidity sensor is constructed of a highly porous material that allows the moisture to move freely in and out of the sensor. As the density changes so does the resistance, and the resistance change is converted to relative humidity. A sealed solid state transducer is used to measure the barometric pressure. These sensor signals may be recorded using a variety of recording devices such as strip chart recorder, analog gauges, digital displays, magnetic tape, or solid state memory.

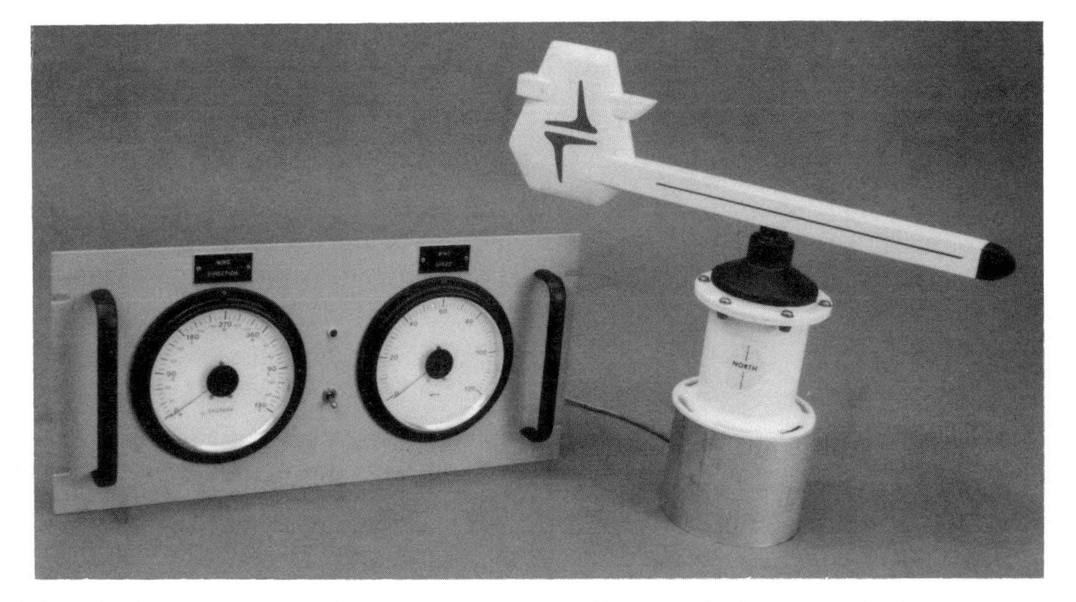

Figure 9-17. Wind speed and direction vortex anemometer (Reprinted with permission from InterOcean Systems Inc.).

WAVES AND TIDES

Many wave and tide gauges measure the pressure variation of the water surface to evaluate the wave and tide climate. One type of wave (pressure) sensor is a silicon semiconductor strain gauge that is temperature compensated and incorporates a low pass hydraulic filter on one side of the differential transducer. The sensor is isolated from the water by an oil filled diaphragm. The installation depth for typical meters is 0 to 60 m, but greater depths are possible. The instruments can be placed on the seafloor, installed in a moored system, or attached to a piling as illustrated in Figure 9-18.

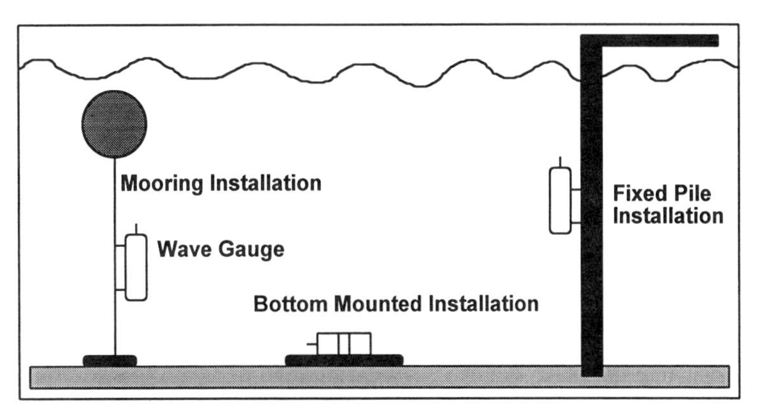

Figure 9-18. Wave and tide gauge installation examples.

The data recording is usually selectable as to rate, duration, and period, and these data are recorded on a magnetic cassette tape or in solid state memory. For example, the gauge may be set to record every 0.5 s and accumulate 1024 data points and then shut itself off. This may be done every three hours and the instrument would last approximately 30 days before having to be replaced. Another approach is to set the instrument to record for 15 min every three hours. When the instrument is recovered, the cassette tape or microprocessor provides the data needed for computer analysis. These instruments are powered by alkaline batteries, and the length of time that the instrument can remain in the water depends on the rate of data collection. An example this type of wave and tide gauge is illustrated in Figure 9-19.

Figure 9-19. Example wave and tide gauge. (Reprinted with permission from InterOcean Systems Inc.).

Another type of wave gauge is a wave orbital following buoy similar to that shown in Figure 9-20. This instrument provides a means for measuring wave direction and heave. The sensors include pitch, roll, compass, and accelerometer instrumentation that are located inside a pressure housing, and the electronics and battery supply are located inside the sphere. The system may be moored or free floating, and the data may be stored internally or transmitted by radio link to shore or satellite. Another type of wave measuring buoy is called the wave rider buoy which is manufactured in The Netherlands, and it also uses an accelerometer to measure the wave height.

Figure 9-20. The Wave Track buoy and receiver. (Reprinted with permission from ENDECO/YSI Inc.)

Tide gauges use floats and pressure transducers to determine the tidal variation. A buoyant float is placed in a stilling well that allows water to rise and fall but doesn't respond to short period waves. The float device is connected mechanically to an ink pen recording system to record the water level variation. These systems have been used in protected waters such as bays and estuaries. Pressure sensors like those used in wave gauges and silicon semi-conductor strain gauges are commonly used in coastal waters as well as bays and estuaries. The tide gauge shown in Figure 9-21 senses the tide through a low pass hydraulic filter that removes the short period waves typically caused by wind and boat wakes. Like the wave gauge previously mentioned, this tide gauge can be mounted in a rigid mooring, bottom mount, or attached to a fixed pile. Data are recorded on a cassette magnetic tape, and the instrument is powered by alkaline batteries with a life expectancy of 60 days.

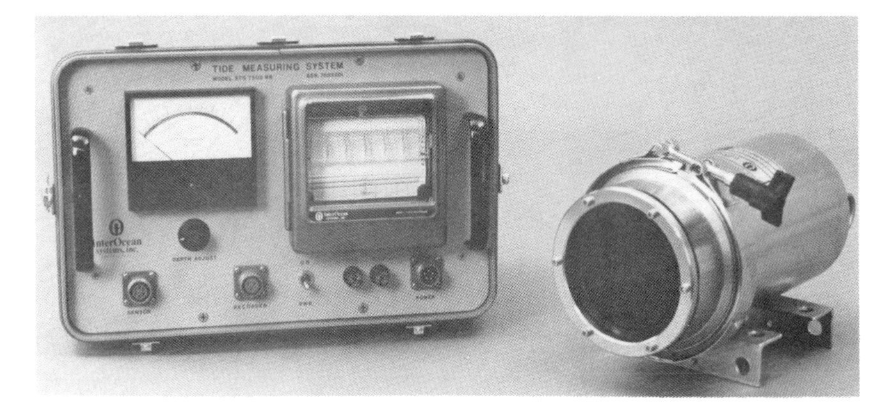

Figure 9-21. Pressure sensor type tide gauge. (Reprinted with permission from InterOcean Systems Inc.)

Wave gauges are important instruments for measurements of water level and wave heights in laboratory experimental facilities. A good reference for laboratory measurements for coastal applications is Hughes (1993). There are two major types of wave gauges used in the laboratory, and these are resistance and capacitance wave gauges. Figure 9-22 illustrates the resistance wave gauge. This type of laboratory wave gauge has two parallel wires separated by a fixed wazzu distance, and the wires are oriented normal to the direction that the wave train is traveling. The conductance between the wires is measured and that conductance is proportional to the length of wire submerged beneath the water surface. The conductance varies with the temperature and conductance of the water, and some resistance gauges have compensating circuits to account for these variations. Otherwise, frequent calibration is necessary. The accuracy is usually ± 1 mm, with a resolution of ± 0.1 mm (Sharp 1981).

The capacitance type wave gauge is also illustrated in Figure 9-22 (Markle and Greer 1992). This sensor consists of a thin insulated wire held taut by the supporting structure. The wire insulation acts as a capacitor between the inside conducting wire and the water that serves as a ground. The capacitance varies linearly as the sea surface elevation changes. The insulation needs to be free of cracks. Electronic circuits determine the variation in capacitance as a voltage output that is subsequently converted to sea surface elevation change after the appropriate calibration constants are applied. The major advantages of the capacitance wave gauges are low cost, good linearity and dynamic response, and good stability.

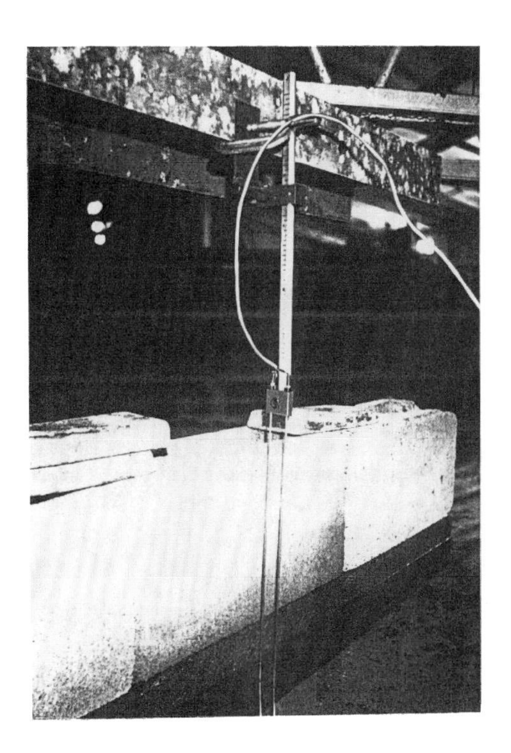

Resistance **Capacitance**

Figure 9-22. Stainless steel, dual rod, resistance-type (Zimmerman 1995) and capacitance type (Markle and Greer 1992) wave gauges for the laboratory. Reprinted with permission from Markle and Greer, 1992, "Crescent City Instrumented Model Dolos Study; Coastal Model Investigation." (Full citing in references) and Zimmerman and Randall, 1995, "Operational Analysis of a Plunger-Type Wave maker and Design of a Diffraction Experiment on a Detached Breakwater." (Full citing in references).

WATER SAMPLERS

The collection of samples of ocean, bay, or lake water is often necessary for use in calibration and actual laboratory analysis of the chemical or biological contents. Water samplers are devices that are attached to a winch line and lowered to a specified depth of water from a surface vessel. The samplers are as small as a liter and as large as 10 liters. Normally, the devices are open at the top and bottom while being lowered to the sampling depth. Then, a small weight (called a messenger) is attached to the winch line and released. The messenger weight slides down the line and contacts a triggering mechanism that closes the bottom and top openings. Several samplers can be attached to the winch line and can be triggered by the same messenger weight when rigged appropriately. The water sampler or samplers are winched to the surface and their contents are emptied into an appropriate containers through a special spout on the sampler. Then, the containers are delivered to the laboratory on board the vessel or ashore for further analysis. Examples of water samplers are illustrated in Figure 9-23.

Roessette Sampler

Water Bottle on Winch Line

Figure 9-23. Examples of water samplers used in ocean engineering and oceanography.

SEDIMENT SAMPLERS

Devices for collecting sediment samples include tube and grab samplers. The core sampler is an open ended tube that is thrust vertically into the sediment to a desired depth. The sampler is then withdrawn from soil while the sediment sample is retained inside the tube. The tube usually has a thin walled plastic tube liner that can be removed with the sample intact.

A common method for driving the samplers into the sediment is by gravity because a large weight that is hydrodynamically shaped is mounted on top of the tube. In ocean applications, the tube and weight are dropped through the water, and the tube penetrates the ocean bottom sediments. A wire rope line is normally attached to the corer and is reeled back to the surface support vessel using a winch. The core samples are removed, refrigerated, and returned to the laboratory for analysis. A well known coring device used in shallow waters to obtain core samples ranging from 0.61 to 1.8 m (2 to 6 ft) is the Phleger corer as shown in Table 9-1.

Table 9-1. Examples of sediment sampling equipment

Device Name	Illustration	Weight (lb)	Description
Peterson Grab		40-90	Collects sample covering area of 1 ft^2 to a depth of 1 ft depending on sediment type.
Shipek		150	Collects sample covering area of 64 in^2 to a depth of 4 in.
Ekman		9	Works only in soft sediments covering area of up to 64 in2 to a depth of approximately 6 in.
Ponar		45-60	Collects sample covering area of 81 in^2 to a depth of 1 ft. Not effective in clay.
Drag Bucket		Varies	Collects shallow sediment slice near the surface and comes in several sizes and shapes.
Phleger Corer		20 -90	Core samples obtained by self-weight penetration or by pushing barrel into the sediment. Depth of penetration depends on sediment with 2 to 6 ft being accomplished in soft sediments.
Piston Corer	See Figure 9-24	200-300	Core samples obtained by self-weight and water pressure. Depth of penetration up to 80 ft.

Grab samplers are sampling instruments that are frequently used for obtaining surface samples of ocean, estuarine, or lake bottom sediments. Several of the more common devices are tabulated in Table 9-1. Essentially these devices scoop or bite into the surface sediment with jaws and then close to contain the sample. This device is normally dropped to the sea bottom similar to the coring devices with a winch line attached. Once the jaws are closed as a result of bottom contact or a tripping mechanism, the sampler is winched to the surface and brought aboard the support vessel. The sediment sample can then be analyzed on board or saved for later analysis at a laboratory after the vessel returns to port. These grab samplers are relatively inexpensive and easy to use.

Deeper sediment samples can be obtained using a piston corer that is capable of penetrating into the sea bottom as much as 25 m (82 ft). A piston corer shown in Figure 9-24 shows the core barrel and top weight on ship prior to being dropped over the side. The corer is allowed to free fall to the sea floor. The corer reaches the seafloor and continues to penetrate the sediment, and the sediment sample is forced into the cylinder. Tension on the cable draws a piston within the core barrel up to the top of the barrel minimizing the disturbance to the sample. The corer with the sediment sample inside is then winched to the surface and retrieved. Very deep cores can be obtained using rotary drilling techniques such as those used in oil drilling ships and drilling platforms. These techniques have been used on oceanographic research vessels such as the Glomar Challenger and JOIDES Resolution to collect cores for the Ocean Drilling Program.

Figure 9-24. Example of piston corer being lowered from the research vessel *Knorr*. (Reprinted with permission from Woods Hole Oceanographic Institute).

OCEANOGRAPHIC WINCHES

Lowering and raising ocean instrumentation from ships, offshore platforms, and fixed piers is commonly accomplished with a winch. These winches have a drum for storing the cable (typically wire rope or electromechanical cable), and the cable is run over sheaves that are attached to an A-frame, or other structure, on the vessel to allow the instruments to be lowered over the side. Manual, electric, and hydraulic power are used to drive the winch. Examples of oceanographic winches are illustrated in Figure 9-25.

Figure 9-25. Typical winches used in ocean research and operations. (Reprinted with permission from InterOcean Systems Inc.).

Winches are an important piece of equipment aboard a vessel and are used for various tasks such as coring, towing, hydrographic casts, water sampling, mid and deep water trawls, buoy laying, and deploying large objects. Deck winches are typically classified as light duty, hydrographic, and heavy duty. Light duty winches carry small diameter single or multiple strand wire or electrical cable on a small single drum. A simple fairlead system is used, and the drive system can be manual, electric, or hydraulic and may be portable. The hydrographic winch is somewhat heavier, more rugged, and carries larger diameter wire that makes it effective for water sampling, plankton trawls, and hydrographic measurements. It is commonly mounted on a welded frame and usually has a variable speed reversing electric motor, a magnetic brake, chain drive, and may have a level-wind device. The magnetic brake is used in case of loss of electric power, and the level-wind insures that the cable is layered properly as it rolls onto the drum. Heavy duty winches are used for heavy trawling, anchoring, and other work requiring substantial line loads. An example is deep sea coring that can expose the winch to loads in excess of 10 tons. Deep sea coring requires a large force to pull out the coring device after it has penetrated deep into the seafloor. High hoisting speeds are necessary to conserve ship time on station while recovering instruments. Useful equations for estimating winch horsepower and drum wire storage length (InterOcean 1985) are

$$P = \frac{V\,W}{33{,}000\,E} \qquad \textbf{9-5}$$

where W is the load in pounds, V is the desired cable speed (ft/min), and E is the drive efficiency. A reasonable approximation for the length of wire that can be stored on the winch drum is

$$L = \frac{w\,h}{4\,d_r^2}\left(d_d + h\right) \qquad \textbf{9-6}$$

where L is line length (ft) that fits on drum, h is height (in) of wire rope on drum when full, d_d is drum diameter (in), w is drum width (in), and d_r is wire rope diameter (in).

UNDERWATER RELEASE DEVICES

In ocean engineering there are many instances requiring an instrument, mooring line, or anchor to be released from the seafloor. In typical deep waters it is impractical to use divers and/or remotely operated vehicles for this type of activity. However, release devices can handle loads of near a million pounds, and this release mechanism is actuated either by an electric or hydraulic motor with a control signal that is hard wired, acoustic, or from a timer.

Acoustic releases have been developed that operate on multiple transponder interrogation frequencies. As illustrated in Figure 9-26, the equipment for an acoustic release system includes a command transmitter, acoustic receiver, a transponder, and range display. The release system uses a secure sequential tone coding technique that eliminates problems resulting from ambient noise and multipath distortion.

Rig anchor releases, Figure 9-26, are used for the release of drillships and semi-submersible drilling vessels from their mooring systems. These are used to quickly abandon a drilling site when unfavorable weather conditions arise. Later, the vessels can return to the site and reconnect to the mooring systems without having to reset anchors.

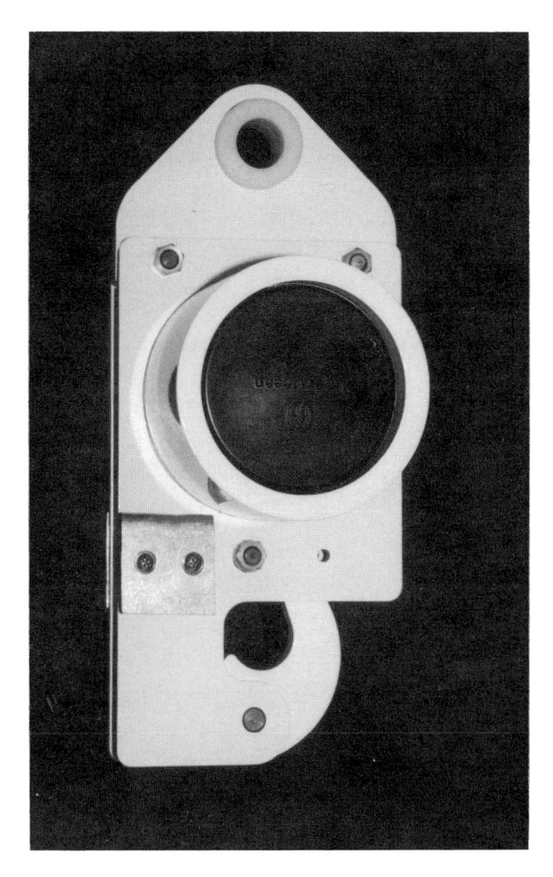

Acoustic Release

Rig Anchor Release

Figure 9-26. Example acoustic and rig anchor release systems. (Reprinted with permission from InterOcean Systems Inc.).

REFERENCES

DANTEC. "Laser Doppler Anemometry." Dantec Elektronik, Publ. No. 3205, Denmark. 1983.

ENDECO/YSI, ENDECO/YSI General Catalog. Marion, 1990.

Hughes, S. A. *Physical Models and Laboratory Techniques in Coastal Engineering*. New Jersey: World Scientific, 1993.

InterOcean Systems Inc. "InterOcean Systems Catalog." San Diego, 1985.

Lewis, E. L. "The Practical Salinity Scale 1978 and its antecedents." *IEEE Journal of Oceanic Engineering*, OE-5(1980): 3-8.

Markle, D. G., and H. C. Greer. "Crescent City Instrumented Model Dolos Study; Coastal Model Investigation." Technical Report CERC-92-15, US Army Engineer Waterways Experiment Station, Coastal Engineering Research Center, Vicksburg, 1992.

Pickard, G. L. *Descriptive Physical Oceanography*, 2nd. Edition. New York: Pergamon Press, 1975.

Pickard, G. L. and W. J. Emery. *Descriptive Physical Oceanography: An Introduction*, 5th Edition. New York: Pergamon Press, 1993.

Sharp, H. H. *Hydraulic Modeling*. London: Butterworths, 1981.
Williams, J. *Oceanographic Instrumentation.* Annapolis: Naval Institute Press, 1973.
Zimmerman, S. T, and R. E. Randall. "Operational Analysis of a Plunger-Type Wavemaker and Design of a Diffraction Experiment on a Detached Breakwater." TEES, Texas A&M University, COE Report No. 348, December 1995.

CHAPTER 10: PHYSICAL MODELING

INTRODUCTION

Physical modeling is used extensively in the study of fluid flow phenomena. The hydraulic engineer uses hydraulic models to study the effects of hydraulic structures (e.g. spillways, river basin changes). The ocean engineer uses models to optimize the design of offshore structures (e.g. oil production platforms, offshore breakwaters) and to evaluate physical processes in bays, estuaries, and along the coastline. Engineers from other disciplines use models in the design and analysis of aircraft, ships, turbines, etc. The objective herein is to briefly discuss the fundamental concepts of dimensions, dimensional analysis, and similarity that are used in physical model testing.

The history of physical modeling dates back to the 1500's when Da Vinci investigated the fluid flow associated with free jets and vortices using visual observation of experiments (Hughes 1993). In the 1700's Newton studied similarity criteria of mechanical processes and Smeaton conducted scale model tests on the performance of water wheels (Hudson et al. 1979). Measurements with ships in a towing tank were conducted by Froude in 1870, and Reynolds conducted movable bed model tests for river beds in 1885. A breakthrough was accomplished by Buckingham in 1914 with the development of the Buckingham Pi theorem for describing dimensionless parameters (Buckingham 1914). A classic text on the subject of dimensional analysis was published by Bridgeman (1922). The subjects of physical modeling and dimensional analysis are discussed in many texts such as Murphy (1950), Langhaar (1951), Sedov (1959), Schuring (1977), Yalin (1971 and 1989), and Sharp (1981).

There are several excellent references related to physical modeling of ocean engineering applications such as found in basic fluid mechanics texts of White (1986) and Munson et al. (1994). Modeling related to coastal processes is described by Hudson et al. (1979), Hughes (1993), Dalrymple (1985 and 1989), Dean (1985), Kamphuis (1991), Keulegan (1966), Le Mehaute (1990), and Svendsen (1985), and the modeling of offshore structures is discussed by Chakrabarti (1989 and 1994).

DIMENSIONS AND UNITS

In ocean engineering and fluid flow processes, the primary dimensions are mass (M), length (L), time (T), and temperature (θ). The secondary dimensions are derived from these primary dimensions. In order to give meaning to a dimension, units are attached to the final result (e.g. pounds, Newtons, centimeters, and inches). The International System (SI) and the British gravitational (BG) units are the more frequently used units in engineering. One of the most important secondary dimensions is that of force (F), which is derived from Newton's second law ($\vec{F} = m\vec{a}$) and dimensionally written as:

$$F = MLT^{-2} \qquad \textbf{10-1}$$

In terms of the SI and BG unit systems, the units of force are

$$\begin{aligned} &\text{SI} \quad 1 \text{ newton of force} - 1\text{ N} = 1\text{kg m s}^{-2} \\ &\text{BG} \quad 1 \text{ pound of force} - 1 \text{ lb} = 1 \text{ slug ft s}^{-2} \end{aligned} \qquad \textbf{10-2}$$

The primary dimensions and some of the important secondary dimensions in the SI and BG systems are tabulated in Table 10-1.

Table 10-1. Dimensions and units in the International (SI) and British Gravitational (BG) systems.

Primary dimensions	**SI Units**	**BG Units**	**Conversion Factor**
Mass (M)	Kilogram (kg)	Slug	1 slug = 14.5939 kg
Length (L)	Meter (m)	Feet (ft)	1 ft = 0.3048 m
Time (T)	Second (s)	Second (s)	1 s = 1 s
Temperature (θ)	Kelvin (K)	Rankine (R)	1 K = 1.8 R
Secondary dimensions	**SI Units**	**BG Units**	**Conversion Factor**
Acceleration (LT^{-2})	m/s^2	ft/s^2	1 ft/s^2 = 0.3048 m/s^2
Angular velocity (T^{-1})	s^{-1}	s^{-1}	1 s^{-1} = 1 s^{-1}
Area (L^2)	m^2	ft^2	1 m^2 = 10.764 ft^2
Density (ML^{-3})	kg/m^3	$slugs/ft^3$	1 $slug/ft^3$ = 515.4 kg/m^3
Dynamic viscosity ($ML^{-1}T^{-1}$)	kg/(m-s)	slugs/(ft-s)	1 slug/(ft-s) = 47.88 kg/(m-s)
Energy, heat, work (ML^2T^{-2})	Joule (J) = N-m	ft-lb	1 ft-lb = 1.3558 J
Force (MLT^{-2})	Newton (N)	Pound (lb)	1 lb = 4.4482 N
Frequency (T^{-1})	s^{-1}	s^{-1}	1 s^{-1} = 1 s^{-1}
Kinematic viscosity ($L^{-2}T$)	m^2/s	ft^2/s	1 m^2/s = 10.76 ft^2/s
Power (ML^2T^{-2})	Watt (W) = J/s	(ft-lb)/s	1 (ft-lb)/s = 1.3558 W
Pressure or stress ($ML^{-1}T^{-2}$)	Pascal (Pa) = N/m^2	lb/ft^2	1 lb/ft^2 = 47.88 Pa
Specific heat ($L^2T^{-2\,-1}$)	$m^2/(s^2\text{-K})$	$ft^2/(s^2\text{-R})$	1 $m^2/(s^2\text{-K})$ = 5.980 $ft^2/(s^2\text{-R})$
Velocity (LT^{-1})	m/s	ft/s	1 ft/s = 0.3048 m/s
Volume (L^3)	m^3	ft^3	1 m^3 = 35.315 ft^3

DIMENSIONAL ANALYSIS

The dimensional analysis method is used to reduce the number and complexity of the experimental variables that govern a particular physical problem. If the problem depends upon "n" dimensional variables, then dimensional analysis reduces the problem to "k" dimensionless variables. Usually n-k equals the number of primary dimensions (mass, length, time, and temperature) that govern the problem. In addition to reducing the number of variables, dimensional analysis saves time and money, assists in the planning of experiments, and provides scaling laws.

The principal of dimensional homogeneity states that an equation expresses the correct relationship between variables for a physical process when each of its additive terms has the same dimensions. In other words the equation is dimensionally homogeneous. For example, the Bernoulli equation for incompressible flow is

$$\frac{p}{\gamma} + \frac{V^2}{2g} + z = \text{constant} \qquad \textbf{10-3}$$

where p is the pressure, γ is the specific weight of the fluid, V is the average velocity, g is gravitational acceleration, and z is the elevation. The additive terms and the constant have the dimensions of length (L). Therefore, the equation is dimensionally homogeneous and gives the correct results for any consistent set of units.

Buckingham Pi Theorem

This theorem, originally developed in the early 1900s (Buckingham 1914), is a well known dimensional analysis method for reducing the number of dimensional variables to a smaller number of dimensionless groups. The number of dimensionless groups are power products defined by $\pi_1, \pi_2, \ldots \pi_k$. When a physical process involves n dimensional variables, it is possible to reduce the number of variables to k dimensionless variables, or π terms. The reduction ($j = n - k$) is the maximum number of variables that do not form a π group among themselves and is less than or equal to the number of primary dimensions describing the variables.

Consider the case of a submerged sphere in a flowing stream. The force on the submerged body depends on the body diameter (D), stream velocity (V), fluid density (ρ), and fluid viscosity (μ), and it may be expressed as

$$F = f(D, V, \rho, \mu) \qquad \textbf{10-4}$$

This equation contains five variables (F, L, V, ρ, μ) that are described by three primary dimensions (M, L, T). Therefore, $n = 5$ and $j \leq 3$. It is estimated that the number of variables can be reduced to two π dimensionless groups ($k = n - j = 5 - 3 = 2$).

Next, the π terms are evaluated one at a time. Select j variables that do not form a π term themselves. Each π term is a power product of these j variables and one additional variable. Assign any convenient non-zero exponent to the additional variable. The exponents of the various dimensions of the variables are equated, and the equations are solved for the exponents of the selected variables for each π term. The steps used in the Buckingham Pi analysis are:

1. Determine the number of variables (n) affecting the problem. If any important variables are missed, the analysis will be invalid.

2. Make a list of dimensions of each variable according to MLTθ. The dimensions of selected fluid mechanics properties are tabulated in Table 10-2.

3. Determine the value of j by evaluating the number of different dimensions included in the variables. Find j variables that do not form a π term in themselves. If this is not possible, reduce j by one and repeat.

4. Make final selection of j variables that do not form a π term, these are called repeating variables. Try for some generality since they appear in each π term.

Table 10-2. Dimensions for selected ocean and fluid properties.

Quantity	Symbol	MLTθ
Angle	θ	none
Angular velocity	ω	T^{-1}
Area	A	L^2
Density	ρ	ML^{-3}
Dynamic viscosity	μ	$ML^{-1}T^{-1}$
Force	F	MLT^{-2}
Kinematic viscosity	$\nu = \mu/\rho$	L^2T^{-1}
Length	L	L
Mass	m	M
Mass flow rate	$\dot{m}$	MT^{-1}
Moment, torque	M	ML^2T^{-2}
Power	P	ML^2T^{-1}
Pressure, stress	p, σ	$ML^{-1}T^{-2}$
Salinity	S	none
Specific heat	c_p, c_v	$L^2T^{-2}\theta^{-1}$
Specific weight	$\gamma = \rho g$	$ML^{-2}T^{-2}$
Speed of sound	c	LT^{-1}
Strain rate	$\dot{\varepsilon}$	T^{-1}
Surface tension	ς	MT^{-2}
Temperature	T	θ
Velocity	V	LT^{-1}
Volume	V	L^3
Volume flow rate	Q	L^3T^{-1}
Water depth	d	L
Wave frequency	f	T^{-1}
Wave height	H	L
Wave length	λ, L	L
Wave number	$k = 2\pi/L$	L^{-1}
Wave period	T	T
Work, energy	W, E	ML^2T^{-2}

5. Add one additional variable from the remaining list of n variables and form a power product. Equate the exponents and solve for the values that make the power product (π term) dimensionless. Arrange for the dependent variables (e.g. force, pressure drop, power, etc.) to be in the numerator if possible for better results. Add one new variable each time to the selected j variables until all ($n - j = k$) desired π terms are found.

6. Check the final π terms and insure each one is dimensionless. Determine the final dimensionless function which is typically one π term set equal to a function of the product of the remaining π terms.

The above steps are best illustrated in detail through an example problem. Therefore, let us consider the example of a fixed submerged sphere in a flowing stream. The force on the submerged sphere depends on the sphere diameter (D), stream velocity (V), fluid density (ρ), and fluid dynamic viscosity (μ) .

Step 1: Determine the general function and count the variables.

$$F = f(D, V, \rho, \mu) \tag{10-5}$$

There are five variables (n = 5).

Step 2: List dimensions of each variable.

F	D	V	ρ	μ
MLT^{-2}	L	LT^{-1}	ML^{-3}	$ML^{-1}T^{-1}$

Step 3: Determine j. No variable contains θ. Therefore, j must be ≤ 3. Inspect the list of variables and find that D, V, and ρ cannot form a π term because only ρ contains M and only V contains T. Therefore, j = 3, and k = n - j = 5 - 3 = 2. Thus, two π terms can be found.

Step 4: Select the j repeating variables. The variables D, V, ρ are good.

Step 5: The repeating variables (D, V, ρ) are combined with one of the remaining variables sequentially to find two π terms. For the first π term, add force (F). Select any exponent to place it in the numerator or denominator to any power. In this problem, F is the dependent variable, so it is selected to appear to the first power in the numerator.

$$\pi_1 = L^a V^b r^c F = M^o L^o T^o$$
$$\pi_1 = (L)^a (LT^{-1})^b (ML^{-3})^c (MLT^{-2}) = M^o L^o T^o \tag{10-6}$$

Equating exponents yields

Length: a + b - 3c + 1 = 0
Mass: c + 1 = 0
Time: - b - 2 = 0

Solving the three simultaneous equations yields a = -2, b = -2, and c = -1. Therefore,

$$\pi_1 = L^{-2} V^{-2} \rho^{-1} F = \frac{F}{L^2 V^2 \rho} \tag{10-7}$$

Now solve for π_2. Add dynamic viscosity (μ) and select any power. By custom the power (-1) is selected to place the dynamic viscosity in the denominator.

$$\pi_2 = L^a V^b \rho^c \mu^{-1} = M^o L^o T^o$$
$$\pi_2 = (L)^a (LT^{-1})^b (ML^{-3})^c (ML^{-1}T^{-1})^{-1} = M^o L^o T^o \tag{10-8}$$

Equating exponents yields

Length: a + b - 3c + 1 = 0
Mass: c - 1 = 0

Time: $-b \quad + 1 = 0$

The solution is a = 1, b = 1, and c = 1. Therefore, the second π term is

$$\pi_2 = LV\rho\mu^{-1} = \frac{\rho VL}{\mu} = Re \qquad \textbf{10-9}$$

which is the Reynolds number (Re).

Step 6: Insure π_1 and π_2 are non-dimensional and determine the final functional relationship as

$$\frac{F}{\rho V^2 L^2} = g\left(\frac{\rho VL}{\mu}\right)$$

or $\qquad \textbf{10-10}$

$$F = g\left(\frac{\rho VL}{\mu}\right)\rho V^2 L^2$$

where "g" indicates the first π parameter is a function of the Reynolds number.

The typical expression for drag force on a submerged body is

$$F = \frac{1}{2} C_D \, A \rho V^2 \qquad \textbf{10-11}$$

where C_D = f(Re), A is the cross-sectional area normal to the flow, and 1/2 is a proportionality constant.

Dimensionless Parameters

The Buckingham Pi theorem may be used to analyze many different problems and find the dimensionless parameters that apply in each problem. Another powerful approach is to use the basic flow equations. These equations and their boundary conditions may be non-dimensionalized to reveal basic dimensionless parameters. The details of this procedure is not undertaken herein but may be found in various fluid mechanics texts such as Munson et al. (1994) and White (1986). Some of the typical dimensionless parameters in ocean engineering are tabulated in Table 10-3.

MODELING

Dimensions, units, dimensional homogeneity, the Buckingham Pi theorem, and dimensionless parameters have been discussed so far. These ideas are straightforward and seem to be easily accomplished. In actuality, the selection of the important variables is difficult and usually requires good judgment and experience. Many difficult decisions must be made such as: Is viscosity unimportant? Can surface tension be neglected? Is wall roughness important? Can some of the pi terms be neglected?

Table 10-3. Typical dimensionless parameters in Ocean Engineering.

Parameter	Definition	Force Ratio	Application
Cavitation number (Euler number)	$Ca = \frac{p - p_v}{\rho V^2}$	$\frac{\text{Pressure}}{\text{Inertia}}$	Cavitation
Drag coefficient	$C_D = \frac{D}{\frac{1}{2}\rho V^2 A}$	$\frac{\text{Drag force}}{\text{Dynamic Force}}$	Hydrodynamics, Aerodynamics
Froude number	$Fr = \frac{V^2}{g L} = \frac{V}{\sqrt{g L}}$	$\frac{\text{Inertia}}{\text{Gravity}}$	Free Surface Flow
Keulegan-Carpenter number	$KC = \frac{VT}{D}$	Period Parameter	Hydrodynamics, Wave Forces
Lift coefficient	$C_L = \frac{L}{\frac{\rho V^2 A}{2}}$	$\frac{\text{Lift force}}{\text{Dynamic Force}}$	Hydrodynamics, Aerodynamics
Mach number	$Ma = \frac{V}{c}$	$\frac{\text{Flow Speed}}{\text{Sound Speed}}$	Compressible Flow
Pressure coefficient	$C_p = \frac{p - p_a}{\frac{1}{2}\rho V^2}$	$\frac{\text{Static Pressure}}{\text{Dynamic Pressure}}$	Hydrodynamics, Aerodynamics
Relative water depth	d/L	$\frac{\text{Water Depth}}{\text{Wave Length}}$	Waves, Hydrodynamics
Reynolds number	$Re = \frac{\rho VL}{\mu}$	$\frac{\text{Inertia}}{\text{Viscous}}$	Viscous Flow
Roughness ratio	$\frac{\varepsilon}{L}$	$\frac{\text{Wall Roughness}}{\text{Body Length}}$	Turbulent, rough walls, pipe flow
Strouhal number	$St = \frac{\omega L}{V}$	$\frac{\text{Oscillation}}{\text{Mean Speed}}$	Oscillating Flow Vortex Shedding
Ursell number	$Ur = \frac{HL^2}{d^3}$	Depth Parameter	Hydrodynamics, Waves
Weber number	$We = \frac{\rho V^2 L}{\varsigma}$	$\frac{\text{Inertia}}{\text{Surface Tension}}$	Free Surface Flow

When the dimensional analysis is completed, the experimenter wants to have similarity between the model and the prototype and to determine the relationship between the π parameters.

$$\pi_1 = f(\pi_2, \pi_3, \ldots \pi_k) \qquad \textbf{10-12}$$

When the above dimensionless parameters have the same corresponding values for the model and prototype, there is complete similarity between the flow conditions in the model (subscript m) and prototype (subscript p). In terms of equations, this means that if $\pi_{2m} = \pi_{2p}$, $\pi_{3m} = \pi_{3p}$, etc., then π_{1m} will equal π_{1p}. However, it is very difficult to attain complete similarity. Generally, there are three types of similarity: geometric, kinematic, and dynamic.

Geometric Similarity

Geometric similarity is attained when all body dimensions in all three coordinates for the model and prototype have the same linear scale ratio. This means that all length scales must be the same. If the model is one-fifth ($L_m/L_p = 1/5$) the size of the prototype, then its length, height, and width are one-fifth as large. All angles are the same. The flow direction and orientation of model and prototype must be identical. If there is any change in these details, geometric similarity is violated and experimental justification is required to show that the change has little or no affect.

Kinematic Similarity

Kinematic similarity is attained when the model and prototype have the same length scale ratio and the same time scale ratio. The fact that the model and prototype have the same length scale ratio implies there is geometric similarity. Having the same time scale ratio may require additional consideration such as Reynolds number equivalency ($Re_m = Re_p$). In the case of the free surface, frictionless flows kinematic similarity is attained if the Froude numbers are equal.

$$Fr_m = \frac{V_m^2}{gL_m} = \frac{V_p^2}{gL_p} = Fr_p \qquad \textbf{10-13}$$

The Froude number contains only length and time dimensions, and thus it implies kinematic similarity that determines the relation between length and time. If the length scale ratio is

$$\alpha = \frac{L_m}{L_p} \qquad \textbf{10-14}$$

and the gravitational acceleration is the same in model and prototype, which is usually the case on the earth's surface, then the velocity scale ratio is

$$\frac{V_m}{V_p} = \sqrt{\frac{L_m}{L_p}} = \sqrt{\alpha} \qquad \textbf{10-15}$$

Since V = L/T,

$$\frac{\frac{L_m}{T_m}}{\frac{L_p}{T_p}} = \sqrt{\alpha}\,, \quad \frac{L_m}{L_p} = \frac{T_m}{T_p}\sqrt{\alpha}\,, \quad \text{and} \quad \frac{T_m}{T_p} = \sqrt{\alpha} \qquad \textbf{10-16}$$

Thus, the time scale ratio is the square root of the length scale ratio (α). For example, in wave modeling the wave height and length are related by the length scale (α) and the wave period, propagation speed, and particle velocities are related by square root of the length scale (α).

Dynamic Similarity

Dynamic similarity is attained when the model and prototype have the same length, time, and force scale ratios. Geometric similarity must be attained first. Dynamic similarity exists at the same time with kinematic similarity when the force ratios are the same in the model and prototype. For an incompressible flow with no free surface, this dynamic similarity is guaranteed if the Reynolds number in the model and prototype is the same.

In the case of a free surface, both the Reynolds and Froude numbers must be equal in the model and prototype

$$\mathrm{Re_m} = \mathrm{Re_p} \text{ and } \mathrm{Fr_m} = \mathrm{Fr_p} \qquad \textbf{10-17}$$

Dynamic similarity insures kinematic similarity. Perfect dynamic similarity is rarely attained because simultaneous equivalence of Reynolds and Froude numbers requires drastic changes in fluid properties. In the case of a free surface hydraulic model, equivalent Reynolds number requires that

$$\frac{\mathrm{V_m\,L_m}}{\nu_m} = \frac{\mathrm{V_p\,L_p}}{\nu_p}$$

$$\frac{\nu_m}{\nu_p} = \frac{\mathrm{V_m\,L_m}}{\mathrm{V_p\,L_p}} \qquad \textbf{10-18}$$

For a length scale ratio (α) the Froude number equivalency requires that

$$\frac{\mathrm{V_m}}{\mathrm{V_p}} = \sqrt{\alpha} \qquad \textbf{10-19}$$

Therefore, the kinematic viscosity ratio is

$$\frac{\nu_\mathrm{m}}{\nu_\mathrm{p}} = \alpha^{\frac{3}{2}} \qquad \textbf{10-20}$$

Most hydraulic prototypes involve water as the fluid. For a 1/10 scale model, the kinematic viscosity in the model must be 3.16 x 10^{-2} times the kinematic viscosity of water. No fluid satisfies this criteria. The closest fluids are mercury or gasoline, and neither of them are desirable to use. An example modeling problem is illustrated in Table 10-4.

Example Problem 10-1

A ship of 400 ft length is to be tested by a model 10 ft long. If the ship travels at 30 knots, what speed must the model be towed for dynamic similitude between model and prototype? If the drag of the model is 2 lb, what prototype drag is to be expected?

Table 10-4. Results of example physical modeling problem 10-1.

Given	$L_p = 400$ ft, $L_m = 10$ ft, $V_p = 30$ kts, $D_m = 2$ lb
Find	V_m, D_p
Solution	$F_p = F_m\ ;\ \frac{V_p}{\sqrt{gL_p}} = \frac{V_m}{\sqrt{gL_m}}$ Since $V_p = 30 \times 1.689 = 50.7 \text{ft/s}\ ;\ V_m = 50.7\frac{\sqrt{32.2(10)}}{\sqrt{32.2(400)}} = 8 \text{ ft/s}$ $\frac{D_m}{\rho_m V_m^2 L_m^2} = \frac{D_p}{\rho_p V_p^2 L_p^2}$ Since $\rho_m = \rho_p\ ;\ \frac{2}{(8)^2(10)^2} = \frac{D_p}{(50.7)^2(400)^2}$ $D_p = 128{,}525$ lb Check Reynolds Number ; $R_p = \frac{VL}{\nu} = \frac{(50.7)(400)}{1.05 \times 10^{-5}} = 1.9 \times 10^9\ ; R_m = \frac{(8)(10)}{1.05 \times 10^{-5}} = 7.6 \times 10^6$ Exact dynamic similarity is not attained and there is a Reynolds number effect.

Fixed Bed Modeling

These models have solid boundaries that do not change as a result of the hydrodynamic process occurring within the model area. Such models are used to study waves, currents, interactions of waves, winds and currents with structures such as breakwaters, offshore fixed and floating platforms, piers, jetties, and harbors. Two dimensional model facilities are usually long and narrow tanks or flumes used to study wind, wave and current interactions, and wave kinematics. Three dimensional facilities are more complex with the length and width of the facility near the same order of magnitude. Thus, directionality effects may be studied in these usually large facilities.

Movable Bed Modeling

In these model facilities, the bed, or bottom, is constructed of material, such as sand, that can move or react to the hydrodynamic forces applied. In most cases, the bed is constructed with sediment material that is similar to that in the bay, lake, estuary, or ocean that is being modeled. Scaling is difficult to satisfy in movable bed models because it is difficult to scale the sediment size and consequently, near prototype size is desired for such modeling. Movable bed modeling is used to model scour around subsea pipelines and coastal structures (jetties, breakwaters), beach profile development, effects of storms on beach fills, seabed ripple formation, littoral drift, and dredged material islands.

Distorted Modeling

In some modeling applications, it is necessary to deviate from the geometric scale in one of the model directions and this is called a distorted model. In other words, the horizontal and vertical scales are different, and as a consequence the geometric similitude is relaxed. An example is the modeling of an estuary that is several kilometers wide and only 4 meters deep. In order to fit the model in a facility a horizontal scale of 1:500 may be used. Applying the same scale to the depth yields a model depth of 8 mm which is not reasonable. Using a scale of 1:50 for the vertical dimension yields a depth of 8 cm which is more reasonable. The construction of a physical model using a 1:500 horizontal and a 1:50 vertical scale is considered a distorted model. Detailed discussion of these distorted models may be found in Langhaar (1951), Hudson et al. (1979), Hughes (1993), and others.

PHYSICAL MODELING FACILITIES

Facilities used for physical modeling in ocean engineering include long narrow wave tanks, towing tanks, variable slope flumes, wind-wave tanks, shallow water wave basins, and deep water wave basins, circulating water tunnels, low speed wind tunnels, centrifugal pump test loops and pipelines. Many of the smaller facilities are located at academic institutions and the larger facilities usually operated by government organizations. Some selected large modeling facilities that are used for modeling of offshore and coastal applications are listed in Table 10-5 and Table 10-6.

According to Hughes (1993) the first hydraulic modeling laboratory in the US was established at Lehigh University in 1887. The 20th century saw the establishment of such facilities expand worldwide. In 1914, the Institute of Hydraulic Engineering was founded in Hanover, Germany, and the Hydraulic Laboratory at the University of Iowa was started in 1918. The Delft Hydraulics Laboratory in the Netherlands began in 1927 to assist that country in protecting the land from coastal flooding. Two years later (1929) the Waterways Experiment Station in Vicksburg, Mississippi was started in the US where a model of the Mississippi River was constructed as well as models of many important ports and harbors in the United States. Other well known modeling facilities worldwide include: Hydraulics Laboratory of the National Research Council of Canada (started 1945), Port and Harbor Research Institute in Japan (started 1946), Laboratoire National D'Hydraulique in France (1946), Hydraulics Research Station in Wallingford, England (1947), and Danish Hydraulic Institute in Denmark (1964).

Many academic institutions such as University of Florida, Massachusetts Institute of Technology, University of Michigan, University of Rhode Island, California Institute of Technology, Texas A&M University, US Naval Academy, University of Michigan, University of Delaware, Oregon State University, Technical University of Delft, New Foundland Memorial University have physical model testing facilities. In 1989, the Offshore Technology Research Center opened the large deep water wave basin, Figure 10-1, that is 45.7 m (150 ft) long, 30.5 m (100 ft) wide and 5.8 m (19 ft) deep with a 9.1 m (30 ft) long by 4.6 m (15 ft) pit that is 16.7 m (55 ft) at its maximum depth. An example of a common two dimensional wave tank facility such as the one located in the Hydromechanics Laboratory at Texas A&M University in College Station, Texas is shown in Figure 10-2. The large towing tank housed at the US Navy's David Taylor Research Center is illustrated in Figure 10-3. A large shallow water basin, Figure 10-4, used to model large harbor facilities such as the Port of Los Angeles in California is a example of the one of the several Coastal Engineering Research Center shallow water wave tank facilities at the US Army Engineer Waterways Experiment Station.

Table 10-5. Selected deep water and towing tank model testing facilities around the world.

Name	Size (length, width, depth) (m)	Wave Maker & Capabilities	Wave Absorbing Type	Carriage Speed (m/s)	Wind and Current	Location
Danish Maritime Institute	240 x 12 x 5.5	Double flap hydraulic, regular & irregular, max height 0.9 m period 0.5-7 s	NA	11	NA	Lyngby, Denmark
David Taylor Research Center (DTRC)	79.3 x 73.2 x 6.1 846 x 15.5 x 6.7 905 x 6.4 x 3	Pneumatic, max height 0.6m	Concrete with fixed bars	7.7 10.2 35.8-51.2		Bethesda, Maryland USA
Institute of Marine Dynamics Towing Tank,	200 x 12 x 7	dual-flap, dry-back max height 1 m	Corrugated plated bolted to rigid framework	10	NA	St Johns, Newfoundland
Marine Research Institute (MARIN)	100 x 24.5 x 2.5 Pit depth of 6 m 60 x 40 x 1.2 Pit depth of 3 m 252 x 10.5 x 5.5 220 x 4 x 4	Regular & Irregular, Max height 0.3 & 0.4 m sig. Period 0.3 - 5 s	Lattice on circular arc plates	4.5 9 15 & 30	Current 0.1 - 0.6 m/s	Netherlands
Norwegian Hydrodynamic Laboratory (MARINTEK)	80 x 50 x 10	Hinged double flap, 144 flaps, hydraulic, regular & irregular max height 0.9 m	NA	NA	Current 0.2 m/s	Trondheim, Norway
Offshore Model Basin	90 x 14.6 x 4.6 circular pit 9 m deep	Single flap hydraulic max height 0.74 m	Metal shavings	6	NA	Escondido, California USA
Offshore Technology Research Center (OTRC)	45.7 x 30.5 x 5.8 Pit 9.1 x 4.6 x 16.7 with adjustable floor in pit.	48 paddles, flap, hydraulic, regular, irregular, long- & short-crested, oblique & focused Period 0.5 - 4.0 s max height 0.8 m	Progressive Expanded Metal Panels	Instrument carriage speed of 0.5	Wind: bank of multiple, variable speed fans (0-12 m/s) Current: multiport jet manifolds (0-0.6 m/s)	Texas A&M Univ. College Station, Texas, USA

Table 10-6. Selected facilities used for modeling coastal processes.

Name	Approximate Floor Space Area (ft^2)	Wave Making	Location
Coastal Engineering Laboratory University of Florida	35,000	Regular, Irregular	Gainesville, Florida, USA
Coastal Engineering Laboratory University of Delaware	NA	Regular, Irregular, Directional	Newark, Delaware, USA
Coastal Engineering Laboratory Dalian University	NA	Regular, Irregular, Directional	China
Coastal Engineering Research Center US Army Corps of Engineers Waterways Experiment Station	450,000	Regular, Irregular, Directional	Vicksburg, Mississippi, USA
Delft Hydraulics Technical University of Delft	NA	Regular, Irregular, Directional	Delft, Netherlands
Fluid Dynamics Laboratory Massachusetts Institute of Technology	NA	Regular, Irregular	Cambridge, Massachusetts, USA
Hydraulic Laboratory Ministry of Transportation	NA	Regular, Irregular, Directional	Japan
Hydraulics Laboratory	NA	Regular, Irregular, Directional	Wallingford, England
Hydraulics Laboratory National Research Council	NA	Regular, Irregular, Directional	Canada
Hydromechanics Laboratory Texas A&M University	15,000	Regular, Irregular	College Station, Texas, USA
National Hydraulic Laboratory	NA	Regular, Irregular, Directional	France

Figure 10-1. Photograph of the Offshore Technology Research Center wave basin showing model of semisubmersible at top, wave maker at middle left, model oil boom at middle right, and large waves and instrument carriage at bottom (Reprinted with permission from Offshore Technology Research Center).

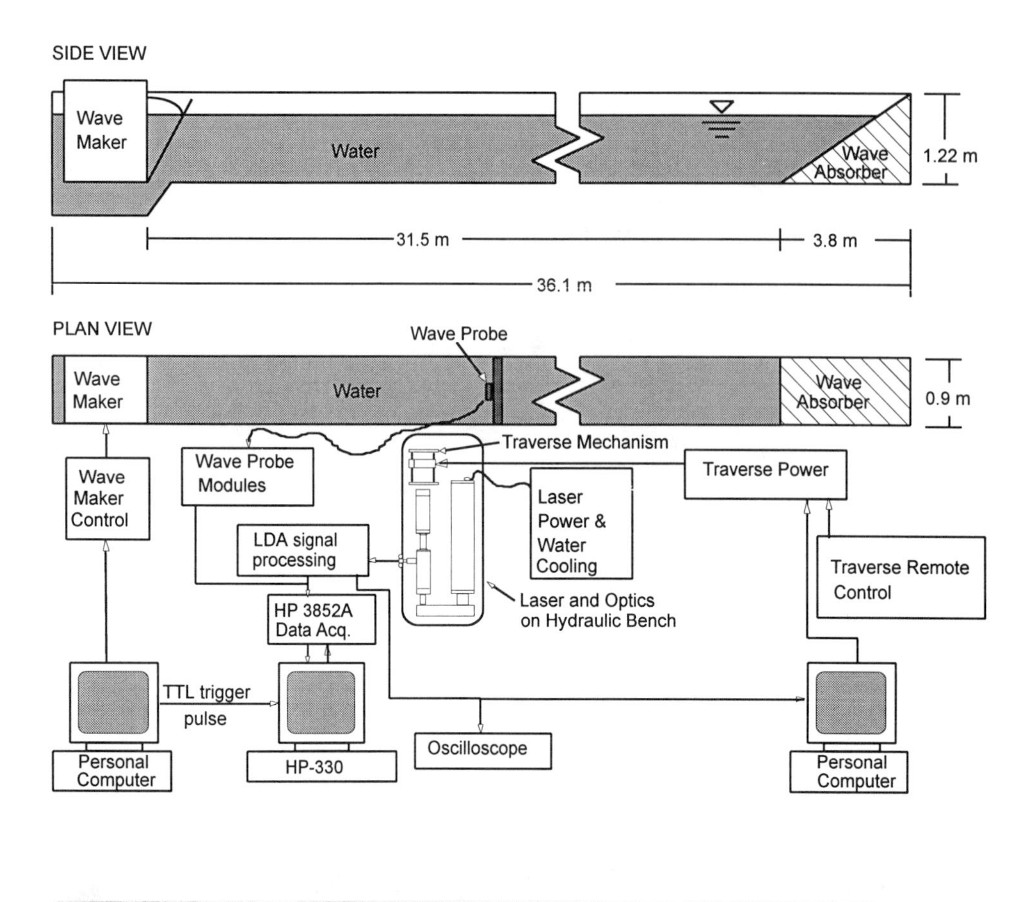

Figure 10-2. Schematic and photograph of glass-walled wave tank at Texas A&M University.

Figure 10-3. Photograph of the towing tank at the David Taylor Research Center.

Figure 10-4. Photograph of shallow water modeling facility for the Los Angeles harbor at the Coastal Engineering Research Center at the US Army Engineer Waterways Experiment Station in Vicksburg, Mississippi.

REFERENCES

Bridgeman, P. W. *Dimensional Analysis.* New Haven: Yale University Press, 1922.

Buckingham, E. "On Physically Similar Systems; Illustrations of the Use of Dimensional Equations." *Physics Review*, 4(1914): 345-376.

Chakrabarti, S. K. *Offshore Structure Modeling.* River Ridge: World Scientific Publishing Co., 1994.

Chakrabarti, S. K. "Modeling of Offshore Structures," in *Application in Coastal Modeling*, V. C. Lakhan and A. S. Trenhaile, Eds., Elsevier Oceanography Series 49(1989).

Dalrymple, R. A. "Introduction to Physical Models in Coastal Engineering," *Physical Modeling in Coastal Engineering.* R. A. Dalrymple, Ed., Balkema, Rotterdam, 1985.

Dalrymple, R. A. "Physical Modeling of Littoral Processes," *Recent Advances in Hydraulic Physical Modeling*, R. Martins, Ed. Dordrecht: Kluwer Academic Publishers, 1989.

Dean, R. G. "Physical Modeling of Littoral Processes," in *Physical Modeling in Coastal Engineering*, R. A. Dalrymple, Ed. Rotterdam: Balkema, 1985.

Hudson, R. Y., F. A. Herrmann, R. A Sager, R. W. Whalin, G. H. Keulegan, C. E. Chatham, and L. Z. Hales. "Coastal Hydraulic Models." Special Report No. 5, US Army Engineer Waterways Experiment Station, Vicksburg, 1979.

Hughes, S. A. *Physical Models and Laboratory Techniques in Coastal Engineering.* River Ridge: World Scientific Publishing Co., 1993.

Kamphuis, J. W. Physical Modeling, *Handbook of Coastal and Ocean Engineering*, J. B. Herbich, Ed., Vol. 2. Houston: Gulf Publishing Company, 1991.

Keulegan, G. H. "Model Laws for Coastal and Estuarine Models," *Estuary and Coastline Hydrodynamics*, A. T. Ippen, Ed. New York: McGraw-Hill Book Company, Inc., 1966.

Langhaar, H. L. *Dimensional Analysis and Theory of Models.* New York: John Wiley & Sons, 1951.

Le Mehaute, B. Similitude, *Ocean Engineering Science*, B. Le Mehaute, Ed., Vol. 9, Part B in the series *The Sea.* New York: John Wiley and Sons, 1990.

Munson, B. R., D. F. Young, and T. H. Okiishi. *Fundamentals of Fluid Mechanics*, 2nd Edition. New York: John Wiley & Sons, Inc., 1994.

Murphy, G. *Similitude in Engineering.* New York: Ronald Press, 1950.

Schuring, D. J. *Scale Models in Engineering.* New York: Pergamon Press, 1977.

Sedov, L. *Similarity and Dimensional Methods in Mechanics.* New York: Academic Press, 1959.

Sharp, J. J. *Hydraulic Modeling.* London: Butterworth. 1981.

Svendsen, I. A. "Physical Modeling of Water Waves," *Physical Modeling in Coastal Engineering*, R. A. Dalrymple, Ed. Rotterdam: Balkema, 1985.

White, F. M. *Fluid Mechanics*, Second Ed. New York: McGraw-Hill Book Company, 1986.

Yalin, M. S. "Fundamentals of Hydraulic Physical Modeling," *Recent Advances in Hydraulic Physical Modeling*, R. Martins, Ed. Dordrecht: Kluwer Academic Publishers, Dordecht, 1989.

Yalin, M. S. *Theory of Hydraulic Models.* London: The MacMillan Press, Ltd, 1971.

PROBLEMS

10-1. The expression for evaluating the drag force on a submerged body is $F = \frac{1}{2}\rho C_D A V^2$ where F is the drag force, ρ is the fluid density, C_D is a dimensionless drag coefficient, A is the projected area normal to the flow direction, and V is the average fluid velocity. Show this equation is dimensionally homogeneous.

10-2. Use the Buckingham Pi theorem to determine the relationship for the drag force on a ship moving through water. The important properties are ship length (L), velocity (V), gravity (g), water density (ρ), and dynamic viscosity (μ) of water.

10-3. The period of heave oscillation (T) of a spar buoy varies with its cross-sectional area (A), gravity (g), mass (m), and the water density (ρ). Use dimensional analysis to determine the relationship for the heave period.

10-4. A prototype ship is 40 m long and designed for a cruising speed of 12 m/s. The drag is to be simulated by a 1 m long model that is towed in a towing tank. Using Froude scaling, determine the tow speed, the ratio of model to prototype drag and power.

10-5. A vertical pile for an offshore platform is expected to be located in the ocean where the wave height is 3.5 m, wave period is 13 s, and the current is 3 m/s. A 1/20 scale model of the pile is placed in a modeling facility. Using Froude scaling, determine the magnitude of the velocity, wave height, and wave period in the modeling facility in order to model the ocean conditions.

10-6. A 1:40 scale model of a submarine propeller is tested in a tow tank at 1500 rpm and exhibits a power of 1.4 ft-lb/s. Using Froude scaling law, determine the rpm and power output of the prototype propeller under dynamically similar conditions.

10-7. The cylindrical piling for an offshore platform is expected to encounter currents of 160 cm/s and waves of a 13 s period and 4 m height. If a 1:60 scale model is tested in the wave tank, what is the current speed, wave period, and wave height?

10-8. The buoyant force (F_b) acting on a body submerged in a fluid is a function of the fluid specific weight (γ), the body volume (V). Show by dimensional analysis that the buoyant force is directly proportional to the specific weight.

CHAPTER 11: ENVIRONMENT, SAFETY, AND ETHICS

INTRODUCTION

Ocean engineers design systems that function in the ocean environment. These systems must be structurally sound so that they can withstand the severity of the ocean environmental forces (e.g. wind, current, storm surges, and waves). Also, ocean systems must be designed to avoid polluting or damaging marine life in coastal and offshore waters. Engineers need to be cognizant of the environmental consequences or effects of their designs and prototype systems. Safety of human life in operating engineering systems or equipment is another concern for ocean engineers. This chapter identifies the applicable environmental and safety laws that apply to engineering in the ocean environment, and the regulatory agencies or organizations that enforce these laws or regulations are also identified. Extensive efforts are often required by ocean engineers to develop environmental impact statements concerning a proposed engineering design for the ocean environment, and the rudiments of these statements are described. Also, the applicable safety laws for ocean operations are discussed. Ethics for engineers are related to professional activities and professional registration and this is briefly described.

OCEAN SYSTEM DESIGN AND SAFETY

The design and safety of ocean systems (e.g. fixed and floating offshore structures, drilling vessels, diving systems, and subsea systems) are checked and assured through procedures known as verification, certification, and classification. These procedures are administered by special regulatory agencies that are independent of the organizations that design and operate ocean systems.

Classification

Classification societies publish standards, rules, and guidelines for the design of ocean systems. Then, design organizations submit their engineering calculations, specifications, and engineering drawings to the responsible classification society, and this society checks for the compliance of the design with the societies' rules and standards. If the design is found to satisfy the society's standards, then it is said to be classed. The classification procedure also is concerned with the integrity and serviceability of the system over the life of the system installation, and therefore, special surveys and inspections are conducted during the operational lifetime of the system. The classification of ocean going vessels is a long standing and internationally accepted procedure.

Certification

Owners or a government may require conformance to specific rules or standards for a design or construction of an ocean system. A limited number of authorized certification agents perform this service. If the ocean system is found to conform to the specified rules or standards then it is said to be certified. Surveys and inspections are normally required after the initial certification to ensure continued compliance with the rules.

Verification

The procedure for verification is similar to that of certification. In the U.S., verification is used to ensure integrity of offshore platforms, and it is administered through the Platform Verification Program (Marine Board 1977). This procedure uses a third party review to assure conformance to government standards. Verification is conducted separately for design, fabrication, and installation, and different inspectors may be used for each of the phases. The list of inspectors is published by the U.S. Government.

REGULATORY BODIES

The need of the shipping industry to assure the structural and mechanical integrity of marine vessels resulted in the development of independent regulatory bodies, or classification societies. Prominent regulatory bodies include the American Bureau of Shipping (U.S.), Det Norske Veritas (Norway), and Lloyd's Register of Shipping (U.K.). The services of these regulatory bodies and others have expanded to other ocean systems such as offshore platforms, mobile drilling units, and floating production systems. These regulatory bodies are typically non-profit organizations that assess fees for their services. Some of the more prominent regulatory bodies and an abbreviated list of their published standards are tabulated in Table 11-1.

Table 11-1. Abbreviated summary of well known regulatory bodies and standards.

Regulatory Body	Rules, Standards, Responsibilities
American Bureau of Ships (ABS)	• Building and Classing Steel Vessels (ABS 1990) • Pressure Vessels • Building and Classing Mobile Offshore Drilling Units (ABS 1990)
American Petroleum Institute (API)	• Recommended Practice for Planning, Designing, and Constructing Fixed Offshore Platforms Working Stress Design, Twentieth Edition (API RP 2A-WSD 1993). • Recommended Practice for Design, Analysis, and Maintenance of Moorings for Floating Production Systems, First Edition (API RP 2FP1 1993)) • Recommended Practice for Analysis, Design, Installation and Testing of Basic Surface Safety Systems for Offshore Production Platforms (API RP 14C 1994) • Recommended Practice for Planning, Designing, and Constructing Tension Leg Platforms, First Edition (API RP 2T 1987)
Det Norske Veritas (DNV)	• Rules for Certification of Diving Systems • Rules for Classification of Mobile Offshore Units • Rules for Design, Construction and Inspection of Fixed Offshore Structures (DNV 1978)
American Society of Mechanical Engineers (ASME)	• Pressure vessel codes • Pressure vessels for human occupancy code
American National Steel Institute (ANSI)	• Piping for gas and oil • Pressure vessels

In the U.S., there are many federal agencies that are involved in regulating various aspects of offshore, or outer continental shelf, operations. Table 11-2 shows the responsibilities of nine of the more active agencies and their areas of responsibility (Cox 1986).

Table 11-2. U. S. federal agencies involved in regulating offshore oil and gas operations (Marine Board 1981).

Operation	EPA[1]	FAA[1]	FWS[1]	MMS[1]	MTB[1]	OSHA[1]	STATE	USACE[1]	USCG[1]
Aircraft		*							
Environmental issues									
Air emissions	*			*			*		
Discharge standards	*								
Oil-spill reports & cleanup	*			*			*		*
Exploration & production operations				*			*		
Hazardous materials	*								*
Installations									
Fire fighting equipment				*		*			*
Equipment	*			*		*			
Life-saving equipment									*
Structures			*	*				*	
Leasing				*			*		
Marine vessels									*
Personnel safety				*		*			*
Pipelines			*	*	*		*	*	

[1]Definitions: Environmental Protection Agency (EPA), Federal Aviation Administration (FAA), Fish and Wildlife Service (FWS), Materials Transportation Bureau, Office of Pipeline Safety (MTB), Minerals Management Service (MMS), Occupational Safety and Health Administration (OSHA), U. S. Army Corps of Engineers (USACE), and U. S. Coast Guard (USCG).

In 1982, the Minerals Management Service (MMS) assumed the responsibility for all leasing and resource management of submerged lands on the continental shelf outside state boundaries. MMS is an agency inside the U.S. Department of Interior. The responsibilities include resource development of nation's energy needs, protection of marine environment, attaining fair market value, and maintaining competition.

ENVIRONMENTAL LAWS

Regulation of the placement of structures, dredging of navigable waterways, and the disposal of dredged material within waters of the United States and ocean waters is a complex issue and is a shared responsibility of the Environmental Protection Agency (EPA) and the US Army Corps of Engineers (USACE). MPRSA, CWA, and NEPA are the major federal statute/laws governing dredging projects, but a number of other Federal Laws, Executive Orders, must also be considered. Jurisdiction of MPRSA and CWA are illustrated in Figure 11-1. Procedures for evaluating dredged material that is proposed for disposal in ocean waters is governed by USEPA/USACE (1991) and for disposal in inland or near coastal waters is governed by USEPA/USACE (1996).

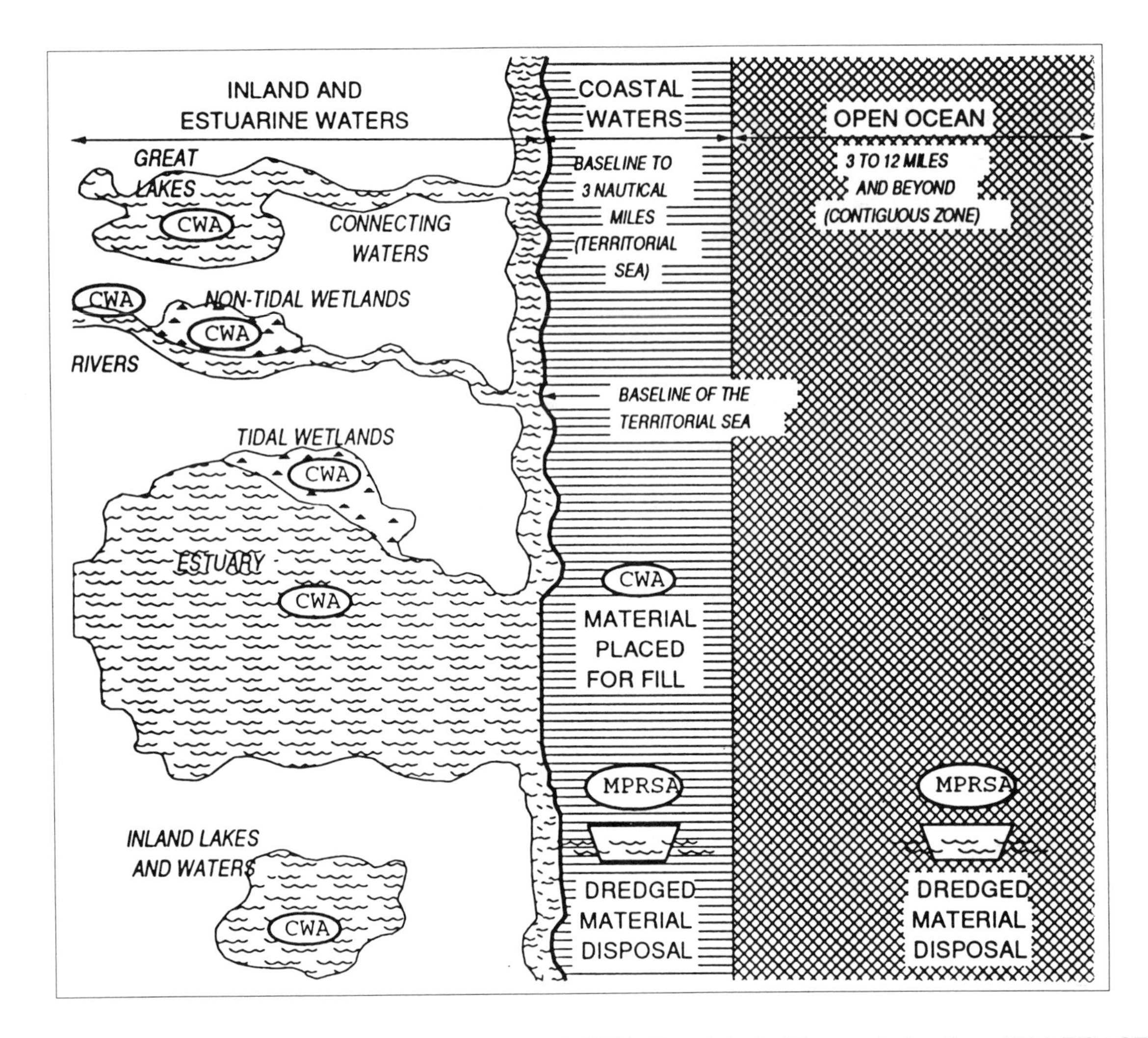

Figure 11-1. Jurisdiction boundaries for MPRSA and CWA. Reprinted with permission from EPA/USACE, 1992, "Evaluating Environmental Effects of Dredged Material Management Alternatives- A Technical Framework." (Full citing in references).

A brief description of these environmental laws are now described and many of which can be found in the Code of Federal Regulations (CFR).

MPRSA - Marine Protection, Research, and Sanctuaries Act - Ocean Dumping Act-London Dumping Convention.

This act is also referred to the London Dumping Convention and the Ocean Dumping Act. Section 102 requires the EPA to develop environmental criteria in consultation with the USACE. Section 103 assigns the USACE responsibility for authorizing the ocean disposal of dredged material. The USACE must apply the criteria developed by the EPA. Section 102 gives authority to the EPA to designate ocean disposal sites. Section 103 authorizes the USACE to select ocean disposal sites for project specific use when an EPA site is not feasible or a site has not been designated.

CWA - Clean Water Act [33 USC ff 1251-1387]

The Clean Water Act is also the Federal Water Pollution Control Act and Amendments of 1972. Section 404 requires the EPA, in conjunction with the USACE, to publish guidelines for the discharge of dredged or fill material such that unacceptable adverse environmental impacts do not occur. Section 404 assigns responsibility to the USACE for authorizing all discharges and requires the application of EPA guidelines. Section 401 provides the States a certification role for project compliance with the applicable State water quality standards.

NEPA - National Environmental Policy Act

Dredged material disposal activities must comply with the applicable NEPA requirements regarding identification and evaluation of alternatives. Section 102(2) requires examination of alternatives to the action proposed, and these alternatives are analyzed in a Environmental Assessment (EA) or Environmental Impact Statement (EIS). For USACE dredging projects, USACE is responsible for developing alternatives for the discharge of dredged material including all facets of the dredging and discharge operation, including cost, technical feasibility, and overall environmental protection. Compliance with environmental criteria of the MPRSA and/or the CWA guidelines is the controlling factor used by the USACE to determine the environmental acceptability of disposal alternatives.

The NEPA process is finalized in one of two ways. First, a Finding of No Significant Impact (FONSI) is the final decision document when an EA finds that the preparation of an EIS is not required. Secondly, an EIS is prepared, and the decision document is called a Record of Decision (ROD) that specifies the recommended action and discusses the alternatives considered (EPA/USACE 1992).

RCRA - Resource Conservation and Recovery Act (1976) [42 USC ff 6901-6992k]

RCRA regulates the collection, generation, transportation, recovery, separation, and disposal of solid wastes that includes liquids, semi-liquids, and contained gases. Under RCRA, a waste is hazardous if it is specifically listed as hazardous or if the waste has a hazardous characteristic. A hazardous characteristic means the waste is ignitable, corrosive, reactive, or toxic. Anyone who generates, stores, treats, processes, or disposes of hazardous wastes must abide by the RCRA hazardous waste management provisions. Some wastes associated with offshore oil and gas exploration and production are exempt from RCRA, and the EPA lists exempt and nonexempt offshore oil and gas wastes (Butler and Binion 1993).

CERCLA - Comprehensive Environmental Response, Compensation, and Liability Act (Superfund) (1980) and SARA- Superfund Amendments and Reauthorization Act (1986) [42 USC ff 9601-9675]

CERCLA provided authority to publish a prioritized list of highly polluted locations or sites. Only hazardous substances are covered by CERCLA and petroleum is excluded from the definition of hazardous substances. Natural gas, natural gas liquids, and refined petroleum products such as gasoline are also excluded. However, substances associated with oil and gas exploration and production are hazardous and these include methanol, caustic soda, and many mud additives. The liability under CERCLA is very formidable because it is strict, retroactive, and joint. Strict means that it is immaterial whether the party involved is at fault or not. Retroactive means that a party may be liable for cleanup even before that party owned the location of the hazardous substance. Joint means each party that is potentially liable may be liable for the total cleanup costs and not just a proportionate amount. Also, a party that did not contribute at all to the pollution is not liable, but the party must prove that it did not contribute to the pollution and did not know of the polluted condition of the site. SARA provides a means and funding for quick, responsible cleanup of locations that threaten the environment or public health.

SDWA - Safe Drinking Water Act (1974) [42 USC ff 300f-300h-26]

SDWA applies to the protection of drinking water from underground sources through underground injection.

OPA-90 - Oil Pollution Control Act (1990) [33 USC ff 2701-2761]

The OPA-90 addresses marine oil spills for both onshore and offshore facilities. This act is relatively new so the ramifications of the act are still evolving.

TSCA - Toxic Substances Control Act (1976) [15 USC ff 2601-2671; 40 CFR Part 761]

TSCA is an act that requires that chemical manufacturers, importers, and processors must supply information related to chemicals handled by each organization. Crude oil and natural gas are naturally occurring substances and are excluded by EPA regulation from the TSCA reporting requirements.

OSHA - Occupational Safety and Health Act [29 USC FF 651 et seq.]

This act has wide applications and requires notification of users of hazardous substances. The notification includes the use of warning labels on containers and the issue of Material Safety Data Sheets (MSDS). The organization is required to evaluate and inventory chemical hazards, properly label on-site containers, make the MSDS available to workers, train workers to protect themselves, and develop programs for communicating procedures for handling hazardous substances.

LDC - London Dumping Convention (1972) [26 UST 2403:TIAS 8165]

The U.S. is a signatory to the International Treaty concerned with marine-waste disposal which is the London Dumping Convention (LDC). LDC jurisdiction includes all waters seaward of the baseline of the territorial sea. MPRSA Section 102 criteria reflect the standards of the LDC.

Coastal Zone Management Act

The Coastal Zone Management Act requires the USACE to coordinate the permit review of all Federal projects with all participating State level coastal zone review agencies.

River and Harbors Act of 1899 [33 USC f 407]

This act requires a USACE permit for any work or structure in navigable waters of the U.S. The act also requires permits for placement of fill material in navigable waters.

Fish and Wildlife Coordination Act of 1958

This act requires USACE to consult with Federal and State fish and wildlife agencies to prevent damages to wildlife and provide for the development and improvement of wildlife resources for any proposed Federal project in a stream or other body of water.

Endangered Species Act of 1988 [16 USC ff 1531-1544]

This act establishes a consultation process between U.S. federal agencies and the Secretaries of the Interior or Commerce for conducting programs for the conservation and protection of endangered species. It protects threatened or endangered species of animal and plant life.

Water Resources Development Act of 1986

This act created a financing arrangement for dredging associated with navigation maintenance and improvement projects. Local sponsors finance one half the cost of improvement and one half the cost for additional maintenance dredging, and the Federal Government finances the other half.

National Historic Preservation Act of 1966 [16 USC ff 470-470w-6]

This act requires the consideration of the effects of the proposed project on any site building, structure, or object that is or may be eligible for inclusion in the National Register of Historic Places.

MBTA - Migratory Bird Treaty Act [16 USC ff 703-711]

Under this act an operator is responsible to place nets or other covers to keep migratory birds out of open pits and storage.

There are other federal statutes that are related to activities in the offshore and coastal waters that include the Comprehensive Environmental Response, Compensation, and Liability Act of 1980, Rivers and Harbors Improvement Act of 1978, Submerged Lands Act of 1953, Rivers and Harbors, Flood Control Acts of 1970, National Fishing Enhancement Act of 1984,

Federal Insecticide, Fungicide, and Rodenticide Act (1972), the Marine Mammal Protection Act (1972), and the Hazardous Material Transportation Act (1990). In addition, there are numerous Executive Orders that affect ocean engineering projects such as Executive Order No. 11988 Flood Plains that requires consideration of alternatives to incompatible development in flood plains. Another order is Executive Order No. 11990 - Wetlands that provides for protection of Federally regulated wetlands. There are many other Executive Orders that may affect ocean engineering applications and these are found in the very voluminous Code of Federal Regulations (CFR) that are typically located in major libraries.

ENVIRONMENTAL IMPACT STATEMENTS

An environmental impact statement is often required to assess the impact of the implementation of a new engineering system or change to an existing system. The cost of these studies is borne by the person, organization, or agency requesting the new or changed system. As an example, the permitting system of the Corps of Engineers is briefly described. Three types of permits exist: individual, nationwide, and general. An individual permit is required for locating a structure, excavating or discharging dredged material in waters of the United States. Nationwide permits are issued for some smaller or minor water bodies, and general permits are issued for certain regions that may require specific notification and reporting procedures. The typical Corps of Engineers review process is illustrated in Figure 11-2. Herbich (1990) discusses the preparation of an EIS and summarizes the permit application process including an example.

SAFETY EQUIPMENT

Offshore platforms, floating production systems, mobile drilling units, and offshore vessels require safety equipment and safety plans to be in place. The safety requirements are outlined in the Code of Federal Regulations (CFR) and the U. S. Coast Guard is responsible for enforcing and approving safety equipment and plans. An international treaty, Safety of Life at Sea (SOLAS) also addresses safety at sea, and the United States is a signatory of that treaty (US Department of State 1974).

Safety equipment includes life saving equipment (e.g. life preservers, life boats, and escape capsules). For example, the Code of Federal Regulations requires that each manned platform provide at least two approved life floats that have enough capacity to accommodate all persons present at one time. The Code of Federal Regulations can be found in most government and university technical libraries in the government documents section.

Fire safety is another critical safety issue for the ocean industry. Detection equipment must be installed to determine the presence of a fire. Fire fighting equipment that includes fire extinguishers, fire sprinkler systems, fire water pumps, fire mains and hose reels are also required. Automatic fire sprinkler systems are required in all accommodation spaces. Fixed fire fighting systems that use halon, carbon dioxide, and dry powder are necessary for control rooms and machinery spaces.

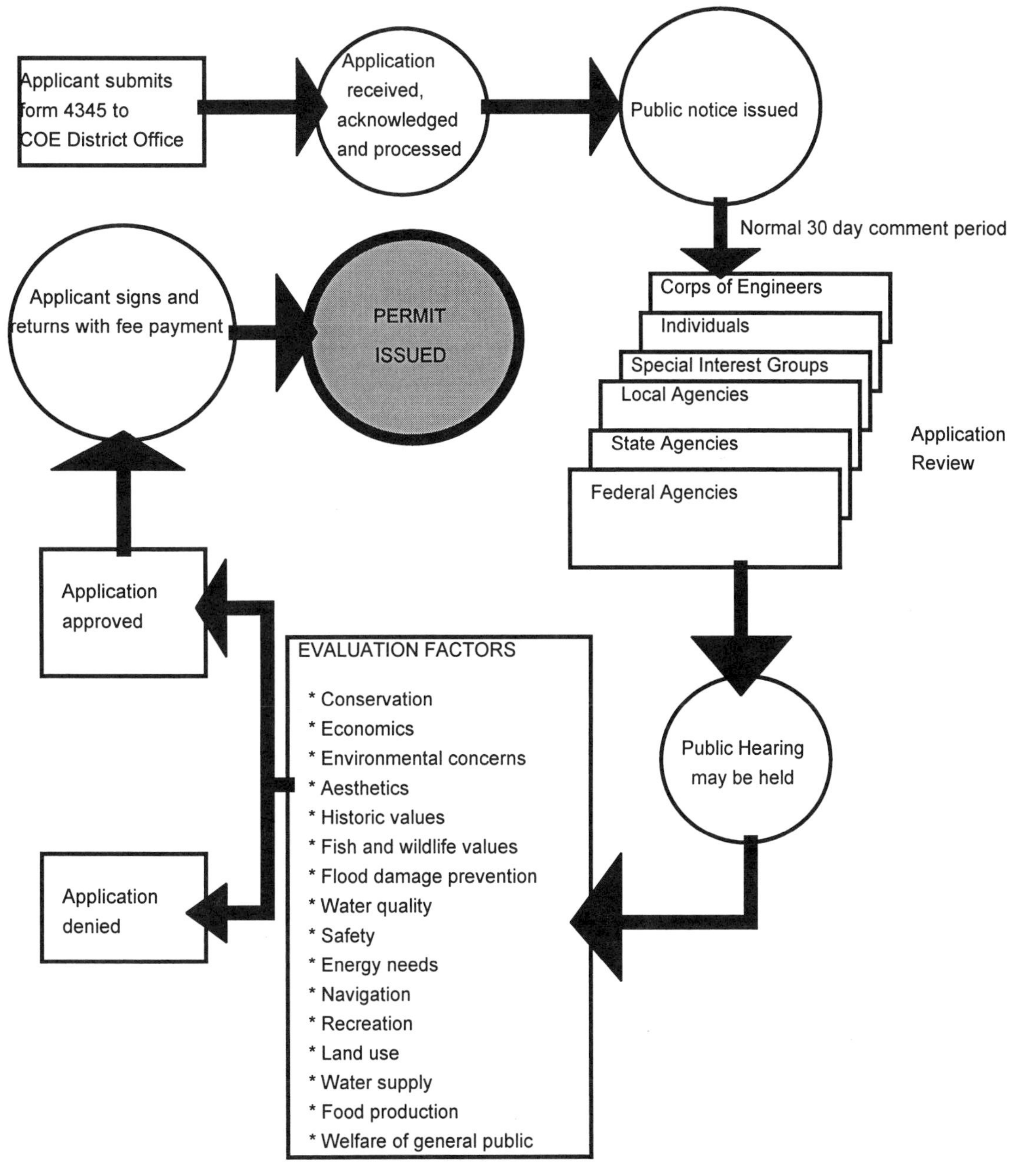

Figure 11-2. Flow chart for Corps of Engineers permit review process.

ETHICS AND PROFESSIONALISM

Ethics is defined in Webster's dictionary as " the discipline dealing with what is good and bad and with moral duty and obligation" or " the principles of conduct governing a individual or a group". The Society of Naval Architects and Marine Engineers (SNAME) is the lead professional society responsible for establishing the accreditation criteria for academic ocean engineering curricula. Other professional societies participating with SNAME are the American Society of Civil Engineers (ASCE) and the Institute of Electrical and Electronics Engineers

(IEEE). SNAME has a code of ethics that are considered applicable to Ocean Engineers. The SNAME code of ethics modified to include Ocean Engineers is contained in Table 11-3.

Table 11-3. Modified Society of Naval Architects and Marine Engineers code of ethics (SNAME 1996)

Code of Ethics

FOREWORD

Engineering work continues to be an increasingly important factor in the progress of civilization and in the welfare of the community. The Engineering Profession is held responsible for the planning, construction and operation of such work, and is entitled to the position and authority that will enable it to discharge this responsibility and to render service to humanity. Honesty, justice and courtesy form a moral philosophy that, associated with the mutual interest among all peoples, constitutes the foundation of ethics. As professionals, naval architects, marine and ocean engineers should recognize such standards, not by passive observance, but as a set of dynamic principles to guide conduct.

FUNDAMENTAL PRINCIPLES

Naval Architects, Marine and Ocean Engineers should maintain and advance the integrity, honor, and dignity of their professions by:

- using their knowledge, experience, and skill for the enhancement of human well-being and as good stewards of the environment
- striving to increase the competence of the professions of naval architecture, marine, and ocean engineering, and
- being honest and impartial, and serving with fidelity the public, their employers, and clients.

SPECIFIC CANONS

Naval Architects, Marine and Ocean Engineers shall:

1. carry on their professional work in a spirit of fairness to employees and contractors, fidelity to clients and employers, loyalty to their country, and devotion to the high ideals of courtesy and personal honor.
2. hold paramount the safety, health and welfare of the public in the performance of their professional duties. They will interest themselves in the public welfare, in behalf of which they will be ready to apply their special knowledge, skill and training for the use and benefit of mankind.
3. refrain from associating themselves with, or allowing the use of their names by, any enterprise of questionable character.
4. advertise only in a dignified manner, being careful to avoid misleading statements.
5. regard as confidential any information obtained by them as to the business affairs and technical methods or procedures of a client or employer.
6. inform a client or employer of any business connections, interests or affiliations that might influence their judgment or impair the disinterested quality or their services.
7. refrain from using any improper or questionable methods of soliciting professional work and will decline to pay or to accept commissions for securing such work.
8. accept compensation, financial or otherwise, for a particular service, from one source only, except with the full knowledge and consent of all interested parties.
9. build their professional reputations on the merits of their services and shall not compete unfairly with others.
10. perform services only in areas of their competence
11. cooperate in advancing the professions of naval architecture and marine engineering by exchanging general information and experience with their fellow naval architects, marine and ocean engineers and students, and also by contributing to the work of technical societies, schools of applied science, and the technical press.
12. continue their professional development throughout their careers and shall provide opportunities for the professional development of those naval architects, marine and ocean engineers under their supervision.

Ocean engineers should strive to become professional engineers in the state or country in which they practice. These states or countries have a governing organization that oversees the practice of engineering within their boundaries of jurisdiction. If an engineer does not abide by the ethical standards of the governing body, then the engineer's license to practice as a professional engineer can be revoked. Table 11-4 is excerpted from the State of Texas Law and Rules Concerning the Practice of Engineering and Professional Engineering Registration (Texas State Board of Registration for Professional Engineers 1988).

Table 11-4. Excerpt from the State of Texas Law and Rules concerning the practice of engineering and professional engineering registration.

PROFESSIONAL RESPONSIBILITY
• The engineer shall not participate in any engineering practice, judgment or decisions that may result in an engineering system that endangers the lives, safety, or welfare of the general public.
INDEPENDENT PROFESSIONAL JUDGMENT
• Avoid all conflicts of interest with clients and employers. • Do not accept financial or other benefits from more than one party for services on the same project or assignment. • Don't solicit or accept financial or other favors from suppliers of materials or any other contractors for projects. • As a public servant, don't make personal use of public property or services for personal use.
ACTION SHALL BE COMPETENT
• Engineer shall not accept any engineering employment or undertake an engineering assignment for which the engineer is not qualified by education or experience to undertake.
CONFIDENCES AND PRIVATE INFORMATION
• The engineer must maintain confidence of information of clients and employers.
PROFESSIONAL PRACTICE AND REPUTATION
• Engineer shall not offer gifts or favors to secure any engineering work or assignments.
RESPONSIBILITY TO THE ENGINEERING PRACTICE
• Engineer shall not engage in illegal acts, circumvent laws, discredit the engineering profession, or conduct fraudulent acts.
PREVENTION OF UNAUTHORIZED PRACTICE
• Do not practice engineering in any way that would violate the laws regulating the practice of professional engineering.

REFERENCES

American Bureau of Shipping (ABS). *Rules for Building and Classing Steel Vessels.* Baltimore: Port City Press, 1990.

American Bureau of Shipping (ABS). *Rules for Building and Classing Mobile Offshore Drilling Units*. Paramus: American Bureau of Shipping, 1990.

American Petroleum Institute (API). *Analysis of Spread Mooring Systems for Floating Drilling Units*, API-RP-2P. American Petroleum Institute, Washington, 1987.

American Petroleum Institute (API). *Recommended Practice for Analysis, Design, Installation and Testing of Basic Surface Safety Systems for Offshore Production Platforms*. API-RP-14C, American Petroleum Institute, Washington, 1986.

American Petroleum Institute (API). *Recommended Practice for Design, Analysis, and Maintenance of Catenary Mooring for Floating Production Systems*. API-RP-2F-P1, American Petroleum Institute, Washington, 1989.

American Petroleum Institute (API). *Recommended Practice for Planning, Designing, and Constructing Fixed Offshore Platforms*. API-RP-2A, American Petroleum Institute, Washington, 1987.

Butler and Binion. *Environmental Law Simplified.* Tulsa: PennWell Publishing Company, 1993.

Cox, J. W. "Standards and Regulations," Chapter 4, *Planning and Design of Fixed Offshore Platforms*, McClelland, B. and Reifel, M. D., Editors. New York: Van Nostrand Reinhold Company, 1986.

Det Norske Veritas (DNV). *Rules for the Design, Construction and Inspection of Fixed Offshore Structures*, 1978.

EPA/USACE. "Evaluating Environmental Effects of Dredged Material Management Alternatives- A Technical Framework." EPA 842-B-92-008. Washington: US Government Printing Office, November 1992.

Herbich, J. B. *Handbook of Dredging Engineering*. New York: McGraw-Hill, 1990.

Marine Board. *Safety and Offshore Oil*. Washington: Assembly of Engineering, National Research Council, 1981.

Marine Board. *Verification of Fixed Offshore Oil and Gas Platforms*. Washington: Assembly of Engineering, National Research Council, 1977.

Society of Naval Architects and Marine Engineers. 1996 SNAME Membership Directory and Information Book. Jersey City: Society of Naval Architects and Marine Engineers, 1996.

U. S. Department of State. "United States Treaties and Other International Agreements," *Safety of Life at Sea*. Washington: U.S. Government Printing Office, 1974.

USEPA/USACE. "Evaluation of Dredged Material Proposed for Discharge in Inland and Near-Coastal Waters - Testing Manual." In preparation, Office of Water, U.S. Environmental Protection Agency, Washington, 1996.

USEPA/USACE. "Evaluation of Dredged Material Proposed for Ocean Disposal (Testing Manual)." EPA-503/8-91/001, Office of Water, U.S. Environmental Protection Agency, Washington, 1991.

CHAPTER 12: OCEAN ENGINEERING DESIGN

INTRODUCTION

Engineering design is difficult to define precisely. Dym (1994) states that: "Engineering design is the systematic, intelligent generation and evaluation of specifications for artifacts whose form and function achieve stated objectives and satisfy specified constraints." Ray (1985) gives the following definition: "Design is the formulation of an inquiry or a plan or a scheme in order to arrive at the required end-product. This will involve a systematic and detailed evaluation of the problems, alternatives, and solutions." Design problems are usually considered open-ended, or ill-defined, in that several solutions may be acceptable, and the engineer must determine the optimum solution. For ocean engineering, design may be defined as the systematic development of a facility that is to be placed in an ocean, or other water body, and satisfy certain constraints and perform in a safe, environmentally sound and reliable way. Some examples of ocean engineering design include: fixed structures (e.g. fixed offshore oil platforms, submerged pipelines, and subsea cables), mooring and berthing facilities, moored systems (e.g. navigation buoys, instrument buoys, floating production systems, tension leg platforms, and floating breakwaters), coastal protection facilities (e.g. breakwaters, jetties, revetments, seawalls), instrumentation (current meters, wave gauges, sonars), dredges, submarines, remotely operated vehicles (tethered and autonomous), floating facilities (semisubmersible drilling rigs, jack-up rigs, pipe-lay vessels, and barges), beach restoration, harbors, ports, and marinas.

The results of engineering design are communicated to the builders, manufacturers, or constructors in such a way that the item can be built for use by the public or organization requesting the new design. The communication of designs is usually accomplished by engineering design drawings that are constructed by mechanical means, drafting, or through the use of computers (e.g. computer-aided drawings, Autocad). These drawings are used to build the item, and consequently, very detailed information (dimensions, tolerances, materials, parts, welding specifications) is placed on the drawings, and detailed codes and conventions that are specified by regulatory agencies and other organizations are used by the draftsperson in the development of the detailed engineering drawings. Drawings are often stored in drawing files on paper, but in current times, the files are commonly stored electronically in computer files.

It is useful to introduce the student to the design process at an early stage (first or second year of their college career) to give them exposure to the exciting design problems on which ocean engineers work. It also prepares them for senior level design courses, assists in maintaining continued interest in their discipline, and provides a view toward the start of an engineering career. The fundamental design process and several example elementary design problems are described. Elementary individual or small group design topics are developed that can be completed in a short (3 - 6 weeks) period that consist of a literature search, elementary (conceptual) design and a brief report. Of course, these topics can easily be expanded to a more detailed and complete design project. Some additional references that address the area of engineering design in more detail include Allmendinger (1990), Cross (1989), Dieter (1983), Dym (1994), Eide et al. (1979), Hill (1970), and Ray (1985).

THE DESIGN PROCESS

There are several models used to describe the process by which engineering design is accomplished (Cross 1989). Figure 12-1 illustrates a descriptive type of model for design. The ellipses indicate end points in the design process and the rectangular shapes show where work or studies are in progress.

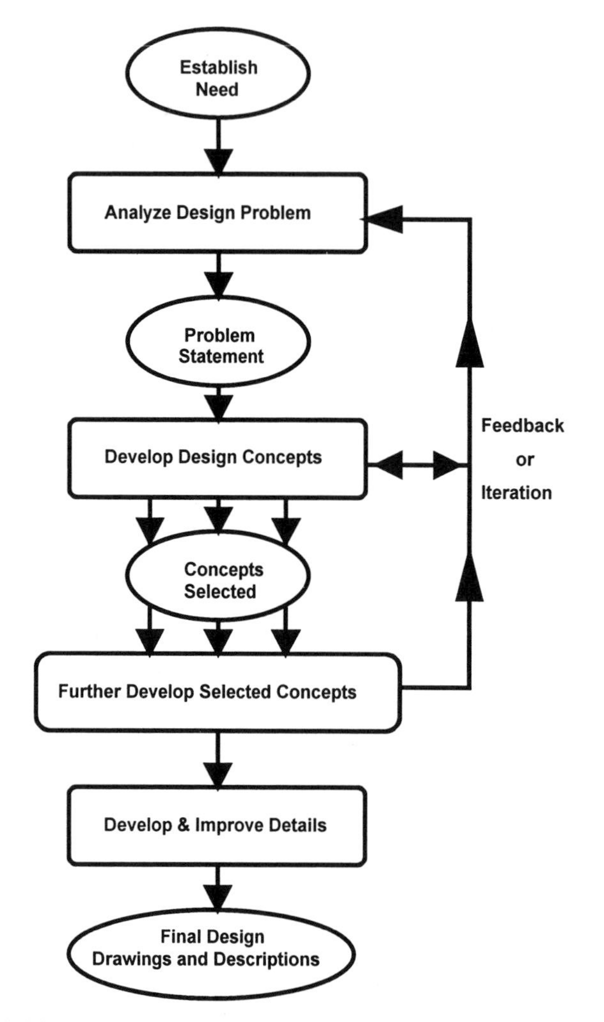

Figure 12-1. A descriptive model for the design process. Reprinted with permission from French, 1985, *Conceptual Design for Engineers*. (Full citing in references).

The first step in the design process is to establish a need. This is usually initiated by the public, industrial organization, commercial business, or federal and state government. For example, an offshore petroleum company establishes the need to place an offshore platform in the Gulf of Mexico or a state decides it needs an offshore breakwater to prevent further erosion of a shoreline. Now that a need has been identified, the next step is to analyze the problem in sufficient detail to arrive at a clear statement of the problem that identifies the criteria the new design must satisfy and the possible materials to be used. Example design criteria include the environmental conditions (waves, currents, and winds) the design must be able to resist, the cost constraints, and the codes of practice [e.g. American Bureau of Ships (ABS), American

Petroleum Institute (API), Det Norske Veritas (DNV), US Army Corps of Engineers (USACE), Environmental Protection Agency (EPA)]. Once the problem statement has been initially established, then the design team works to identify several conceptual designs. Several methods for developing these conceptual designs include brainstorming, literature searches, reviewing similar past design projects, interviewing prospective users, questionnaires, and simple testing. Upon the selection of several candidate conceptual designs, the design team must further develop the concepts by applying their engineering knowledge and experience to analyze and determine the feasibility of the design concepts.

After the design team develops the initial conceptual designs, then there is an opportunity to receive feedback and return to the initial design steps and refine the problem statement. Also, the designers may decide to eliminate one or more of the concepts, and perhaps, they may decide to initiate a new design concept that must be further analyzed and developed. This feedback or iterative process can occur many times and result in several repeats of the analyses of the design concepts. This iterative process has also been called the design spiral and is illustrated in Figure 12-2 for a submersible (Allmendinger 1990).

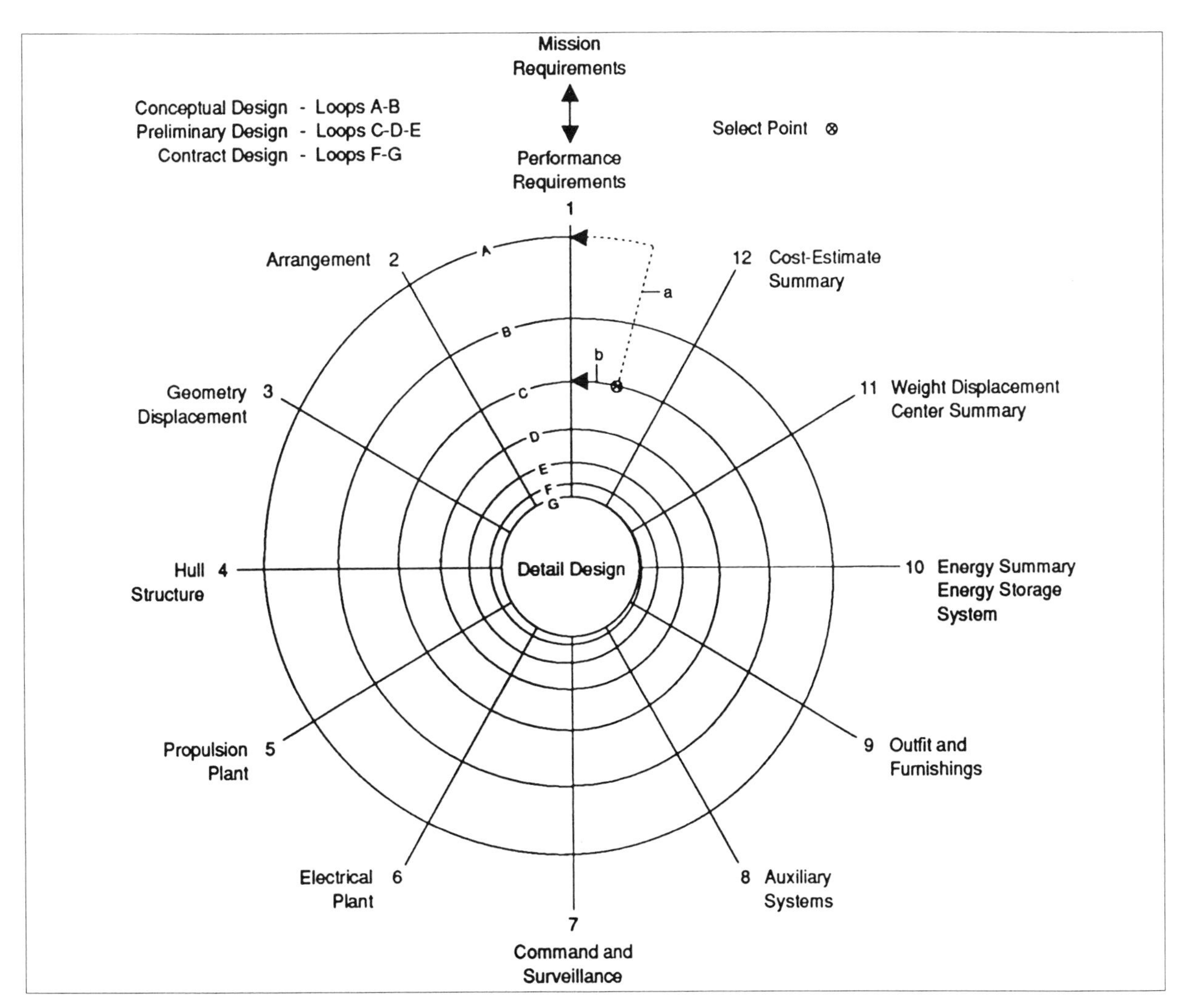

Figure 12-2. The design spiral for the design of a submersible. Reprinted with permission from Allmendinger, 1990, *Submersible Vehicle Systems Design*. (Full citing in references).

After the iterations have been completed, the conceptual designs are evaluated by some method to select the best concept that is to be further developed in detail with a final design description and detailed design drawings produced. These final descriptive products are used to manufacture, build, or construct the final design. A rational method for evaluating the different concepts is often called the method of weighted objectives (Cross 1989). The objective and criteria identified in the problem statement are given weights and an evaluation number from 1 to 10, 1 to 5, or other system is assigned based on the engineers best judgment. Numerical values are assigned and either the highest or lowest score determines the best concept that is to be developed further. An example is the evaluation of three different automobile design concepts that are about the same price. The objectives of low fuel consumption, low cost of spare parts, maintenance, and comfort are used in the evaluation. Weighting values of 0.5, 0.2, 0.1, and 0.2 are assigned to the previously mentioned objectives. The results are tabulated in Table 12-1, and show that automobile number 2 is the best overall design concept.

Table 12-1. Example of weighted objectives method applied to automobile designs (Cross 1989).

Objective	Weight	Parameter	Automobile #1			Automobile #2			Automobile #3		
			Mag.	Score	Value	Mag.	Score	Value	Mag.	Score	Value
Low fuel consumption	0.5	miles per gallon	33	2	1.0	40	4	2.0	36	3	1.5
Low cost of spare parts	0.2	Cost of 5 typical parts	$36	7	1.4	$44	5	1.0	$56	2	0.4
Maintenance	0.1	Simplicity of service	Very simple	5	0.5	Compli -cated	2	0.2	Ave.	3	0.3
Comfort	0.2	Comfort rating	Poor	2	0.4	Very good	5	1.0	Good	4	0.8
Totals					3.3			4.2			3.0

ELEMENTARY CASE STUDIES

Simple Offshore Platform

Need

Environmental data needs to be collected for a ten year period in offshore waters to support offshore oil installations. A schematic of the platform is illustrated in Figure 12-3.

Problem Statement

An oil company desires to place a single pile platform in 30.5 m (100 ft) of water to collect long term (10 year) environmental data in the Gulf of Mexico. The platform is to be located where a piled foundation can be placed in order to secure the platform to the sea bottom that is a soft sediment. The platform must support two people and ocean instrumentation to measure wind speed, water current, ambient pressure, tidal elevation, wave height, dissolved oxygen, and salinity. The surface area of the housing instrumentation equipment and power supplies is 2 m^2 (21.5 ft^2) and the estimate topside weight is 44.5 kN (10,000 lb). The company requests a preliminary design for the size of a single cylindrical tubular supported platform.

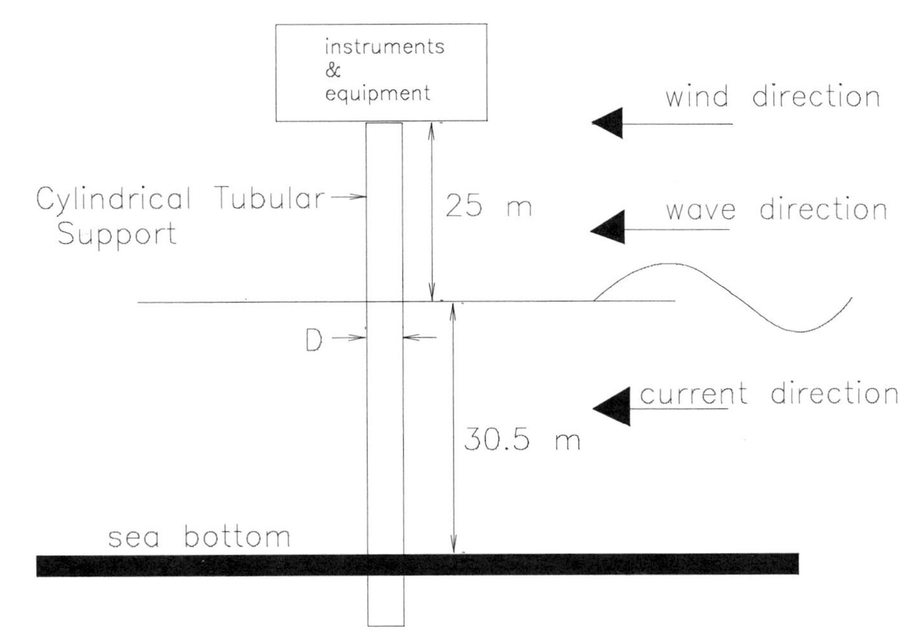

Figure 12-3. Schematic of single pile platform.

Design Criteria

After consultation with the hypothetical oil company, the design team established a set of design criteria that are shown in Table 12-2.

Table 12-2. Design criteria for single leg instrument platform.

Parameter	Value
Water depth (mean low water)	30.5m (100 ft)
Maximum current	1.03 m/s (2 kts)
Maximum deep water wave height (API 1987)	17.4 m (57 ft)
Return period of maximum wave (API 1987)	100 yr
Maximum deep water wave period	13 s
Topside exposed surface area	2 m^2 (21.5 ft^2)
Maximum topside weight	44.5 kN (10000 lb or 5 short tons)
Maximum wind speed	44 m/s (98.4 mph)
Material for cylindrical tubular support	steel

Design Computations

A major part of designing a fixed platform is determining the wave, wind, and current forces and moments on the platform cylinder. Wave forces and moments are determined using linear wave theory and the Morison equation as discussed in Chapter 3. Comparing the results of these to the allowable bending moment for different diameters, wall thickness, and yield strengths, the proper diameter for the platform cylindrical pile can then be determined. The results of the calculations are shown in Table 12-3 for a cylinder diameter of 1.067 m (42 in).

Table 12-3. Calculation of forces and moments on cylinder supporting instrument platform.

Given:	Wave Period (T) s						13
	Gravity (g) m/s^2						9.81
	Depth (d) m						30.5
	Pi						3.14159
	Wave Height (H) m						17.4
	Diameter (D) m						1.524
	Density (ρ) kg/m^3						1030
	Kinematic viscosity m^2/s						1.17E-06
	Wind speed (U) m/s						44
	Current speed (U) m/s						1.03
	Area normal to the wind (Aw) m^2						40.1
	Topside surface area normal to wind (At) m^2						2
	Distance to centroid of topside area (D2) m						0.5
	Height of platform from SWL to top deck (H2) m						25
	Air density (ρair) kg/m^3						1.23
Find:	Fi, Fd, Ft, Mi, Md, Mt on one leg over one wave period and plot						
Solution:	Pipe area normal to the current flow (Ac) m^2						46.482
	Deep water wave length (Lo) m			Lo = 1.56*T^2			263.64
	Relative water depth d/Lo.			This is an intermediate water depth.			0.115688
	Wave length for intermediate water depth L using Equation 2-40						207.896
	Wave Number (k) k=2*Pi/L						0.030223
	Maximum Horizontal Velocity (umax) m/s using Equation 2-47						5.785938
	Kuelegan-Carpenter Number (K=umax*T/D)						49.35512
	Reynolds Number (R=umax*D/kinematic viscosity)						7536555
	Drag coefficient Cd						0.62
	Inertia coefficient Cm						1.8
	Evaluate forces & moments using Morison Equation (Equation 3-4&3-5)						
	A2=(2kd +sinh 2kd)/(16 sinh^2(kd))						0.274938
	A1=Pi(a/2H)						0.06879
	C1=2*pi*rho*a*H^2*L/T^2						1836664
	A3=pi*r(1+2kdsinh(kd)-cosh(kd))/(4*H*sinh(kd))						0.016887
	A4=-(2*k^2d^2+2kdsinh(2kd)+1-cosh(2kd))/64*sinh^2(kd))						-0.07175
	C2=2*pi*rho*a*H^2*L^2/T^2=C1*L						3.82E+08
	Evaluate forces and moments due to wind and current(Eq. 3-9)						
	Results from Morison Equation for theta varying between 0 and 2Pi (one wave period)						
Theta	Finertia	Fdrag	Ftotal	Minertia	Mdrag	Mtotal	
0	0	313081	313081	0	-1.7E+07	-1.7E+07	
0.2	45181.25	300723.8	345905.1	2305879	-1.6E+07	-1.4E+07	
0.4	88561.27	265603.3	354164.6	4519830	-1.4E+07	-9889460	
0.6	128410.6	213264.1	341674.8	6553590	-1.2E+07	-5016240	
0.8	163140.7	151969.6	315110.2	8326079	-8244528	81550.52	

Table 12-3. continued

	Results from Morison Equation for theta varying between 0 and 2Pi (one wave period)							
Theta	Finertia	Fdrag	Ftotal	Minertia	Mdrag	Mtotal		
1.0	191366.8	91396.66	282763.5	9766633	-4958376	4808257		
1.2	211963.7	41108.52	253072.2	10817822	-2230185	8587637		
1.4	224110.3	9044.543	233154.9	11437739	-490677	10947062		
1.570796	227419.4	3.34E-08	227419.4	11606619	-1.8E-06	11606619		
1.8	221471.8	-16161.5	205310.3	11303078	876779.7	12179858		
2	206791.8	-54218.8	152573	10553869	2941433	13495302		
2.2	183867.7	-108430	75437.28	9383910	5882479	15266389		
2.4	153613.4	-170238	-16624.2	7839844	9235591	17075435		
2.6	117235	-229882	-112647	5983228	12471386	18454614		
2.8	76182.79	-277948	-201765	3888080	15079003	18967082		
3	32093.42	-306846	-274753	1637926	16646757	18284683		
3.141593	-0.07878	-313081	-313081	-4.02065	16985011	16985007		
3.2	-13275.4	-312014	-325290	-677526	16927133	16249607		
3.4	-58115	-292636	-350751	-2965968	15875868	12909900		
3.6	-100638	-251772	-352410	-5136166	13658932	8522766		
3.8	-139148	-195873	-335021	-7101601	10626331	3524730		
4	-172112	-133764	-305875	-8783918	7256845	-1527073		
4.2	-198213	-75250.8	-273464	-1E+07	4082444	-6033604		
4.4	-216413	-29571.6	-245984	-1.1E+07	1604294	-9440589		
4.6	-225985	-3937.99	-229923	-1.2E+07	213640.7	-1.1E+07		
4.712389	-227419	1.2E-10	-227419	-1.2E+07	-6.5E-09	-1.2E+07		
4.8	-226547	2396.971	-224150	-1.2E+07	-130038	-1.2E+07		
5	-218078	25191.82	-192886	-1.1E+07	-1366686	-1.2E+07		
5.2	-200915	68723.74	-132191	-1E+07	-3728344	-1.4E+07		
5.4	-175742	126120	-49621.6	-8969183	-6842158	-1.6E+07		
5.6	-143562	188319	44756.72	-7326871	-1E+07	-1.8E+07		
5.8	-105660	245500.8	139841.3	-5392460	-1.3E+07	-1.9E+07		
6	-63544.5	288637.8	225093.3	-3243069	-1.6E+07	-1.9E+07		
6.2	-18896.1	310919.5	292023.4	-964387	-1.7E+07	-1.8E+07		
6.283185	0.000641	313081	313081	0.032735	-1.7E+07	-1.7E+07		
Evaluate the force on cylinder due to current								
	Fcurrent= 0.5*Cd*rho*Ac*U^2							15745.56
	Mcurrent=Fcurrent*d/2							240119.8
Evaluate the force of wind on platform above water								
	Fwind= 0.5*Cd*rho*Aw*U^2							29601.69
	Mwind=Fwind*(d+h2)							1272873
Evaluate the maximum force and moment of wave on cylinder								
	Fwave max							313081
	Mwave max							-1.7E+07
Evaluate the total maximum force and maximum moment about bottom								
	Ftotal=Fwave+Fwind+Fcurrent (N)							358428.2
	Mmax=Mwind+Mwave+Mcurrent (N-m)							18498003

The wind, waves, and current are assumed to be all in the same direction. The current is also assumed to be constant over the water depth. The total maximum moment is determined by summing the contributions from the wind, wave, and current. Calculations of the total maximum moment were also accomplished for cylinder diameters of 0.762 m (30 in), 0.9144 m (36 in), 1.219 m (48 in), 1.372 m (54 in), and 1.524 m (60 in). Information on allowable bending moments for different size cylinder pipes, for different wall thickness, and two different yield strengths of steel pipe was obtained from Hsu (1984), and the results are tabulated in Table 12-4.

Table 12-4. Comparison of allowable bending moments for three cylindrical pipe wall thickness with computed moments (M) due to wind, wave, and current loads.

248 MPa (36 ksi) steel pipe of wall thickness (t)

Cylindrical Pile Support Diameter (m)	Maximum Moment (M) Due to Environmental Forces (kN-m)	Allowable Moment for Wall Thickness t=0.0254 (m) (kN-m)	Allowable Moment for Wall Thickness t=0.0381 (m) (kN-m)	Allowable Moment for Wall Thickness t=0.0508 (m) (kN-m)
0.762	9280	2034	2983	3797
0.9144	11124	2848	4475	5831
1.067	12970	4204	6509	8272
1.219	14809	4746	7458	9899
1.372	16659	6102	9628	12882
1.524	18498	9492	14238	18713
1.829	22187	14238	20340	25764

290 MPa (42 ksi) steel pipe of wall thickness (t)

Cylindrical Pile Support Diameter (m)	Maximum Moment (M) Due to Environmental Forces (kN-m)	Allowable Moment for Wall Thickness t=0.0254 (m) (kN-m)	Allowable Moment for Wall Thickness t=0.0381 (m) (kN-m)	Allowable Moment for Wall Thickness t=0.0508 (m) (kN-m)
Dia. (m)	M (kN-m)	t=0.0254 (m)	t-0.0381 (m)	t=0.0508 (m)
0.762	9280	2576	3729	4746
0.9144	11124	3932	5560	7051
1.067	12970	5288	7865	9763
1.219	14809	5560	8814	11390
1.372	16659	7322	11255	14238
1.524	18498	10848	16272	21018
1.829	22187	16272	23730	29832

Summary

The tabular results from Table 12-4 are also illustrated in Figure 12-5 for 248 MPa (36 ksi) steel pipe and Figure 12-5 for 290 MPa (42 ksi). Steel pipe with wall thickness of 0.0254, 0.0381, and 0.0508 m (1, 1.5, and 2 in) are shown. After evaluation of the forces and moments on the single cylindrical pile, the 290 MPa (42 ksi) working stress steel pile with a leg diameter of 1.829 m (72 in) and a wall thickness of 0.0508 m (2 in) was chosen. This is the diameter where the bending moment at the bottom is 22,187 kN-m, and the allowable moment is estimated as 29,832 kN-m. Since the allowable moment is greater than the moment computed for the moment due to the environmental forces, then a margin of safety exists that is typically called the safety factor. In this design, the safety factor is 1.34, and this is within the range establish in the design criteria. With this size of wave, the calculations made are in the drag dominated regime and inertia is an order of magnitude less. The forces and moments due to inertia, wind, and current are included to show that they are small. The shear stress due to the lateral forces caused by the same environmental loads was neglected in this design problem. Normally the bending stresses near the mudline caused by the maximum bending moment controls the design. If a greater safety factor is desired then a larger working stress steel is required.

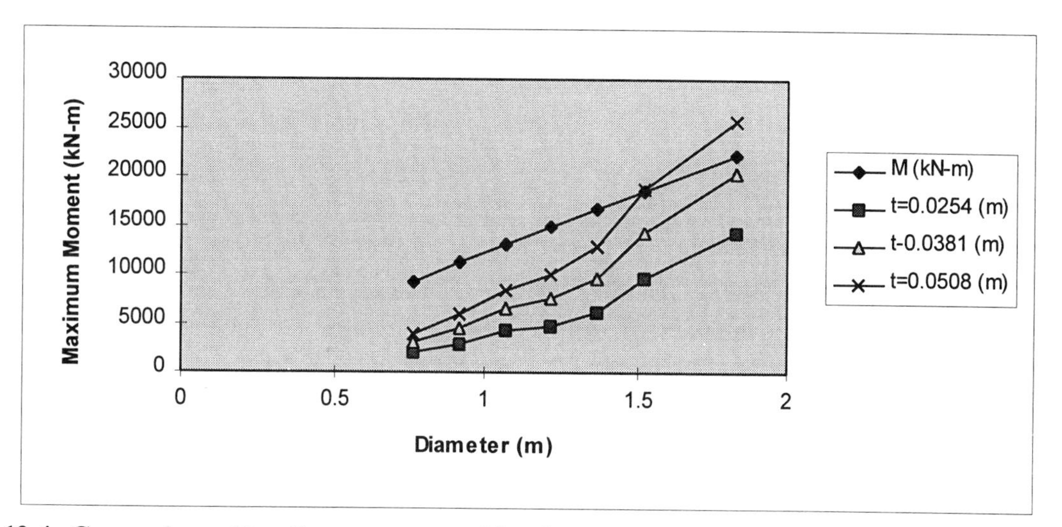

Figure 12-4. Comparison of bending moment resulting from wind, wave and current forces with the allowable bending moment for a selected steel pipe with 248 MPa (36 ksi) yield strength and wall thickness (t).

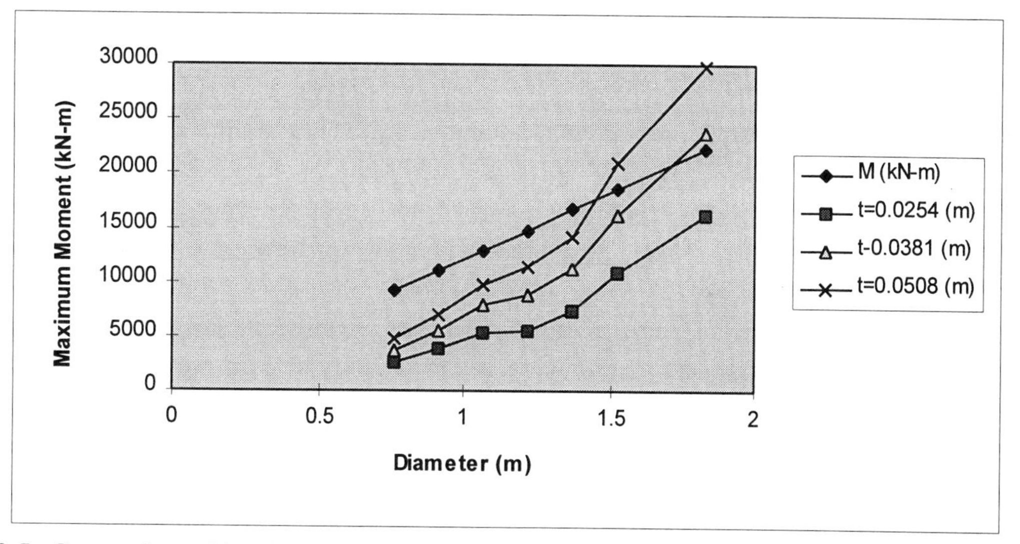

Figure 12-5. Comparison of bending moment resulting from wind, wave and current forces with the allowable bending moment for a selected steel pipe with 290 MPa (42 ksi) yield strength and wall thickness (t).

Remotely Operated Vehicle

Need

A small tethered remotely operated vehicle is needed for a wave basin facility (Figure 12-6) to retrieve tools, inspect equipment and connections, and to assist scuba divers in the installation of models and instrumentation.

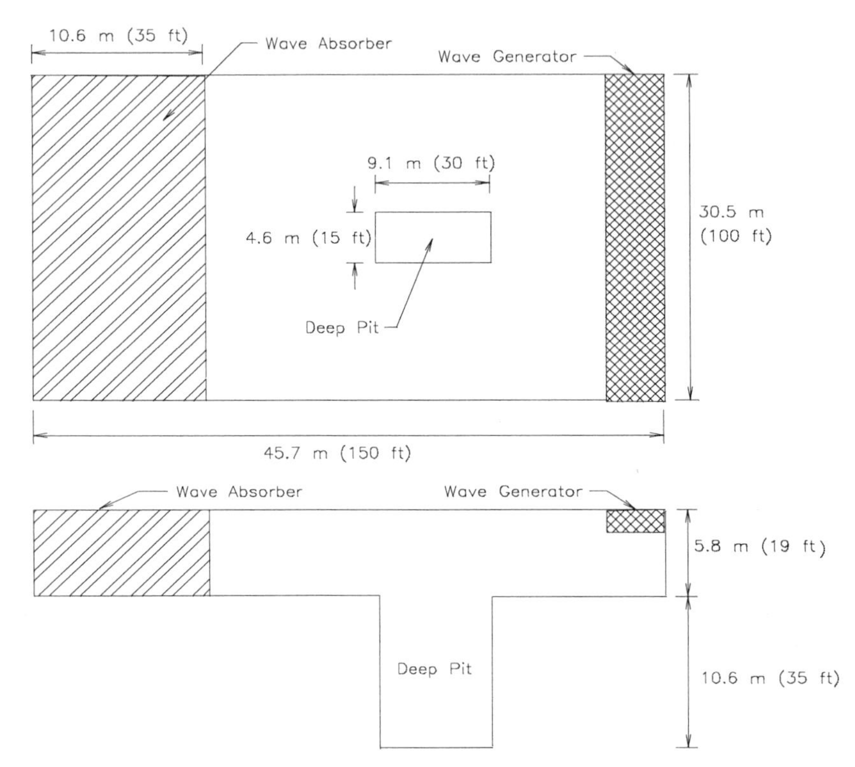

Figure 12-6. Schematic of wave basin in which remotely operated vehicle is to operate.

Problem Statement

Design a small remotely operated vehicle that can work in the wave basin shown in Figure 12-5 and be capable of retrieving an object weighing 25 lb in water. The wave basin contains fresh water at 70 °F. The maximum speed of the ROV is to be 1.5 kts. The ROV must have a mechanical arm capable of gripping and twisting. An underwater video camera and lights are necessary to view the work. A video monitor and controls for the mechanical arm and thruster are to be located on the deck at the wave basin for observation and use by the ROV operator. Power and control signals for the ROV are to be delivered through an umbilical.

Design Criteria

The design criteria are summarized in Table 12-3. Remotely operated vehicles are near neutrally buoyant submersibles that can be used for a number of operations in an underwater environment. As with any design, there are a number of factors that must be determined in order to proceed with the design process. Key parameters for determining the operating characteristics of a remotely operated vehicle include the determination of drag, thrust, and hydrostatics.

Table 12-5. Design criteria for remotely operated vehicle.

Parameter	Criteria
Maximum Speed	1.5 kts (2.6 ft/s, 0.8 m/s)
Lifting Capability	25 lb (111.2 N)
Maximum Depth	55 ft (16.8 m)
Maximum Footprint	2 ft x 1 ft (0.61 m x 0.31 m)
Viewing	Video Camera only
Propellers	Electric, Caged

Drag Force

The drag force is important in estimating the speed of the vehicle. It can be determined from the following equation:

$$F_D = \frac{1}{2}\rho C_D A_D V_C^2 \qquad \textbf{12-1}$$

where ρ is the mass density of water, C_D is the drag coefficient, A_D is the projected area normal to the flow direction, and V_C is the average magnitude of the current velocity or speed of the vehicle.

Thrust

The amount of available thrust can be determined theoretically by computing the change in momentum of the water that is accelerated through the thruster using the following equation

$$F_T = \rho A_T V_T^2 \qquad \textbf{12-2}$$

where A_T is the cross-sectional area of thruster's stream of water and V_T is the average velocity of the water through the thruster. V_T is computed using

$$V_T = \frac{NPE}{60} \qquad \textbf{12-3}$$

where N is the number of revolutions per minute of the propeller, P is the propeller pitch, and E is the overall efficiency of the propeller. A_T is computed using

$$A_T = (\pi/4)\left(d_p^2 - d_h^2\right) \qquad \textbf{12-4}$$

where d_p is the diameter of propeller and d_h is the hub diameter.

Center of Gravity and Buoyancy

To determine the longitudinal and vertical center of gravity (CG), it is necessary to itemize each piece of equipment by weight and location according to fixed axes. Then, the moments of each item are calculated and the CG's are determined. An example of these calculations are shown in Table 12-6. The results show the vertical center of gravity (VCG) is below the vertical center of buoyancy (VCB) by 0.14 ft, and this shows the vehicle is stable. A larger GB can be obtained by increasing the ballast and amount of buoyancy material. The vehicle weighs 233 lb in air, and it is 1 lb positively buoyant in the water. The longitudinal center of buoyancy and center of gravity are nearly the same, so only a small pitch angle is expected. The transverse CB and CG are the same so no list angle is anticipated. A smaller vehicle may be desirable, and the designer would have to reduce the overall vehicle weight and size.

Table 12-6. Center of gravity and buoyancy for remotely operated vehicle.

				From Volume							From Weight			
ITEM	W(air)	Volume	LCB	L-Mom	TCB	T-Mom	VCB	V-Mom	LCG	L-Mom	TCG	T-Mom	VCG	V-Mom
	(lbs)	(ft^3)	(ft)	(ft^4)	(ft)	(ft^4)	(ft)	(ft^4)	(ft)	(lb-ft)	(ft)	(lb-ft)	(ft)	(lb-ft)
	*	*	*		*		*		*		*		*	
Aluminum Frame	15.00	0.50	1.00	0.50	0.00	0.00	0.50	0.25	1.00	15.00	0.00	0.00	0.50	7.50
Port Vertical Thruster	8.00	0.15	1.00	0.15	-0.38	-0.06	0.75	0.11	1.00	8.00	-0.38	-3.00	0.75	6.00
Azimuthing Horizontal Thruster	15.00	0.30	1.00	0.30	0.00	0.00	1.25	0.38	1.00	15.00	0.00	0.00	1.25	18.75
Starboard Vertical Thruster	8.00	0.15	1.00	0.15	0.38	0.06	0.75	0.11	1.00	8.00	0.38	3.00	0.75	6.00
Pan Video Camera	12.00	0.30	0.75	0.23	0.00	0.00	0.50	0.15	0.75	9.00	0.00	0.00	0.50	6.00
Electronics Housing & Equipment	20.00	0.50	1.75	0.88	0.00	0.00	0.25	0.13	1.75	35.00	0.00	0.00	0.15	3.00
Buoyancy Top Cover	45.00	1.00	1.00	1.00	0.00	0.00	0.75	0.75	1.00	45.00	0.00	0.00	0.75	33.75
Port Light	4.00	0.10	0.00	0.00	-0.38	-0.04	0.25	0.03	0.00	0.00	-0.38	-1.50	0.25	1.00
Starboard Light	4.00	0.10	0.00	0.00	0.38	0.04	0.25	0.03	0.00	0.00	0.38	1.50	0.25	1.00
Mechanical Arm	32.00	0.40	0.50	0.20	0.00	0.00	0.00	0.00	0.75	24.00	0.00	0.00	0.00	0.00
Ballast	60.00	0.10	1.10	0.11	0.00	0.00	0.10	0.01	1.10	66.00	0.00	0.00	0.10	6.00
Movable ballast	5.00	0.05	0.10	0.01	0.00	0.00	0.00	0.00	0.00	0.00	0.00	0.00	0.00	0.00
Movable buoyancy	5.00	0.10	2.00	0.20	0.00	0.00	0.75	0.08	2.00	10.00	0.00	0.00	0.75	3.75
TOTALS	233.00	3.75	0.99	3.72	0.00	0.00	0.54	2.01	1.01	235.00	0.00	0.00	0.40	92.75
Weight Water Displaced	234													
VCG (ft)	0.39807		Note	Assume origin of axis is at the bow of the vehicle and even with the keel										
VCB (ft)	0.536			x-axis runs longitudinal along vehicle keel from bow to stern										
V GB (ft)	0.13793			y-axis is transverse running perpendicular to keel (+ starboard & - port)										
LCG (ft)	1.00858			z-axis is vertical and perpendicular to keel and positive vertically upward										
LCB (ft)	0.99067													
TCG (ft)	0.00		*	Input values										
TCB (ft)	0													

Summary

Once the weight and the drag force are determined, then the amount of thrust needed can be computed. To change the amount of thrust produced factors such as RPMs, propeller pitch, and propeller size must be adjusted. The calculations of drag and thrust for the ROV are tabulated in Table 12-7. The total drag (F_{total}) increases to 72 lb at a vehicle speed of 3.4 ft/s. Propeller thrust is also estimated for different RPM, propeller pitch, hub diameter, propeller diameter, and propeller efficiency. The values of developed propeller thrust are plotted in Figure 12-7 as a function of RPM for each propeller type (propeller id numbers 1-8). To develop a thrust of 72 lb, Figure 12-7 shows that propeller id number 7 running at 2250 RPM is a possible choice. To obtain a vertical lift of 25 lb, the two vertical thrusters can be estimated by using Figure 12-7. Entering the ordinate axis at 12.5 lb shows that propeller id 5 operating a 1500 RPM would develop the necessary thrust and have additional capability by increasing the RPM to 1600 to bring the load up through the water column. A schematic drawing of the design of the remotely operated vehicle is illustrated in Figure 12-8 showing the overall dimensions of the vehicle to be 2 ft long, 1.5 ft wide, and 1 ft high. The aluminum tubular frame holds three thrusters, electronic equipment housing, video camera, lights, and a mechanical arm.

Table 12-7. Calculations of ROV drag and thrust forces.

Drag and Thrust Calculations								
ROV Projected area (ft^2) =			2					
Tether Diameter (ft)=			0.0625					
ROV Drag Coefficient (Cd)=			1.5					
ROV tether drag coefficient =			1.2					
ROV characteristic length (ft)=			2					
Water kinematic viscosity (ft^2/s)=			1.21E-05					
Water density (slugs/ft^3)			1.94					
Drag Force Calculations								
Velocity	**Re (ROV)**	**Re (tether)**	**F(tether)**	**F(ROV)**	**F(total)**			
(ft/s)			(lbs)	(lbs)	(lbs)			
0.20	3.31E+04	1.03E+03	0.13	0.12	0.25			
0.60	9.92E+04	3.10E+03	1.20	1.05	2.25			
1.00	1.65E+05	5.17E+03	3.33	2.91	6.24			
1.40	2.31E+05	7.23E+03	6.54	5.70	12.24			
1.80	2.98E+05	9.30E+03	10.80	9.43	20.23			
2.20	3.64E+05	1.14E+04	16.14	14.08	30.22			
2.60	4.30E+05	1.34E+04	22.54	19.67	42.21			
3.00	4.96E+05	1.55E+04	30.01	26.19	56.20			
3.40	5.62E+05	1.76E+04	38.55	33.64	72.18			
Velocity of Water Through Thruster and Thruster Force								
RPM	Pitch	Efficiency	Prop Diam	Hub Diam	Area (At)	V(t)	Force	Propeller
	(ft)		ft	ft	ft^2	(ft/s)	lb	id
1000	0.5	0.5	0.5	0.125	0.18	4.17	6.20	5
1500	0.5	0.5	0.5	0.125	0.18	6.25	13.95	5
2000	0.5	0.5	0.5	0.125	0.18	8.33	24.80	5
2500	0.5	0.5	0.5	0.125	0.18	10.42	38.75	5
1000	0.5	0.6	0.5	0.125	0.18	5.00	8.93	6
1500	0.5	0.6	0.5	0.125	0.18	7.50	20.09	6
2000	0.5	0.6	0.5	0.125	0.18	10.00	35.71	6
2500	0.5	0.6	0.5	0.125	0.18	12.50	55.80	6
1000	0.33	0.5	0.33	0.08	0.08	2.75	1.18	1
1500	0.33	0.5	0.33	0.08	0.08	4.13	2.66	1
2000	0.33	0.5	0.33	0.08	0.08	5.50	4.72	1
2500	0.33	0.5	0.33	0.08	0.08	6.88	7.38	1
1000	0.33	0.6	0.33	0.08	0.08	3.30	1.70	2
1500	0.33	0.6	0.33	0.08	0.08	4.95	3.83	2
2000	0.33	0.6	0.33	0.08	0.08	6.60	6.80	2
2500	0.33	0.6	0.33	0.08	0.08	8.25	10.63	2
1000	0.75	0.5	0.5	0.125	0.18	6.25	13.95	7
1500	0.75	0.5	0.5	0.125	0.18	9.38	31.39	7
2000	0.75	0.5	0.5	0.125	0.18	12.50	55.80	7
2500	0.75	0.5	0.5	0.125	0.18	15.63	87.18	7
1000	0.75	0.6	0.5	0.125	0.18	7.50	20.09	8
1500	0.75	0.6	0.5	0.125	0.18	11.25	45.20	8
2000	0.75	0.6	0.5	0.125	0.18	15.00	80.35	8
2500	0.75	0.6	0.5	0.125	0.18	18.75	125.55	8
1000	0.5	0.5	0.33	0.08	0.08	4.17	2.71	3
1500	0.5	0.5	0.33	0.08	0.08	6.25	6.10	3
2000	0.5	0.5	0.33	0.08	0.08	8.33	10.85	3
2500	0.5	0.5	0.33	0.08	0.08	10.42	16.95	3
1000	0.5	0.6	0.33	0.08	0.08	5.00	3.90	4
1500	0.5	0.6	0.33	0.08	0.08	7.50	8.78	4
2000	0.5	0.6	0.33	0.08	0.08	10.00	15.62	4
2500	0.5	0.6	0.33	0.08	0.08	12.50	24.40	4

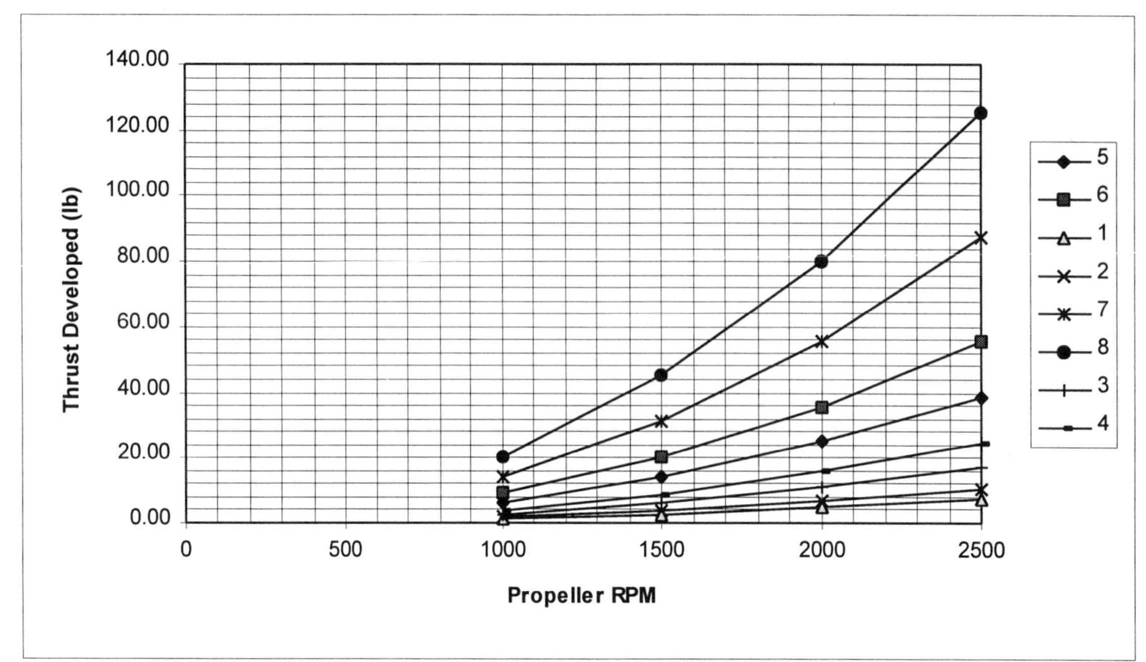

Figure 12-7. Effects of ROV velocity and cross-sectional area on drag force.

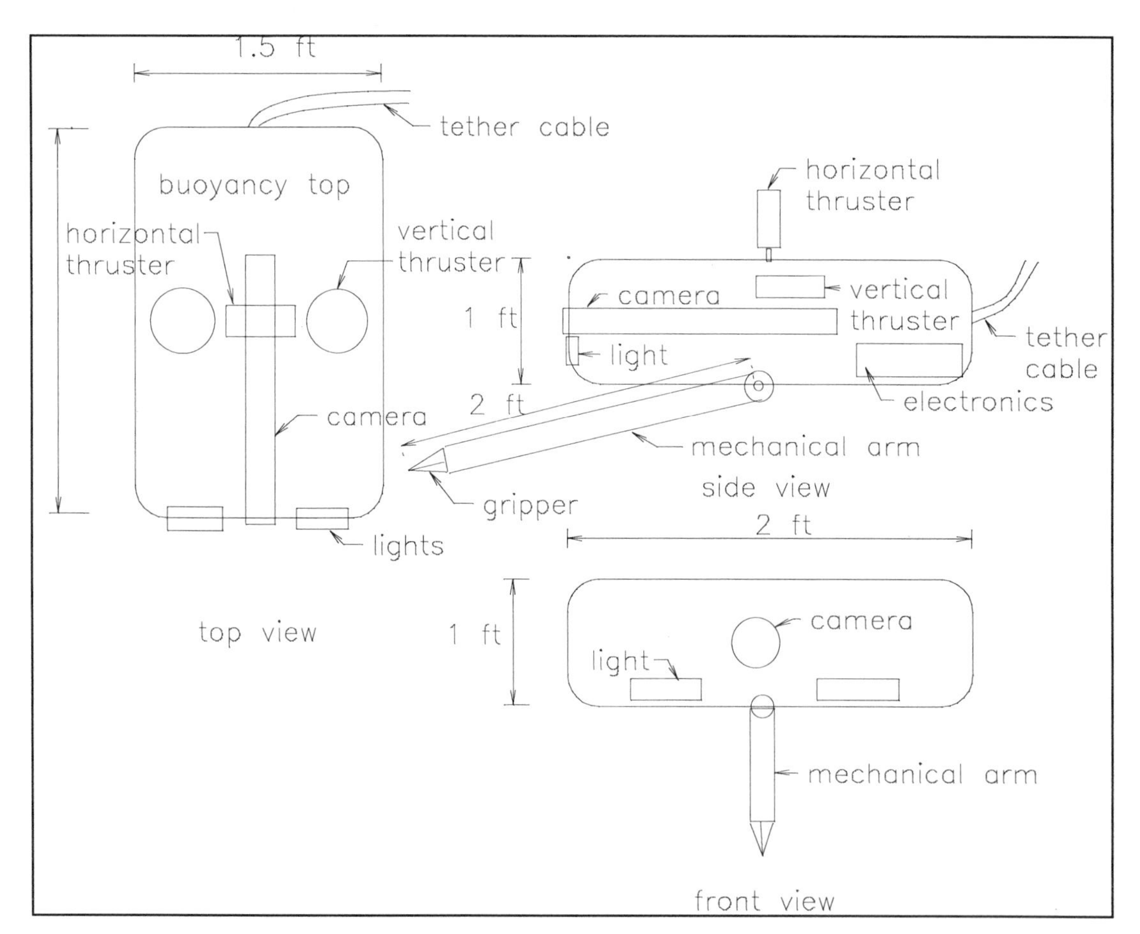

Figure 12-8. Remotely operated vehicle drawing.

Breakwater

Need

A hypothetical port needs a breakwater designed to protect ships during berthing. A schematic of the breakwater location is shown in Figure 12-9.

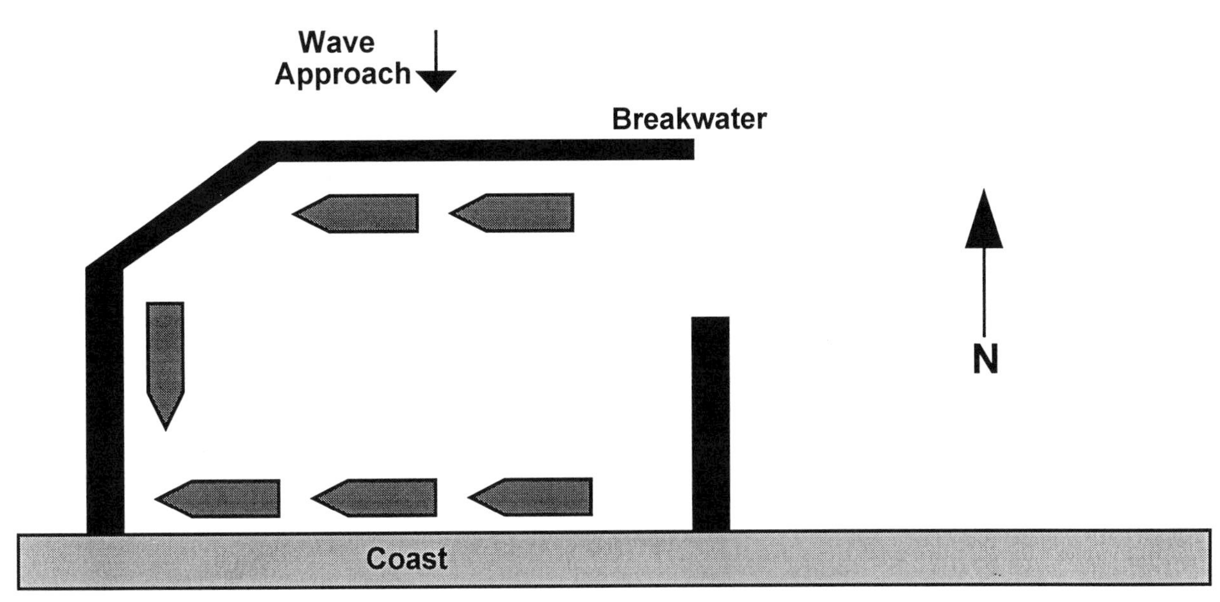

Figure 12-9. Schematic for hypothetical port requiring breakwater design.

Problem Statement

Design the North side breakwater for the hypothetical port. The breakwater design should allow minor overtopping and 25 % damage is acceptable. Consider the design wave height on the ocean side of the breakwater only. The water depth in the port is to be dredged to 50 ft and the mean low water depth at the North side breakwater is 55 ft. The length of the breakwater is to be 3000 ft, and the maximum tidal variation is 3 ft.

Design Criteria

In consultation with the client, the design team established the design criteria that are tabulated in Table 12-8.

Table 12-8. Design criteria for breakwater design.

Parameter	Criteria
Wave Height and Direction	6 ft, 180°
Water depth	40 ft min, 46 ft max
Type of wave	Non-breaking
Type of construction	Random placement from barge
Types of armor units available	Rough quarrystone, tribar, and dolos
Wave location	Offshore side only

Design Considerations

The design of a rubble mound breakwater involves the determination of three major variables that include the geometry of the structure, methods of construction, and selection of materials. There is no particular order in which these must be determined as they are all interrelated. The following information is an outline of the necessary equations and methods for determining these variables. Information regarding the design of this breakwater may be obtained from Chapter 4 and the US Army Corps of Engineers Shore Protection Manual (USACE 1984).

Geometry of Structure

The geometry of a rubble mound breakwater depends on the two main considerations of crest elevation and width. The determination of crest elevation is mainly dependent on whether or not the breakwater is meant to be overtopped, and this depends on the wave conditions at that particular site. A typical recommended breakwater cross section is shown in Figure 12-10 for the common situation when the design waves are expected only on one side (USACE 1984). The structure consists of a bed layer of quarry stone whose weight is between 200 and 6000 times smaller than the armor or cover layer. The middle layer is 10 to 15 times smaller and the cover layer consists of large quarry stone or large manmade concrete armor units such as tetrapods, dolos, tribar, accupode, core-lock, and quadripods. Quarry stone is usually less expensive if it can be obtained close to the breakwater location. A disadvantage to the manmade units is that the units sometimes break.

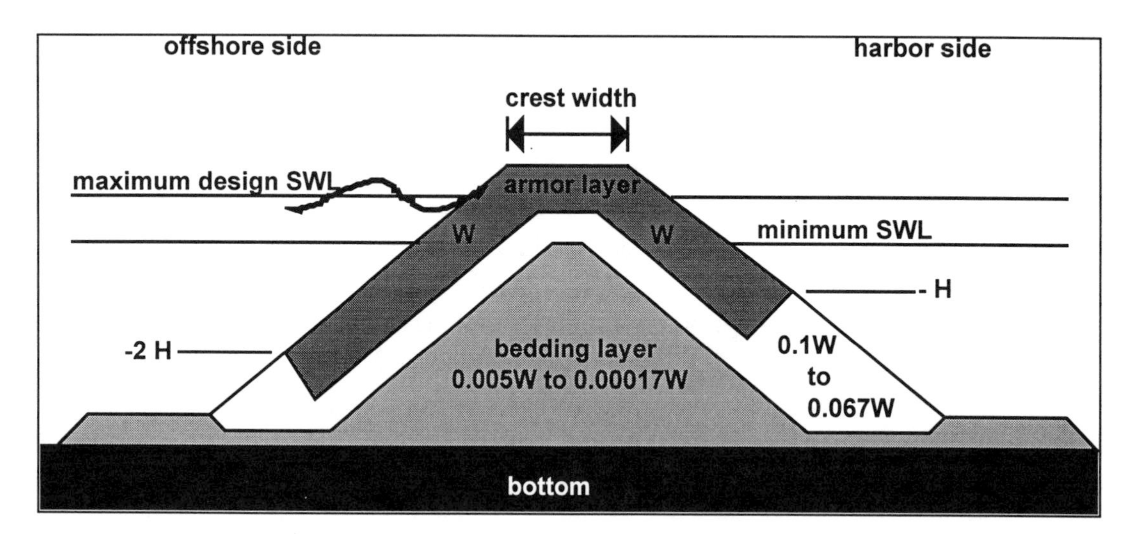

Figure 12-10. Cross section of final breakwater design with dimensions.

The crest elevation is determined for no overtopping and depends on the height of the crest above the maximum water depth. For linear waves, the wave amplitude is 3 ft and to account for the nonlinear waves and runup, a wave amplitude of 4 ft is used. Therefore, the crest elevation is 50 ft. The breakwater crest width is governed by

$$B = nk_{\Delta}\left(\frac{W}{\gamma_a}\right)^{1/3} \qquad \textbf{12-5}$$

where B is the crest width, n is the number of stones, k_Δ is the layer coefficient, W is mass of armor unit in primary cover layer, and γ_a is the unit weight of armor unit. This design is to consider only minor overtopping, and it is recommended that the minimum number of stones or armor units be three. The results for the three armor units available for this design are contained in Table 12-9.

Methods of Construction

Typical construction methods that may be considered include dumping from barge, using a trestle, using a movable platform, dumping off the structure, or combinations of these methods. Some factors that may influence the decision are site conditions, wave climate at the site, final cross section of the structure, materials to be used, and equipment available. For this design, placement from a barge was selected.

Selection of Materials

Points to consider in the selection of armor units are availability, sizes (forms) available, transportation to site, stockpiling, amount needed, and cost. These factors may be applied whether manmade or natural armor stone is to be used. The major variable in selecting armor stone is the weight of the stone, and this can be determined using the Hudson equation

$$W = \frac{\gamma_a H^3}{K_D (S_a - 1)^3 \cot\alpha} \qquad \textbf{12-6}$$

where W is weight of individual armor unit in the primary layer, γ_a is the unit weight of armor unit, H is the design wave height, S_a is the ratio of the armor unit weight to that of the water at the structure, α is the angle of structure slope, and K_D is the stability coefficient.

The thickness of each armor layer and underlayers can then be calculated using

$$r = nk_\Delta \left(\frac{W}{\gamma_a}\right)^{1/3} \qquad \textbf{12-7}$$

where n is the number of armor units that form the armor layer thickness. The final step is to determine the number of individual armor units (N_r) needed and this can be determined by

$$N_r = nAk_\Delta \left(1 - \frac{P}{100}\right)\left(\frac{\gamma_a}{W}\right)^{2/3} \qquad \textbf{12-8}$$

where A is the surface area, k_Δ is the layer coefficient, and P is the average porosity of the cover layer in percent.

Design Results

A spread sheet is assembled to illustrate the design calculations. Formulas are used in the spread sheet (Figure 12-9) to calculate the weight of individual armor units, crest width, cover layer thickness, number of individual armor units per 1000 ft^2 of surface area.

Table 12-9. Spread sheet calculations for rubble mound breakwater design.

Given:
Design Wave Height, H = 6 ft (non-breaking)
Slope (cot theta)= 2
Surface Area = 1000 ft^2
Crest Elevation Safety Factor= 1 ft
Minimum Water Depth= 40 ft
Maximum Water Depth= 46 ft
Breakwater Length= 3000 ft

Find: W, B, r, Nr, estimated cost

Solution:

Crest Elevation of Breakwater

Crest Elevation ft
50

Weight of Individual Armor Units

Unit	w(r) (lb/ft^3)	H (ft)	S(r)	Cot theta	K(D)	Weight of Individual Armor Unit, W (lb)
Quarrystone	140	6	2.1875	2	4	2257.31
Tribar	140	6	2.1875	2	10	902.92
Dolos	140	6	2.1875	2	31.8	283.94

Crest Width

Unit	k(delta)	n	W (lb)	w(r) lb/ft^3	Crest Width, B (ft)
Quarrystone	1	3	2257.31	140	7.57
Tribar	1.02	3	902.92	140	5.69
Dolos	0.94	3	283.94	140	3.57

Number of Individual Armor Units

Unit	k(delta)	n	W (lb)	gamma (a) lb/ft^3	P (%)	A (ft^2)	Req. # of Armor Units
Quarrystone	1	2	2257.31	140	40	1000	188
Tribar	1.02	2	902.92	140	54	1000	271
Dolos	0.94	2	283.94	140	56	1000	516

H/H(D=0) as Function of Damage and Type of Armo

Unit		Damage in Percent						
		0 to 5	5 to 10	10 to 15	15 to 20	20 to 30	30 to 40	40 to 50
Quarrystone	H/H(D=0)	1	1.08	1.14	1.2	1.29	1.41	1.54
Tribar	H/H(D=0)	1	1.11	1.25	1.36	1.5	1.59	1.64
Dolos	H/H(D=0)	1	1.1	1.14	1.17	1.2	1.24	1.27

Armor Layer Thickness

Unit	k(delta)	n	W (lb)	gamma (a) lb/ft^3	Thickness, r (ft)
Quarrystone	1	3	2257.31	140	7.57
Tribar	1.02	3	902.92	140	5.69
Dolos	0.94	3	283.94	140	3.57

Estimate of Cost for Armor Layer

Unit	Surface Area/Foot offshore side ft	Surface Area/Foot harbor side ft	Crest Width ft	Total Surface Area ft^2	Total Weight of Armor Layer tons	Cost /ton 1996 $	Total Cost of Armor Layer 1996 $
Quarrystone	49.19	35.78	7.57	277627.53	58860.91	50	2943045
Tribar	49.19	35.78	5.69	271988.84	33235.96	65	2160337
Dolos	49.19	35.78	3.57	265617.90	19463.85	65	1265150

Summary

The breakwater design calculations shows that the least cost design uses the dolos armor unit. The cost shown is for the armor material only. Additional cost considerations need to be considered also, such shipping to construction site and time to place the armor units. The dolos armor units are man made and are subject to breakage and this needs to be considered in the design selection. The large stability coefficient, K_D, is the main reason for the lower cost of the dolos armor units in the breakwater construction. The schematic and characteristic breakwater dimensions are illustrated in Figure 12-11.

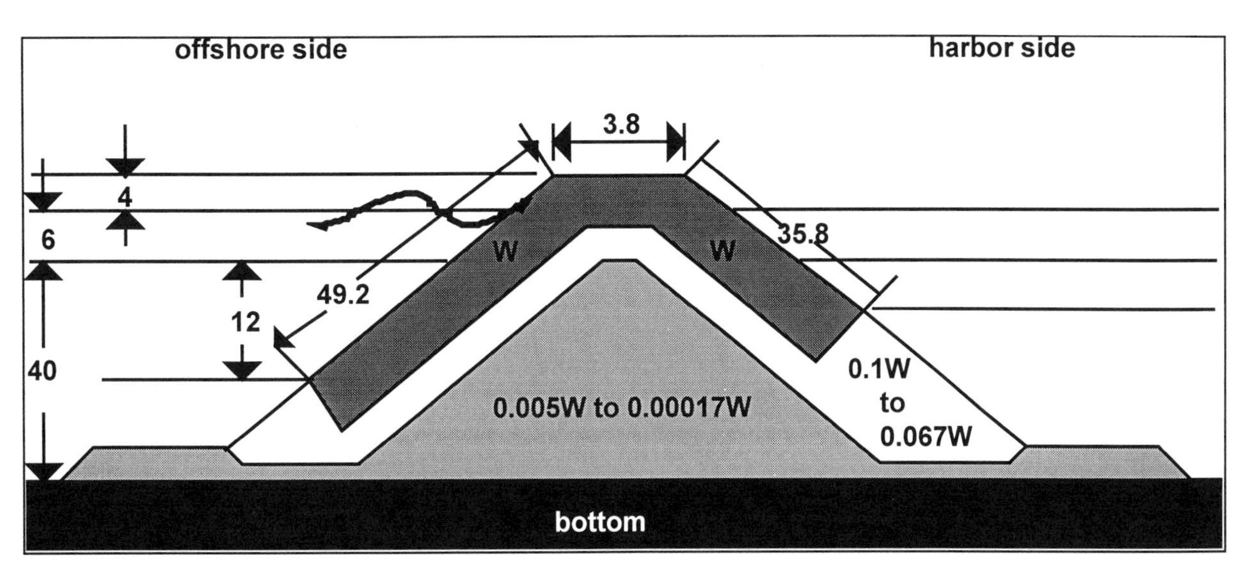

Figure 12-11. Schematic of cross section of final breakwater design using dolos with dimensions in feet.

REFERENCES

Allmendinger, E. E., Editor. *Submersible Vehicle Systems Design.* Jersey City: The Society of Naval Architects and Marine Engineers, 1990.

Cross, N. *Engineering Design Methods.* New York: John Wiley & Sons, Inc., 1989.

Dieter, G. E. *Engineering Design: A Materials and Processing Approach.* New York: McGraw-Hill Book Company, 1983.

Dym, C. L. *Engineering Design: A Synthesis of Views.* New York: Cambridge University Press, 1994.

Eide, A. R., R. D. Jenison, L. H. Mashaw, and L. L. Northup. *Engineering Fundamentals and Problem Solving.* New York: McGraw-Hill Book Company, 1979.

French, M. J. *Conceptual Design for Engineers.* London: Design Council, 1985.

Hill, P. H. *The Science of Engineering Design.* New York: Holt, Rinehart and Winston, Inc., 1970.

Hsu, T. H. *Applied Offshore Structural Engineering.* Houston: Gulf Publishing Co., 1984.

Ray, M. S. *Elements of Engineering Design: An Integrated Approach.* New York: Prentice/Hall International, 1985.

US Army Corps of Engineers (USACE). *Shore Protection Manual*, Vol. I and II, Fourth Edition. Coastal Engineering Research Center, US Army Engineer Waterways Experiment Station, Washington: US Government Printing Office, 1984.

PROBLEMS

The following preliminary design problems are meant to be conceptual in nature. The design results should include a review of appropriate literature, development of design criteria, preliminary engineering analysis and calculations, development of engineering drawings, and production of a final written design report.

12-1. Design a small steel jacketed offshore platform for 150 ft water depth in the Gulf of Mexico.

12-2. Design a remotely operated vehicle for surveying open water disposal sites in less than 100 ft of water.

12-3. Design a taut line submerged mooring for a three current meter array in 300 ft of water.

12-4. Design a buoy system for monitoring waves, currents, wind speed, atmospheric pressure, salinity and dissolved oxygen.

12-5. Design a small boat marina for boats less than 100 ft long.

12-6. Design an offshore breakwater for a local coastal location.

12-7. Design an underwater habitat for a local location in 30 ft of water and where the water temperature is 75 ^{o}F.

12-8. Design a fishing dock for a shallow water lake.

12-9. Design a two person excursion submarine to operate in 90 ft of water.

12-10. Design a floating dock for small sail boats (less than 25 ft) for an inland lake or protected bay.

12-11. Design a small semisubmersible floating platform.

12-12. Design a model semisubmersible for testing in a wave basin/towing tank.

12-13. Design a jetty for new entrance channel at a local coastal location.

APPENDIX A: PROPERTIES AND CONVERSIONS

Table A-1. Physical properties of selected common gases at standard atmospheric pressure.

International System Units (SI)

Gas	Temperature T (°C)	Density ρ (kg/m³)	Specific Weight γ (N/m³)	Dynamic Viscosity μ (N-s/m²)	Kinematic Viscosity ν (m²/s)	Specific Heat Ratio k	Gas Constant R (J/Kg-K)
Air(standard)	15	1.23E+0	1.20E+!	1.79E-5	1.46E-5	1.40	2.869E+2
Carbon Dioxide (CO_2)	20	1.83E+0	1.80E+1	1.47E-5	8.03E-6	1.30	1.889E+2
Helium (He)	20	1.66E-1	1.63E+0	1.94E-5	1.15E-4	1.66	2.077E+3
Hydrogen (H)	20	8.38E-2	8.22E-1	8.84E-6	1.05E-4	1.41	4.124E+3
Methane (MH_3)	20	6.67E-1	6.54E+0	1.10E-5	1.65E-5	1.31	5.183E+2
Nitrogen (N)	20	1.16E+0	1.14E+1	1.76E-5	1.52E-5	1.40	2.968E+2
Oxygen (O_2)	20	1.33E+0	1.30E+1	2.04E-5	1.53E-5	1.40	2.598E+2

British Gravitational Units (BG)

Gas	Temperature T (°F)	Density ρ (slug/ft³)	Specific Weight γ (lb/ft³)	Dynamic Viscosity μ (lb-s/ft²)	Kinematic Viscosity ν (ft²/s)	Specific Heat Ratio k	Gas Constant R (ft-lb/slug-°R)
Standard Air	59	2.38E-3	7.65E-2	3.74E-7	1.57E-4	1.40	1.716E+3
Carbon Dioxide (CO_2)	68	3.55E-3	1.14E-1	3.07E-7	8.65E-5	1.30	1.130 E+3
Helium (He)	68	3.23E-4	1.04E-2	4.09E-7	1.27E-4	1.66	1.242 E+4
Hydrogen (H)	68	1.63E-4	5.25E-3	1.85E-7	1.13E-4	1.41	2.466 E+4
Methane (MH_3)	68	1.29E-3	4.15E-2	2.29E-7	1.78E-4	1.31	3.099 E+3
Nitrogen (N)	68	2.26E-3	7.28E-2	3.68E-7	1.63E-4	1.40	1.775 E+3
Oxygen (O_2)	68	2.58E-3	8.31E-2	4.25E-7	1.65E-4	1.40	1.554 E+3

Note: Values of specific heat (k) vary only slightly with temperature and values of the gas constant R do not vary with temperature.

Table A-2. Physical properties of selected common liquids.

International System Units (SI)

Liquid	Temperature T (°C)	Density ρ (kg/m³)	Specific Weight γ (N/m³)	Dynamic Viscosity μ (N-s/m²)	Kinematic Viscosity ν (m²/s)	Surface Tension σ (N/m)	Vapor Pressure p_v (N/m² (abs))	Bulk Modulus E_v (N/m²)
Gasoline	15.6	6.80E+2	6.67E+0	3.10E-4	4.60E-7	2.20E-2	5.50E+4	1.30E+9
Mercury	20	1.36E+4	1.33E+2	1.57E-3	1.15E-7	4.66E-1	1.60E-1	2.85E+10
SAE 30 oil	15.6	9.12E+2	8.95E+0	3.80E-1	4.20E-4	3.60E-2	-	1.50E+9
Seawater	15.6	1.03E+3	1.01E+1	1.20E-3	1.17E-6	7.34E-2	1.77E+3	2.34E+9
Water	15.6	9.99E+2	9.80E+0	1.12E-3	1.12E-6	7.34E-2	1.77E+3	2.15E+9

British Gravitational Units (BG)

Liquid	Temperature T (°C)	Density ρ (slugs/ft³)	Specific Weight γ (lb/ft³)	Dynamic Viscosity μ (lb-s/ft²)	Kinematic Viscosity ν (ft²/s)	Surface Tension σ (lb/ft)	Vapor Pressure p_v (lb/in² abs))	Bulk Modulus E_v (lb/in²)
Gasoline	60	1.32E+0	4.25E+1	6.50E-6	4.90E-6	1.50E-3	8.00E+0	1.90E+5
Mercury	68	2.63E+1	8.47E+2	3.28E-5	1.25E-6	3.19E-2	2.30E-5	4.14E+6
SAE 30 oil	60	1.77E+0	5.70E+1	8.00E-3	4.50E-3	2.50E-3	-	2.20E+5
Seawater	60	1.99E+0	6.40E+1	2.51E-5	1.26E-5	5.03E-3	2.56E-1	3.39E+5
Water	60	1.94E+0	6.24E+1	2.34E-5	1.21E-5	5.03E-3	2.56E-1	3.12E+5

Table A-3. Converting British Gravitational Units (BG) to International System Units (SI).

Parameter	To Convert from BG Units	Multiply by	To Get SI Units
Acceleration	ft/s^2	3.048E-1	m/s^2
Area	ft^2	9.290E-2	m^2
	acres	4.047E+3	m^2
	$mile^2$	2.590E+6	m^2
Density	lb_m/ft^3	1.602E+1	kg/m^3
	$slugs/ft^3$	5.154E+2	kg/m^3
Energy	BTU (British thermal units)	1.055E+3	Joule
	ft-lb	1.356E+0	Joule (N-m)
	cal	4.186E+0	Joule
Force	lb	4.448E+0	N
	long tons (2240 lb)	9.964E+0	kN (kilonewtons)
	short tons (2000 lb)	8.896E+0	kN
	kip (1000 lb)	4.448E+0	kN
Length	fathom	1.829E+0	m
	ft	3.048E-1	m
	in	2.540E+0	cm
	mile	1.609E+0	km
	nautical mile	1.852E+0	km
Mass	lb_m	4.536E-1	kg
	slug	1.459E+1	kg
Power	ft-lb/s	1.356E+0	watts (N-m/s)
	hp	7.457E+2	watts
Pressure	in of Hg (60 °F)	2.532E+1	mm of Hg (60 °F)
	in of Hg (60 °F)	3.377E+3	N/m^2 Pascal)
	lb/ft^2 (psf)	4.788E+1	N/m^2 (Pascal)
	lb/in^2 (psi)	6.895E+3	N/m^2 (Pascal)
	ata	1.013E+5	N/m^2 (Pascal)
	ft of seawater	1.548E+4	N/m^2 (Pascal)
Specific Weight	lb/ft^3	1.571E+2	N/m^3
Temperature	°R	5.556E-1	°K
Velocity	ft/s	3.048E-1	m/s
	knots	5.144E-1	m/s
	mi/hr (MPH)	4.470E-1	m/s
Viscosity	$lb\text{-}s/ft^2$ (dynamic)	4.788E+1	$N\text{-}s/m^2$ (dynamic)
	ft^2/s (kinematic)	9.290E-2	m^2/s (kinematic)
Volume flowrate	ft^3/s	2.832E-2	m^3/s
	gal/min (GPM)	6.309E-5	gal/min (GPM)
	ft^3/min	28.32	liters/min
Volume	in^3	1.639E-5	m^3
	ft^3	2.832E-2	m^3
	yd^3	7.646E-1	m^3
	barrel (oil, 42 gallons)	1.590E+2	liters
	barrel (oil, 42 gallons)	1.590E-1	m^3
	gal (US)	3.785E+0	liters
	ft^3	2.832E+1	liters

Table A-4. Converting International System Units (SI) to British Gravitational Units (BG).

Parameter	To Convert from SI Units	Multiply by	To Get BG Units
Acceleration	m/s^2	3.281E+0	ft/s^2
Area	m^2	1.706E+1	ft^2
	m^2	2.471E-1	acres
	m^2	3.681E-7	mile^2
Density	kg/m^3	6.243E-2	lbm/ft^3
	kg/m^3	1.940E-3	slugs/ft^3
Energy	Joule	9.478E-4	BTU (British thermal units)
	N-m	7.376E-1	ft-lb
Force	N	2.248E-1	lb
	kN (kilonewtons)	1.004E-1	long tons (2240 lb)
	kN	1.124E-1	short tons (2000 lb)
	kN	2.248E-1	kips
	metric tons	2.205E+3	lb
Length	m	3.281E+0	fathom
	m	5.468E-1	ft
	cm	3.937E-1	in
	km	6.214E-1	mile
	km	5.400E-1	nautical mile
Mass	kg	2.205E+0	lbm
	kg	6.852E-2	slug
Power	N-m/s	7.376E-1	ft-lb/s
	watts	1.341E-3	hp (horsepower)
Pressure	mm of Hg (60 °F)	3.95E-2	in of Hg (60 °F)
	N/m^2 (Pascal)	2.961E-4	in of Hg (60 °F)
	N/m^2 (Pascal)	2.089E-2	lb/ft^2 (psf)
	N/m^2 (Pascal)	1.450E-4	lb/in^2 (psi)
Specific Weight	N/m^2	6.366E-3	lb/ft^2
Temperature	°K	1.800E+0	°R
Velocity	m/s	3.281E+0	ft/s
	m/s	2.237E+0	mph
	m/s	1.944E+0	knots (nautical mile/hr)
Viscosity	N-s/m^2 (dynamic)	2.089E-2	lb-s/ft^2 (dynamic)
	m^2/s (kinematic)	1.076E+1	ft^2/s (kinematic)
Volume flowrate	m^3/s	3.531E+1	ft^3/s
	m^3/s	1.585E+4	gal/min (GPM)
	liters/min	3.531E-2	ft^3/min (CFM)
Volume	cm^3	6.102E-2	in^3
	m^3	3.531E+1	ft^3
	m^3	1.308E+0	yd^3
	liters	6.290E-3	barrels (oil)
	liters	2.642E-1	gal (US)
	liters	3.531E-2	ft^3

e A-5. Properties of sinusoidal waves (Svendsen and Jonsson 1976)

d/L_o	tanh (kd)	d/L	kd	sinh (kd)	cosh (kd)	2kd/sinh (2kd)	H/H_o
0.000	0.000	0.0000	0.000	0.000	1.00	1.000	∞
0.002	0.112	0.0179	0.112	0.113	1.01	0.992	2.12
0.004	0.158	0.0253	0.159	0.160	1.01	0.983	1.79
0.006	0.193	0.0311	0.195	0.197	1.02	0.975	1.62
0.008	0.222	0.0360	0.226	0.228	1.03	0.967	1.51
0.010	0.248	0.0403	0.253	0.256	1.03	0.958	1.43
0.015	0.302	0.0496	0.312	0.317	1.05	0.938	1.31
0.020	0.347	0.0576	0.362	0.370	1.07	0.918	1.23
0.025	0.386	0.0648	0.407	0.418	1.08	0.898	1.17
0.030	0.420	0.0713	0.448	0.463	1.10	0.878	1.13
0.035	0.452	0.0775	0.487	0.506	1.12	0.858	1.09
0.040	0.480	0.0833	0.523	0.548	1.14	0.838	1.06
0.045	0.507	0.0888	0.558	0.588	1.16	0.819	1.04
0.050	0.531	0.0942	0.592	0.627	1.18	0.800	1.02
0.055	0.554	0.0993	0.624	0.665	1.20	0.781	1.01
0.060	0.575	0.104	0.655	0.703	1.22	0.762	0.993
0.065	0.595	0.109	0.686	0.741	1.24	0.744	0.981
0.070	0.614	0.114	0.716	0.779	1.27	0.725	0.971
0.075	0.632	0.119	0.745	0.816	1.29	0.707	0.962
0.080	0.649	0.123	0.774	0.854	1.31	0.690	0.955
0.085	0.665	0.128	0.803	0.892	1.34	0.672	0.948
0.090	0.681	0.132	0.831	0.929	1.37	0.655	0.942
0.095	0.695	0.137	0.858	0.968	1.39	0.637	0.937
0.100	0.709	0.141	0.886	1.010	1.42	0.620	0.933
0.110	0.735	0.150	0.940	1.080	1.48	0.587	0.926
0.120	0.759	0.158	0.994	1.170	1.54	0.555	0.920
0.130	0.780	0.167	1.050	1.250	1.60	0.524	0.917
0.140	0.800	0.175	1.100	1.330	1.67	0.494	0.915
0.150	0.818	0.183	1.15	1.42	1.74	0.465	0.913
0.160	0.835	0.192	1.20	1.52	1.82	0.437	0.913
0.170	0.850	0.200	1.26	1.61	1.90	0.410	0.913
0.180	0.864	0.208	1.31	1.72	1.99	0.384	0.914
0.19	0.877	0.217	1.36	1.82	2.08	0.359	0.916
0.20	0.888	0.225	1.41	1.94	2.18	0.335	0.918

Definitions: d = water depth; L_o = deep water wave length, L = water wave length, k = wave number ($2\pi/L$), H = water wave height, and H_o = deep water wave height.

Table A-5. Properties of sinusoidal waves (Svendsen and Jonsson 1976) (continued).

d/L_0	tanh (kd)	d/L	kd	sinh (kd)	cosh (kd)	2kd/sinh (2kd)	H/H_0
0.20	0.888	0.225	1.41	1.94	2.18	0.335	0.918
0.21	0.899	0.234	1.47	2.05	2.28	0.313	0.920
0.22	0.909	0.242	1.52	2.18	2.40	0.291	0.923
0.23	0.918	0.251	1.57	2.31	2.52	0.271	0.926
0.24	0.926	0.259	1.63	2.45	2.65	0.251	0.929
0.25	0.933	0.268	1.65	2.60	2.78	0.233	0.932
0.26	0.940	0.277	1.74	2.75	2.93	0.215	0.936
0.27	0.946	0.285	1.79	2.92	3.09	0.199	0.939
0.28	0.952	0.294	1.85	3.10	3.25	0.183	0.942
0.29	0.957	0.303	1.90	3.28	3.43	0.169	0.946
0.30	0.961	0.312	1.96	3.48	3.62	0.155	0.949
0.31	0.965	0.321	2.02	3.69	3.83	0.143	0.952
0.32	0.969	0.330	2.08	3.92	4.05	0.131	0.955
0.33	0.972	0.339	2.13	4.16	4.28	0.120	0.958
0.34	0.975	0.349	2.19	4.41	4.53	0.110	0.961
0.35	0.978	0.358	2.25	4.68	4.79	0.100	0.964
0.36	0.980	0.367	2.31	4.97	5.07	0.091	0.967
0.37	0.983	0.377	2.37	5.28	5.37	0.083	0.969
0.38	0.984	0.386	2.43	5.61	5.70	0.076	0.972
0.39	0.986	0.395	2.48	5.96	6.04	0.069	0.974
0.40	0.988	0.405	2.54	6.33	6.41	0.063	0.976
0.41	0.989	0.415	2.60	6.72	6.80	0.057	0.978
0.42	0.990	0.424	2.66	7.15	7.22	0.052	0.980
0.43	0.991	0.434	2.73	7.60	7.66	0.047	0.982
0.44	0.992	0.443	2.79	8.07	8.14	0.042	0.983
0.45	0.993	0.453	2.85	8.59	8.64	0.038	0.985
0.46	0.994	0.463	2.91	9.13	9.18	0.035	0.986
0.47	0.995	0.472	2.97	9.71	9.76	0.031	0.987
0.48	0.995	0.482	3.03	10.3	10.4	0.028	0.988
0.49	0.996	0.492	3.09	11.0	11.0	0.026	0.990
0.50	0.996	0.502	3.15	11.7	11.7	0.123	0.990
∞	1.000	∞	∞	∞	∞	0.000	1.000

Definitions: d = water depth; L_0 = deep water wave length, L = water wave length, k = wave number ($2\pi/L$), H = water wave height, and H_0 = deep water wave height.

Table A-6. Common unit prefixes.

Multiples and submultiples	Prefixes	Symbols
$1{,}000{,}000{,}000 = 10^9$	giga	G
$1{,}000{,}000 = 10^6$	mega	M
$1{,}000 = 10^3$	kilo	k
$0.01 = 10^{-2}$	centi	c
$0.001 = 10^{-3}$	milli	m
$0.000001 = 10^{-6}$	micro	μ
$0.00000001 = 10^{-9}$	nano	n

APPENDIX B: NOMENCLATURE

Table B-1. List of nomenclature.

SYMBOL	DEFINITION
A	frontal area facing flow velocity direction
A	anode surface area
ABS	American Bureau of Ships
A_d	projected area normal to flow direction
ANSI	American National Steel Institute
A_p	pile end area
AP	After perpendicular
API	American Petroleum Institute
A_s	pile side surface area
ASCE	American Society of Civil Engineers
ASME	American Society of Mechanical Engineers
A_T	cross sectional area of the thruster flow stream
a_x	water particle acceleration in x direction
a_z	water particle acceleration in z direction
B	buoyancy index
B	vessel breadth
B, CB	Center of buoyancy
BM	Distance between center of buoyancy and metacenter
b_o	distance between orthogonals in deep water
C	wave celerity
C	undrained shear strength of soil
C	anode current capacity
c	sound propagation velocity or sound speed
Ca	cavitation number
C_B	block coefficient
C_d	drag coefficient
C_{DN}	normal drag coefficient
C_{DT}	tangential drag coefficient for cable
CERCLA	Comprehensive Environmental Response, Compensation, and Liability Act
C_f	friction coefficient
C_F	skin friction coefficient
CFM	cubic feet per minute
CFR	Code of Federal Regulations
C_g	wave group velocity or wave group celerity
CG, G	Center of gravity
C_l, C_L	lift coefficient
C_m	inertia coefficient
C_M	midship coefficient
C_o	deep water wave celerity
C_P	prismatic coefficient
C_P	pressure coefficient
c_p, c_v	specific heat at constant pressure, specific heat at constant volume
C_r	wave reflection coefficient

Table B-1. List of nomenclature (continued).

C_T	resistance coefficient
C_V	volumetric coefficient
c_v	ray vertex velocity
C_{VP}	vertical prismatic coefficient
C_W	wave resistance coefficient
CWA	Clean Water Act
C_{WP}	waterplane coefficient
D	depth of frictional influence
d	water depth
D	diameter
d	sediment grain diameter
D	cable diameter
D	water depth
D	inside diameter of tank
d_{50}	median grain diameter
dB	decibel
d_g	grain diameter
d_h	propeller hub diameter
DI	directivity index
DI_T	transmitting directivity index
DNV	Det Norske Veritas
d_p	propeller diameter
DT	detection threshold
DWL	design load water line
E	total average energy per unit width of wave crest
E	potential
E	propeller efficiency
E	efficiency
E	energy
EHP	effective horsepower
E_k	kinetic energy per unit width of wave crest over the length of the wave
E_L	total energy in one wave length per unit width of wave crest
E_p	potential energy per unit width of wave crest over the length of the wave
EPA	Environmental Protection Agency
f	Coriolis parameter
F	force
f	pile unit skin friction
F	bursting force
f	frequency
FAA	Federal Aviation Administration
F_B	buoyant force
F_c	centrifugal force
F_{cf}	coriolis force
F_d	force due to drag
F_D	drag force
F_D	normal component of force on cable
F_f	force due to friction

Table B-1. List of nomenclature (continued).

F_g	gravity force
F_i	force due to inertia
F_l	force due to lift
F_L	lift force
F_n	force in normal direction
FP	forward perpendicular
F_{pg}	pressure gradient force
Fr	Froude number
F_T	tangential component of force on cable
F_T	ROV thruster force
F_w	submerged unit weight of pipe
FWS	Fish and Wildlife Service
g	acceleration of gravity
GM	distance between CG and metacenter
GZ	righting arm
H	wave height
H_{10}	one tenth wave height
H_{ave}	average wave height
H_b	breaker wave height
H_i	incident wave height
H_{max}	maximum wave height
H_{mo}	wave height based upon energy methods
H_o	deep water wave height
H_r	reflected wave height
H_{rms}	root mean square wave height
H_s, $H_{1/3}$	significant wave height
I	average initial current at anode
I	initial structure current density
I	Moment of inertia
I	instantaneous acoustic intensity
IL	intensity level
i_m	maintenance current density
I_{total}	total current
J	advance coefficient
k	wave number ($k=2\pi/L$)
K	Keulegan-Carpenter number
K	coefficient of lateral earth pressure
K	keel
KB	distance between keel and center of buoyancy
KC	Keulegan-Carpenter number
K_D	breakwater armor unit stability coefficient
KG	distance between keel and center of gravity
KM	distance between keel and metacenter
K_Q	torque coefficient
K_R	wave refraction coefficient
K_S	wave shoaling coefficient
K_T	thrust coefficient
K_x, K_y, K_z	eddy viscosity coefficients
L	wave length

Table B-1. List of nomenclature (continued)

L	length of cylindrical anode
L	vessel length
L	length of tank
L_b	length of basin
LCB, LCF	longitudinal center of buoyancy, flotation
LCG	longitudinal center of gravity
LDC	London Dumping Convention
L_o	deep water wavelength
m	mass
$\dot{m}$	mass flow rate
M	moment
m	beach slope
M	transverse metacenter
m	mass
M	molecular weight
M	moment, torque
M_1	defined by Equation 4-33
Ma	Mach number
MMS	Minerals Management Service
MPRSA	Marine Protection, Research and Sanctuaries Act
N	total number of anodes
n	number of moles
N	number of revolutions per minute
N	level
NEPA	National Environmental Policy Act
NL	noise level
N_q	dimensionless bearing capacity factor
N_R	Reynolds number
^oC	degrees Centigrade
^oF	degrees Fahrenheit
^oK	degrees Kelvin
^oR	degrees Rankine
OSHA	Occupational Safety and Health Administration
O_{slm}	oxygen consumption
$\overline{P}_o$	average wave energy flux in deep water
P	resistivity of the electrolyte
p	pressure
p	pressure inside tank
P	pitch of propeller
p	acoustic pressure
P	total acoustic radiated power
P_{atm}	atmospheric pressure
P_{CO2}	partial pressure of carbon dioxide
P_e	input electric power to transducer
P_g	gauge pressure
PHP	propeller horsepower
p_o	effective overburden pressure at pile tip
P_r	tank rated pressure

Table B-1. List of nomenclature (continued)

q	unit end bearing capacity
Q	volume flow rate
Q_d	ultimate pile bearing capacity
Q_f	pile skin friction resistance
Q_p	pile total end bearing capacity
r	radius
R	runup height
R	anode resistance
r	radius of cylindrical anode
R	gas constant
R	respiratory coefficient (vol. of carbon dioxide produced/vol. of oxygen consumed)
r	range
RCRA	Resource Conservation and Recovery Act
Re	Reynolds number
R_F	skin friction resistance
RL	reverberation level
R_P	pressure resistance
R_T	vessel total resistance
R_u	universal gas constant
R_W	wave making resistance
S	salinity
s	sieve size
S	safety factor
S	tank duration
S	salinity
S_a	specific gravity of armor unit material
SCFM	standard cubic feet per minute
SHP	shaft horsepower
S_l	longitudinal stress
SL	source level
SLM	standard liters per minute
SM	Simpson multiplier
SNAME	Society of Naval Architects and Marine Engineers
SOLAS	Safety of Life at Sea
SPL	sound pressure level
S_t	hoop, circumferential, girth, or tangential stress
St	Strouhal number
SWL	still water level
T	temperature
t	time
T	wave period
T	anode lifetime
T	vessel mean molded draft
T	vessel period of roll
T	propeller thrust
t	thrust deduction factor
T	cable tension
T	temperature
T	absolute temperature

Table B-1. List of nomenclature (continued)

TCG	Transverse center of gravity
T_H	horizontal cable tension
THP	thrust horsepower
T_i	period of oscillation in wave basin
TL	transmission loss
T_m	peak wave period
T_o	cable tension at the origin
TS	target strength
T_s , $T_{1/3}$	significant wave period
T_V	vertical cable tension
u	velocity component in x direction
U	vessel velocity
u	particle velocity
U_{10}	wind speed at height of 10 meter above the water surface
U_A	wind stress factor
U_e	effective velocity
U_m	peak velocity
U_o	surface current
Ur	Ursell number
USACE	US Army Corps of Engineers
USCG	US Coast Guard
U_z	wind speed at height (z) above the water surface
v	velocity component in y direction
$\dot{V}$	volume rate of flow
V	volume of displacement
v	specific volume
V	volume
$\dot{V}_{air}$	volumetric flow rate of air at depth
V_a	velocity of advance
v_b	longshore current at breaker position
V_c	current speed or vehicle speed
VCG, VCB	Vertical center of gravity, buoyancy
V_f	sediment grain fall velocity
V_g	geostrophic current
V_o	freestream velocity
V_r	tank rated capacity
V_t	volume of chamber
V_T	average velocity through thruster
w	velocity component in z direction
W	weight of armor units
w	wake fraction
W	work
Waterplane	horizontal planes parallel to designed load waterplane
We	Weber number
W_I	cable immersed weight
WL	intersection of waterplane with vessel's form
WL	waterline
W_{total}	total anode weight

Table B-1. List of nomenclature (continued)

x	coordinate axis
y	coordinate axis
z	coordinate axis
Z	gas compressibility factor
Greek	
α	slope angle of armor layer on breakwater structure
α	scale ratio
α	absorption coefficient
α_b	angle between the breaker crest and the shoreline
β	slope angle of seabed
β	depth to height ratio of breaking waves in shallow water
Γ	mixing coefficient
γ	fluid specific weight
γ	ratio of normal to tangential drag coefficients for cable
γ_a	specific weight of armor unit
γ_s	solid specific weight
∇	displacement
Δ	mass displacement
δ	friction angle between soil and pile wall
ε	vertical water particle displacement
ε	roughness
ε	roughness
$\dot{\varepsilon}$	strain rate
η	water surface elevation
η_H	hull efficiency
η_i	propeller ideal efficiency
η_o	open water efficiency
η_R	relative rotative efficiency
η_T	total propeller efficiency
θ	phase of wave (θ= kx-ωt)
θ	structure slope angle for runup evaluation
θ	angle of pitch when picking up load
θ_c	critical angle
λ	wave length
μ	dynamic viscosity
μ	coefficient of friction
ν	kinematic viscosity
ξ	surf similarity parameter
ξ	horizontal water particle displacement
ρ	density
σ_N	cavitation index
σ_t	sigma-t
τ	shear stress
ϕ	angle of heel
ϕ	angle between cable and horizontal
ϕ	velocity potential
φ	latitude

Table B-1. List of nomenclature (continued)

Φ	grain size parameter
ψ	ratio of undrained shear strength to effective overburden pressure
ς	surface tension
Ω	Earth's angular velocity
ω	angular velocity
ω	wave angular frequency ($\omega=2\pi/T$)

APPENDIX C: EVEN NUMBERED PROBLEM ANSWERS

Table C-1. List of answers to the even numbered problems.

CHAP	PROB	ANSWER
1		No Problems
2	2	L_o = 25, 56.2, 99.8, 224.6 m; C_o = 6.2, 9.4, 12.5, 18.7 m/s
	4	u = 0.13 ft/s; w = 0; a_x= 0; a_z= -0.85 ft/s^2
	6	$\rho_{s,t,o}$ = 1019.4 kg/m^3 σ_t = 19.4 kg/m^3
	8	L = 2.25 m, C= 1.88 m/s, C_g = 0.94 m/s, E_L= 441 (N-m)/m, P = 414.5 (N-m)/s, u = 0.054 m/s, w = 0.034 m/s, p=9911.8N/m^2
	10	L= 91.1 m, H = 2.29 m
3	2	F = 1403 N
	4	Fdrag =43.6 lb Finertia = 37.2 lb
	6	γ =1064.6 N/m^3
	8	H = 25.2 ft
	10	Finertia = 28765 N
4	2	H_s =13.8 ft, T_s = 8.3 s; H_s = 11 ft, T_s = 7.3 s; F_z = 1050 mi, D = 42 hr
	4	V_f = 15.9 cm/s
	6	C_r = 0.3, H_r = 1.05 ft
	8	V = 2.4 ft/s
	10	H = 56.56 ft, E = 1.14e^7 ft-lb/ft
	12	T_i= 5951.3 s
	14	α_b = 33.9 ^{o}
	16	V_f= 6.84 cm/s
5	2	Discussion, No solution given
	4	Discussion, No solution given
	6	N = 32

CHAP	PROB	ANSWER
6	2	a) weight = 227 lb b) weight = 3486 lb
	4	F_d = 144.6 lb
	6	T_A = 331 lb, Length of chain = 115.4 ft, Distance downstream = 80.8 ft, T_b =571.2 lb
	8	Displaced volume = 312.5 ft^3
7	2	17.4 ft^3/min at depth 492.8 liters/min at depth
	4	20.9 SCFM
	6	P_{co2} = 0.12 ata, Yes
	8	13,684 psi
	10	P_{co2}=0.005 ata, oxygen range = 6 to 40 %
8	2	
	4	TL = 122 dB
	6	SL = 212 dB re 1 μPa
	8	c = 1530 m/s
10	2	D = $[L^2V^2\rho]$ * $F[\rho VL/\mu]$ * $F[V^2/gL]$where "F" indicates the first Π term is a function of the Reynolds Number and the Froude Number.
	4	V_m = 1.90 m/s, D_m/D_p = 1.57x10^{-5}, P_m/P_p = 2.49x10^{-6}
	6	rpm_m = 9487, P_p = 567903 ft-lb/s = 1033 hp
	8	Since neither γ or V had to be raised to an exponent, both γ and V are directly proportional to F_b
11		No Problems
12		No solutions to preliminary design problems

APPENDIX D: INDEX

A

B

C

D

E

F

G

H

I

J

K

L

M

N

O

P

R

S

T

U

V

W

ABOUT THE AUTHOR

Dr. Robert E. Randall graduated from Ohio State University with a Bachelor of Mechanical Engineering degree in 1963. Afterwards, he served in the U.S. Navy as a submarine officer from 1963 to 1967 on the diesel submarines, USS *Grenadier* and *Grouper.* He entered graduate studies in ocean engineering at the University of Rhode Island in 1967 where he received a master of science degree in 1969 and a doctor of philosophy degree in 1972. His first engineering position was with the Naval Underwater Systems Center in Newport, Rhode Island, where he worked as an ocean engineer researching submarine weapon systems. In 1973, he worked as an ocean engineer at the Harbor Branch Foundation in Fort Pierce, Florida, where he conducted engineering design and analysis related to undersea research and exploration with small submersibles. The opportunity to teach and conduct research in the new Ocean Engineering Program at Texas A&M University attracted Dr. Randall to Texas in 1975, and he has been teaching and conducting research in ocean engineering ever since. Dr. Randall is a registered professional engineer in Texas. As a member of the ocean engineering faculty in the Civil Engineering Department at Texas A&M University, he teaches courses in ocean engineering, underwater acoustics, diving and life support engineering, marine dredging, and fluid mechanics. Dr. Randall has been the undergraduate and graduate advisor, associate head and head of the Ocean Engineering Program, associate program manager for the Strategic Petroleum Reserve Project, and is currently the director of the Center for Dredging Studies. He serves as the faculty advisor to the student chapter of The Society of Naval Architects and Marine Engineers and Marine Technology Society. He is a member of the Education Committee for The Society of Naval Architects and Marine Engineers and the Technical Committee on Ocean Engineering for the American Society of Civil Engineers. He serves on the board of directors for the Western Dredging Association. He served as the vice chair and chair of the Ocean and Marine Engineering Division of the American Society of Engineering Education. Dr. Randall received the Texas A&M Association of Former Students distinguished teaching award in 1991, and the Birdwell Endowed Teaching Award in 1997. His students received the ASME Ocean Engineering Division Best Student Team Design Project in 1992. As faculty advisor for the ocean engineering students' human powered submarine project, he guided the students to a Best Safety Design Award for the *Aggie Ray* in 1991, a fifth place finish with *SubMaroon* in 1996, and third place with *Tamu* in 1997. His research accomplishments include development of a tracking system for monitoring negatively buoyant plumes in offshore waters, development of laser techniques for measure kinematics in the steep crested waves, co-development of a simulation for predicting the fate of multiple disposals of dredged material in open water, and co-development of guidelines for capping of contaminated dredged material.

Although *Elements of Ocean Engineering* is the first book written by Dr. Randall, he has published and presented extensively in his field. In addition to the many awards, honors, and citations he has received, he has published fifteen papers in refereed journals, three papers at refereed conferences, 27 contributions to conference proceedings and books, and 72 significant reports.

Dr. Randall is married to Barbara A. Randall, and they have two sons, Brian and Neil, who are graduates of Texas A&M University. Dr. Randall's parents, Peg and Jack Randall reside in Columbus, Ohio, and he has a brother, Jerry W. Randall, in Columbus, Ohio and a sister, Sally Drahn, in Madison, Wisconsin.